Vega Prime
虚拟现实开发技术

王孝平　董秀成　古世甫　著

西南交通大学出版社
·成　都·

内容简介

本书系统地介绍了 Vega Prime 虚拟现实开发技术的相关知识及应用开发的流程和步骤，通过实例深入分析了其中的技术要点和难点，同时提供了一个应用开发框架，使开发者更为容易入门使用 Vega Prime，进阶掌握其核心技术，能够熟练灵活使用 Vega Prime 进行实际虚拟现实项目的开发。对于熟练的开发者，可以大大加快项目的开发进程。

本书内容主要包括 Vega Prime 核心模块的介绍与使用、LynX Prime 的使用、Vega Prime 的 MFC 改造、VSG 自绘图形的使用、OOBB 碰撞检测的实现、液体纹理仿真的实现、鼠标点选效果的实现、通道图形文字显示设计、通道模板效果的实现和 Vega Prime 编程框架设计等。本书是作者根据多年实际项目开发经验编写而成的，可作为大学计算机专业和非计算机专业的虚拟现实开发课程教材，也可作为其他虚拟现实开发人员的参考用书。

图书在版编目（CIP）数据

Vega Prime 虚拟现实开发技术 / 王孝平，董秀成，古世甫著. —成都：西南交通大学出版社，2018.8
ISBN 978-7-5643-6414-4

Ⅰ. ①V… Ⅱ. ①王… ②董… ③古… Ⅲ. ①虚拟现实 – 程序设计 Ⅳ. ①TP391.98

中国版本图书馆 CIP 数据核字（2018）第 207337 号

Vega Prime 虚拟现实开发技术

王孝平　董秀成　古世甫　著

责任编辑	黄庆斌
助理编辑	李华宇
封面设计	何东琳设计工作室

出版发行	西南交通大学出版社
	（四川省成都市二环路北一段 111 号
	西南交通大学创新大厦 21 楼）
邮政编码	610031
发行部电话	028-87600564　028-87600533
网址	http://www.xnjdcbs.com
印刷	四川森林印务有限责任公司

成品尺寸	185 mm × 260 mm
印张	23
字数	573 千
版次	2018 年 8 月第 1 版
印次	2018 年 8 月第 1 次
定价	68.00 元
书号	ISBN 978-7-5643-6414-4

图书如有印装质量问题　本社负责退换
版权所有　盗版必究　举报电话：028-87600562

前　言

　　Vega Prime 是 Presagis 公司推出的最新虚拟现实开发工具,具有面向对象、功能强大、界面友好、平台兼容性好等特点。Vega Prime 集成了最新的虚拟现实研究成果,同时具备开发 32 位和 64 位应用程序的强大功能,被广泛应用于航空航天飞行器试验、工业机械装置实时模拟仿真、房地产视景楼盘呈现、数字化城市等。随着 4G 时代的纵深发展,虚拟技术将会得到更加广泛的应用。

　　Vega Prime 最底层是 OpenGL,OpenGL 之上是 Presagis 公司自己的跨平台场景渲染引擎 VSG,VSG 之上就是 Vega Prime。Vega Prime 提供的 LynX Prime 是一个可视化的图形用户接口编辑工具,能够让开发者不写一行代码而实现虚拟现实仿真;同时提供的 Vega Prime 库,采用 C++方式,提供面向对象编程接口。这样,Vega Prime 既拥有了 OpenGL 的强大功能,又提供了界面友好的开发环境,极大提高了图形开发的效率,满足了虚拟现实仿真这种大型图形开发的要求。

　　本书是在 2012 年出版的国内第一本 Vega Prime 方面图书的基础上,依据作者项目组近几年的实际项目开发经验和总结 Vega Prime 应用开发项目的步骤和方法,通过纠正错误、补充内容、完善结构编写而成的。本书系统地介绍了 Vega Prime 相关知识及应用开发的流程和步骤,通过实例深入分析了其中的技术要点和难点,同时提供了一个应用开发框架,使开发者更为容易入门使用 Vega Prime,进阶掌握其核心技术,能够熟练灵活使用 Vega Prime 进行实际虚拟现实项目的开发。对于熟练的开发者,可以大大加快项目的开发进程。本书内容主要包括 Vega Prime 核心模块介绍与使用、LynX Prime 的使用、Vega Prime 的 MFC 改造、VSG 自绘图形的使用、OOBB 碰撞检测的实现、液体纹理仿真的实现、鼠标点选效果的实现、通道图形文字显示设计、通道模板效果的实现和 Vega Prime 编程框架设计等。

　　本书由西华大学电气信息学院王孝平和古世甫共同完成,其中王孝平负责第 1,5,6,9,10 章,古世甫老师负责第 2,3,4,7,8 章。西华大学电气信息学院董秀成对本书进行了审核。本书得到了董秀成教授工作室的大力支持,包含郑海春和张帆的大力协助,工作室的部分研究生同学也提供了帮助,在此表示感谢。本书还得到了技研新阳集团-西华大学机器人联合实验室和四川省信号与信息处理实验室的大力支持,在此一并表示感谢。

　　由于编者水平有限,书中难免存在不足之处,恳请广大读者批评指正。

作　者

2018 年 8 月

目 录

第 1 章　认知 Vega Prime ··· 1
　1.1　认识 Vega Prime ··· 1
　　　1.1.1　Vega Prime 的特性和基本模块 ··· 2
　　　1.1.2　Vega Prime 的增强模块选项 ·· 3
　　　1.1.3　Vega Prime 的第三方工具 ·· 6
　　　1.1.4　Vega Prime 的应用领域 ·· 9
　1.2　Vega Prime 入门指引 ·· 9
　　　1.2.1　Vega Prime 的版本问题 ·· 9
　　　1.2.2　Vega Prime 入门指引 ·· 10
　1.3　Vega Prime 应用开发 ·· 11
　　　1.3.1　Vega Prime 的结构及资源 ·· 11
　　　1.3.2　Vega Prime 系统结构及应用组成 ·· 13

第 2 章　Vega Prime 主要功能模块 ··· 16
　2.1　Vega Prime 基本概念 ·· 16
　　　2.1.1　Vega Prime 中的六自由度 ·· 16
　　　2.1.2　Vega Prime 中的基本概念 ·· 17
　2.2　Vega Prime 主要功能模块 ·· 18
　　　2.2.1　应用 vpApp ·· 19
　　　2.2.2　内核 vpKernel ·· 20
　　　2.2.3　管道 vpPipeline ·· 21
　　　2.2.4　窗口 vpWindow ·· 21
　　　2.2.5　通道 vpChannel ·· 22
　　　2.2.6　场景 vpScene ·· 23
　　　2.2.7　观察者 vpObserver ··· 24
　　　2.2.8　对象 vpObject ·· 24
　　　2.2.9　自由度 vsDOF ·· 25
　　　2.2.10　转换 vpTransform ·· 26
　　　2.2.11　运动模式 vpMotion ·· 27
　　　2.2.12　碰撞 vpIsector ·· 27

第 3 章 LynX Prime 的使用 29
3.1 LynX Prime 的界面组成 29
3.1.1 LynX Prime 的启动与退出 29
3.1.2 LynX Prime 的界面构成 30
3.2 LynX Prime 的使用 32
3.2.1 创建场景 vpScene 33
3.2.2 操作对象 vpObject 35
3.2.3 设置观察者 vpObserver 37
3.2.4 建立转换 vpTransform 37
3.2.5 创建运动模式 vpMotion 39
3.2.6 应用碰撞检测 vpIsector 40
3.2.7 建立特效 vpFx 43
3.2.8 设置窗口 vpWindow 50
3.2.9 建立环境 vpEnv 51
3.2.10 设置通道 vpChannel 54

第 4 章 运行 Vega Prime 应用 59
4.1 VC++编程基础 59
4.1.1 安装 VC++ 59
4.1.2 进程优先级 60
4.1.3 使用多线程 61
4.1.4 创建控制台程序 63
4.1.5 创建 MFC 对话框程序 64
4.2 配置 Vega Prime 应用程序编译运行环境 65
4.2.1 配置运行控制台仿真程序 66
4.2.2 配置 MFC 对话框运行环境 68
4.3 导出 ACF 文件 68
4.3.1 导出 ACF 文件 68
4.3.2 Vega Prime 的最小应用程序 68
4.4 剖析 Vega Prime 应用程序组成 71
4.4.1 基本组成 72
4.4.2 创建容器 73
4.4.3 初始化模块 73
4.4.4 建立内核 74
4.4.5 建立路径 74
4.4.6 建立管道 74
4.4.7 建立窗口 75
4.4.8 建立通道 76

	4.4.9	建立观察者	77
	4.4.10	建立场景	77
	4.4.11	建立转换	78
	4.4.12	建立对象	78
	4.4.13	建立碰撞检测	80
	4.4.14	建立碰撞服务	81
	4.4.15	建立循环服务	81
	4.4.16	建立环境	81
	4.4.17	建立太阳	82
	4.4.18	建立运动模式	83
	4.4.19	建立特效	83
	4.4.20	配置应用	84

第 5 章 建立基于 MFC 对话框的 Vega Prime 应用程序 87

5.1 配置 MFC 对话框 Vega Prime 应用程序的编译环境 87
5.1.1 MFC 对话框 Vega Prime 应用程序的理解 87
5.1.2 配置 MFC 对话框 Vega Prime 应用程序的编译环境 88

5.2 建立 MFC 对话框 Vega Prime 应用程序 89
5.2.1 MFC 对话框程序界面改造 90
5.2.2 添加 Vega Prime 应用公共类 PublicMember 91
5.2.3 启动 Vega Prime 应用主线程 109
5.2.4 启动包含 ACF 文件的主线程 111

第 6 章 Vega Prime 编程对象的实例使用 113
6.1 建立内核实例 114
6.2 建立管道对象 114
6.3 建立窗口 115
6.4 建立场景 116
6.5 改造路径搜索对象 116
6.6 加载物体函数设计 117
6.7 设置运动模式 120
6.8 建立转换 121
6.9 控制观察者 122
6.10 配置键盘函数 123
6.11 控制物体缩放比例及透明 126
6.12 控制碰撞检测 128
6.13 控制特效 130

6.14	配置灯光效果	131
6.15	制造幻影效果	133
6.16	控制声音	134
6.17	控制父子关系	137
6.18	操作 Switch	140
6.19	操作 DOF	142
6.20	获取 DOF 的坐标	148
6.21	配置多通道	149
6.22	物体平面影子效果	154
6.23	物体颜色控制	155
6.24	雨雪天气控制	159
6.25	场景能见度控制	164
6.26	纯色场景控制	165
6.27	仿真场景全屏设计	167
6.28	加快仿真场景物体加载速度	169

第 7 章 Vega Prime 自绘图形设计 ... 172

7.1	认识 VSG	172
	7.1.1 VSG 的特点	172
	7.1.2 VSG 的功能模块	173
	7.1.3 VSG 的图形绘制过程	173
7.2	VSG 图形绘制	174
	7.2.1 简单几何体绘制	175
	7.2.2 箱体绘制	178
	7.2.3 梯形平台绘制	180
	7.2.4 平面绘制	182
	7.2.5 球体绘制	184
	7.2.6 字符输出	186
7.3	图形纹理控制	189
7.4	图形材质控制	192
7.5	VSG 在场景中显示中文	196
	7.5.1 图片产生类设计	196
	7.5.2 纹理切换控制设计	201

第 8 章 Vega Prime 和 OpenGL 混合编程 ... 206

8.1	OpenGL 基础	206
	8.1.1 OpenGL 的特点	206

8.1.2　OpenGL 开发环境配置 207
　　　8.1.3　OpenGL 程序构成 207
　　　8.1.4　OpenGL 绘制几何图形 209
　　　8.1.5　OpenGL 的颜色模式 215
　　　8.1.6　OpenGL 视图变换 218
　8.2　OpenGL 在 Vega Prime 应用中绘制图形 220
　　　8.2.1　理解 Vega Prime 与 OpenGL 混合编程 220
　　　8.2.2　定义订阅者类 221
　　　8.2.3　Vega Prime 中使用 OpenGL 225

第 9 章　Vega Prime 中的实用功能实现 230
　9.1　Vega Prime 中的重叠效果 230
　　　9.1.1　点到点重叠效果 231
　　　9.1.2　闭环重叠效果 232
　　　9.1.3　二维字体重叠效果 233
　　　9.1.4　二维图片重叠效果 234
　9.2　自定义碰撞检测类 235
　　　9.2.1　Vega Prime 的碰撞检测 235
　　　9.2.2　自定义查找物体顶点类 237
　　　9.2.3　自定义碰撞检测类 239
　　　9.2.4　自定义碰撞检测类的使用 242
　9.3　窗口鼠标控制的完整实现 247
　　　9.3.1　窗口鼠标函数的认识 247
　　　9.3.2　窗口鼠标类的设计实现 248
　　　9.3.3　窗口鼠标类的配置 251
　9.4　场景通道屏幕文字和图形显示 254
　　　9.4.1　通道屏幕文字显示类的设计实现 254
　　　9.4.2　通道屏幕文字显示类的配置调用 263
　　　9.4.3　通道屏幕绘图功能设计实现 265
　9.5　鼠标点选通道对象功能设计 267
　　　9.5.1　通道挑选 Picker 类的设计实现 268
　　　9.5.2　窗口鼠标类的辅助设计实现 272
　　　9.5.3　通道挑选 Picker 类的配置使用 275
　9.6　仿真通道的屏幕图片抓取 278
　　　9.6.1　通道屏幕图片抓取功能的设计 278
　　　9.6.2　通道屏幕图片抓取功能的配置使用 281
　9.7　仿真通道的视频录制 283
　　　9.7.1　通道屏幕视频录制类的设计 283

  9.7.2 通道屏幕视频录制功能的配置使用 289
 9.8 虚拟仿真中的半透明处理和纹理运动仿真 293
  9.8.1 虚拟仿真中的半透明处理 293
  9.8.2 虚拟仿真中的纹理运动仿真 295
 9.9 虚拟仿真中的聚光灯光源使用 299
  9.9.1 聚光灯光源 vpLightLobe 的理解 300
  9.9.2 聚光灯光源 vpLightLobe 的使用 300
 9.10 渲染策略的使用 302
  9.10.1 物体渲染策略的使用 302
  9.10.2 通道渲染的使用 305
 9.11 模板效果的设计与使用 309
  9.11.1 模板效果的基本要素 309
  9.11.2 单通道模板效果设计 310
  9.11.3 双通道模板效果设计 318
 9.12 仿真辅助线程设计 331
  9.12.1 C++中的线程 331
  9.12.2 线程函数 331
  9.12.3 线程控制 332
  9.12.4 数据辅助线程设计 333

第 10 章 Vega Prime 编程框架设计 335

 10.1 MFC 下的框架总体设计 335
 10.2 具体窗口功能设计实现 336
  10.2.1 预备设计 336
  10.2.2 主窗口背景色彩控制 339
  10.2.3 主窗口背景图片布局 341
  10.2.4 主窗口全屏幕自适应设计 342
  10.2.5 功能窗口初步设计 344
  10.2.6 TabControl 初始化设计 345
  10.2.7 TabControl 功能切换设计 347
  10.2.8 功能窗口再次设计 348
 10.3 运行效果设计 354

参考文献 356

附件 虚拟现实开发实例 357

第 1 章　认知 Vega Prime

Vega Prime 是 Presagis 公司推出的最新虚拟现实开发工具，具有面向对象、功能强大、界面友好、平台兼容性好等特点。Presagis 推出的虚拟现实开发工具一直受到市场广泛的欢迎，Vega Prime 就是该公司推出的最新版本。Vega Prime 是一个应用程序编程接口(API)，它大大扩展了 Vega Scene Graph，也是一个跨平台的可视化模拟实时开发工具。Vega Prime 是一个进行实时仿真和虚拟现实开发的高性能软件环境和良好工具，它由以下 3 部分组成：图形用户接口，LynX Prime; Vega Prime 库；C++头文件，可调用的函数。Vega Prime 的功能还被其他特殊功能模块所扩展，这些模块扩展了用户接口的同时，也为应用开发提供了功能库。

【本章重点】

- Vega Prime 的特性；
- Vega Prime 的基本模块；
- Vega Prime 的增强模块；
- Vega Prime 的应用领域；
- Vega Prime 的入门指引；
- Vega Prime 的学习资源；
- Vega Prime 的配置；
- Vega Prime 的组成；
- Vega Prime 的特点。

1.1　认识 Vega Prime

Vega Prime 在提供高级仿真功能的同时，还具有简单易用的优点，使用户能快速准确地开发出满足要求的视景仿真应用程序。它将易用的工具和高级视景仿真功能巧妙地结合起来，从而使用户能够简单迅速地创建、编辑、运行复杂的实时三维仿真应用。由于它大幅减少了源代码的编写，软件的进一步维护和实时性能的优化变得更容易，从而大大提高了开发效率。同时，它还拥有一些特定的功能模块，可以满足特定的仿真要求，如特殊效果、红外和大面积地形管理等。

此外，Vega Prime 包括了许多有利于减少开发时间的特性，包括自动的异步数据库调用、碰撞检测与处理、对延时更新的控制和代码的自动生成。Vega Prime 还具有可扩展可定制的文件加载机制、对平面或球体的地球坐标系统的支持、对应用中每个对象进行优化定位与更新的能力、星象模型、各种运动模式、环境效果、模板、多角度观察对象的能力、上下文相关帮助和设备输入/输出支持等。

1.1.1 Vega Prime 的特性和基本模块

1. Vega Prime 的特性

- 跨平台性：它支持 Microsoft Windows、SGI IRIX、Linux 和 Sun Microsystems Solaris 等操作系统。同时，用户的应用程序也具有跨平台特性，用户可以在任意一种平台上开发应用程序，且无须修改就能在另一个平台上运行。
- 与 C++ STL(Standard Template Library)兼容。
- 支持双精度浮点数。
- 可定制用户界面和可扩展模块：Vega Prime 可扩展的插件式体系结构技术复杂但使用简单。用户可以根据自己的需求调整三维应用程序，能快速设计并实现视景仿真应用程序，以最低的硬件配置获得高性能的运行效果。此外，用户还可以开发自己的模块，并生成定制的类。
- 同时支持 OpenGL 和 Direct3D。
- 高效的生产率：Vega Prime 提供了许多高级功能，能满足绝大部分视景仿真应用的需要，同时还具有简单易用的特性、高效的生产率。
- 支持 MetaFlight 文件格式：MetaFlight 是 Presagis 公司基于 XML 的数据描述规范，它使运行数据库能与简单或复杂的场景数据库相关联，MetaFlight 极大地扩展了 OpenFlight 的应用范围。

2. Vega Prime 的基本模块

Vega Prime 包括 Lynx Prime 图形用户界面配置工具和 Vega Prime 的基础 VSG(Vega Scene Graph)高级跨平台场景渲染 API。此外，Vega Prime 还提供了多个针对不同应用领域的可选模块，使其能满足特殊行业的仿真需要，同时支持用户自主开发自己的模块。

LynX Prime 是一种可扩展的、跨平台的、单一的 GUI 工具，为用户提供了一个简单明了的开发界面。Lynx Prime 基本上继承了 Lynx 的功能，同时又增加了一些新功能：向导功能可以对 Vega Prime 的应用程序进行快速创建、修改和配置，大大提高了生产效率;基于工业标准的 XML 数据交换格式,能与其他应用领域进行最大限度的数据交换;它可以把 ACF(Application Configuration File)自动转换为 C++代码。

VSG(Vega Scene Graph)是跨平台的场景渲染 API，是 Vega Prime 的基础。Vega Prime 包括了 VSG 提供的所有功能，并在易用性和生产效率上做了相应的改进。在为视景仿真和可视化应用提供的各种低成本商业开发软件中，VSG 具有最强大的功能，它为仿真、训练和可视化等高级三维应用开发人员提供了最佳的、可扩展的基础。VSG 具有以下特性：

- 帧频率控制;
- 内存分配;

- 内存泄漏跟踪；
- 基于帧的纹理调用；
- 异步光线点处理；
- (优化的)分布式渲染；
- 跨平台可扩展的开发环境，支持 Windows、Irix、Linux 和 Solaris；
- 与 C++ STL 相兼容的体系结构；
- 强大的可扩展性，允许最大限度的定制，使得用户可调整 VSG 来满足应用需求，而不是根据产品的限制来调整应用需求；
- 支持多处理器多线程的定制与配置；
- 应用程序也具有跨平台性，用户在任意一种平台上开发的应用程序无须进行修改就可以在另一个平台上运行；
- 支持 OpenGL 和 Direct3D 的优化渲染功能，应用程序能基于 OpenGL 或 Direct3D 运行，其间无须改动程序代码；
- 支持双精度浮点数，使几何物体和地形在场景中能够精确地放置并表示；
- 支持虚拟纹理、软件实现图像的动态查阅，使高级功能与平台无关。

3. Vega Prime 的可选模块

Vega Prime 为了满足特定应用开发的需求，除了上述基本模块之外，还提供了功能丰富的可选模块。Vega Prime 的可选模块基本上覆盖了 Vega 的可选模块，其中包括：

- Vega Prime FX：爆炸、烟雾、弹道轨迹和转轮等；
- Vega Prime Distributed Rendering：分布式渲染；
- Vega Prime LADBM：非常大的数据库支持；
- DIS/HLA：分布交互仿真；
- Blueberry：3D 开发环境；
- DI-GUY：三维人体；
- GL-Studio：仪表；
- Vega Prime IR Scene：传感器图像仿真；
- Vega Prime IR Sensor：传感器图像实际效果仿真；
- Vega Prime RadarWorks：基于物理机制的雷达图像仿真；
- Vega Prime Vortex：刚体动力学模拟；
- Vega Prime marine：三维动态海洋。

1.1.2　Vega Prime 的增强模块选项

1. Vega Prime Marine

Vega Prime Marine 为在实时 3D 仿真应用中创建极具真实感的海洋、湖泊、海岸线水流表面提供了理想的解决方案。该选项使用户能够很方便地在任何 Vega Prime 应用中添加动态真实的水流表面效果。图 1.1.1 所示为海洋效果图片。

效果呈现

图 1.1.1　海洋效果图片

提供必要的真实感仿真海洋表面效果以及与之动态交互的船体效果，充分满足了交互式实时 3D 仿真与训练中对综合动态海洋表面的真实性和准确性要求。该模块选项提供的高性能浪花模型，使用户可以轻松控制浪花的形态，包括在风力影响下浪花的方向、高度、长度和形式分布。同时，还可塑造 13 种由不同 Beaufort 标度描述的海洋状态及由 9 种不同海浪模型描述的海洋状态。

开发者能够定义船体特征和参数，以控制船首、船尾和船体外观。浪花的大小和形状完全吻合船体的大小、形状和速度，并且与周围的浪花和船只相交互。该特征使用户能够对仿真环境下船体的速度、机动性和转向进行控制。此外，Vega Prime Marine 还支持多洋面和多观察者效果，并支持正确的真实感海岸线浅水动态仿真，包括海浪冲击效果、水深变化效果和沙滩效果。

2. Vega Prime Camera

Vega Prime Camera 模块，能够模拟出用于任何类型的监视工具、闭路电视系统视频或光学设备的彩色及黑白效果。支持全套效果，Vega Prime Camera 为本土安全、操纵仿真、UAV/UGV、安全演练以及突发事件响应等多种应用提供了理想的工具。各种效果能通过 LynX Prime GUI 接口或 Vega Prime API 进行组合，并简单添加到任何 Vega Prime 场景中。同时，Vega Prime Camera 还提供了最多种类的镜头特效。Vega Prime Camera 支持对每一个摄像效果产生最佳真实感效果，支持对快速原型进行创建和改进的同时预览效果。图 1.1.2 所示为监视效果图片。

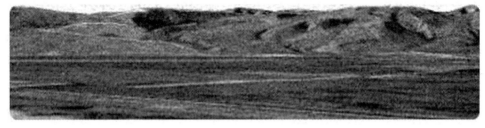

效果呈现

图 1.1.2　监视效果图片

3. Vega Prime LADBM(大范围地景数据库管理)

Vega Prime LADBM 模块，专为应用大规模和复杂的地景数据库创建和调度提供跨平台、扩展性良好的开发环境。高性能的 Vega Prime LADBM 模块能够在动态页面调用和用户自定义页面调用时确保大规模数据库装载与组织的最优化。图 1.1.3 所示为大范围地景图片。

效果呈现

图 1.1.3　大范围地景图片

Vega Prime LADBM 提供了最佳的渲染性能，充分满足定制与扩展性需求，能够最大化地利用现有资源。基于其 MetaFlight XML 文件规格和数据库格式，Vega Prime LADBM 确保大规模数据库的组成和关联能够以一种最有效的新型方式完成。MetaFlight 文件的分级式数据结构确保了运行时场景图像得到最佳性能。利用 Vega Prime 核心特性(包括双精度和多线程特性)，Vega Prime LADBM 为大规模视景仿真应用提供了理想的解决方案。同时，结合 GUI 配置工具（包括易用的向导工具），先进的 API 功能提供了完全符合实时 3D 应用开发的基础构造。

4. Vega Prime FX

Vega Prime FX 模块为实时 3D 应用中大量特殊效果的仿真提供了跨平台且扩展性良好的开发环境。所有的效果都能够采用 LynX Prime GUI 配置工具或直接通过 API 进行访问、修改，并添加到具体应用中。同时，采用 Vega Prime FX，用户只需对某些视觉属性进行预定义或调整，就能够定制场景中效果的显示、时间、触发以及性能特征。图 1.1.4 所示为特效效果图片。

效果呈现

图 1.1.4　特效效果图片

Vega Prime FX 提供了可完全定制和升级的粒子系统，使用户能够极其方便地进行粒子特效的定制和构建。配置属性包括速度、重力、颗粒大小和颗粒生命周期。除了可创建定制的特殊效果外，用户还能够直接访问任意 Vega Prime 应用中的预定义和优化效果。同时，联合 GUI 配置工具(如向导工具和 API 功能)，能为简单快速地创建和展开实时 3D 应用提供理想的特殊效果。

5. Vega Prime Distributed Rendering

Vega Prime Distributed Rendering 模块是实现完全同步的、多通道应用的开发和调度的理想工具，能够在多台图形节点上进行连续一致的渲染。利用 Vega Prime Distributed Rendering 提供的优化渲染性能，主机系统和客户端系统以同一种配置进行互联。直观的接口结构充分满足跨平台实时 3D 应用的开发与调度的需求。图 1.1.5 所示为分布式渲染效果图片。

通常，分布式渲染可以满足多通道连续或非连续显示的应用。任何 Vega Prime 应用均能够通过在图形界面简单添加一些设置进行分布式渲染。Vega Prime Distributed Rendering 模块包括通过局域网对多通道应用进行简单设置和配置的工具。因此，用户能够利用一个 GUI 接口使多通道应用高效运行，允许用户在适当的硬件上对应用进行设置、测试、处理和配置。

效果呈现

图 1.1.5　分布式渲染效果图片

6. Vega Prime LightLobes

Vega Prime LightLobes 模块为 Vega Prime 应用提供了极具真实感的照明效果。能够创建真实的场景照明且避免产生错误的贴图效果。支持实时帧率下的大量移动光源模拟和用户自定义光照类型。Vega Prime LightLobes 模块为照明光源的观察(如针对飞机驾驶员)应用提供理想解决方案。移动光源渲染技术适用于任何支持 OpenGL1.2 或更高版本的硬件平台。照明程度根据光源与地面距离的扩大而减退，或根据地面与观察者的距离变化。这项创新的技术使用户能够在一个应用中使用大量的移动光源，并通过优化绘制时间以实现最佳表现性能。图 1.1.6 所示为光源效果图片。

效果呈现

图 1.1.6　光源效果图片

1.1.3　Vega Prime 的第三方工具

1. Blueberry 3D Dev Environment

Blueberry3D 模块用来在 VegaPrime 中加入基于分形的程序几何体，以创建高度复杂、充满细节的虚拟地理环境。因为表示地形和文化特征的几何体都是动态生成的，它能够在保证帧率的同时达到前所未有的复杂度。同样，它具有 API 及 LynX Prime 界面。图 1.1.7 所示为 Blueberry 3D 效果图片。

效果呈现

图 1.1.7　Blueberry 3D 效果图片

使用Blueberry 3D开发环境，几何形体是在程序运行时根据需要实时生成的。地形和文化特征只是在观察者感兴趣区域内动态生成，细节部分也是在观察者靠近的时候才加入，越走近，细节就越多，包括高精度的污垢、树枝和丰富的植被。另外，植物、树木等还会对一些因素产生反应，如随风摇摆。而细节能达到的程度和数量，取决于用户定义的帧率，或者说硬件越快，场景中的细节就可以越多。

用分形算法，Blueberry 3D开发环境能将多种土壤类型和特性自然地融合在一起、真实地分布植被，每个分形物体都是不同的。但同时，又保证用户每次走近一个地方时，看到的场景和以前保持一致。

2. DIS/HLA for Vega Prime

DIS/HLA for Vega Prime模块，能够非常简单地通过LynX Prime对Vega Prime应用进行互联，无须任何规划即可进行DIS和HLA操作，实现HLA互联，或在多台机器/多参与者之间开发分布式Vega Prime仿真。

DIS/HLA for Vega Prime基于MAK公司的网络工具集VR-Link，包括了VR-Link的所有功能，提供来自MAK产品的灵活和专业的互联技术。该模块能创建一个仿真应用，并使它在多个不同的模块之间进行灵活转换。

用户能够使用DIS协议(DistributedInteractive Simulation，分布式交互仿真)或HLA(高层体系结构)对Vega Prime应用进行网络化拓展。用户可以用Lynx Prime界面进行基本的分布式仿真设定，而无须任何编程。

3. GLStudio for Vega Prime

GL Studio模块由DiSTI开发，使得用户能在VegaPrime场景中方便地加入由GL Stuido创建的交互式对象，而无须编写任何代码。另外，创建好的GL Studio对象能够与用户和其他Vega Prime对象进行交互。图1.1.8所示为GL Studio效果图片。

图1.1.8　GL Studio效果图片

效果呈现

GL Studio(DiSTI的独立产品)创建高质量的、具有照片级真实感的仪器仪表图形显示及人机界面，并生成优化的OpenGL C/C++代码。

GL Studio模块为照片级图像显示提供快速原型创建、设计和调度环境(如仪器和设备模型)，尤其适用于实时3D仿真和训练应用。

4. Immersive for Vega Prime

Immersive for Vega Prime模块提供了Immersive虚拟外设驱动接口，可配置用于几乎所有的Vega Prime应用中，包括walls、tiles等各种类型的应用。同时，还能够配置运行在非立体、主动立体和被动立体显示系统中。Immersive for Vega Prime 提供了与 VRCO Trackd 的连接，可将 Vega Prime 应用于任意基于上述驱动的 Immersive 虚拟外设连接，用以增强应用的可交互性。Immersive for Vega Prime 完全支持多节点的分布式渲染。图 1.1.9 所示为 Immersive for Vega Prime 效果图片。

效果呈现

图 1.1.9　Immersive for Vega Prime 效果图片

5. SpeedTree for Vega Prime

SpeedTree 模块能够在实时帧率下进行真实感植被景观的定义与渲染。该模块集成了来自 IDV 公司的获奖产品 SpeedTree 的技术，此技术目前已经成为 US DoD 训练系统和大多数视景游戏的特定特征。图 1.10 所示为 SpeedTree 效果图片。

效果呈现

图 1.1.10　SpeedTree 效果图片

SpeedTree 模块能够对 Vega Prime 应用中高密度植被进行定义和渲染，并能在达到最佳视觉效果的同时保持原有的渲染效率不变。SpeedTree 模块能生成具备碰撞映射、阴影和精细纹理的植被效果，提供了包括 200 多种树和植物种类的模型库(如阔叶树、针叶树、棕榈树、仙人掌和灌木等)，并允许对现有树型进行修改以及创建新的树型。SpeedTree 植被还能被方便地添加到现有的 OpenFlight 和 MetaFlight 数据库中。

6. Vortex for Vega Prime

Vortex 模块为在实时仿真应用中创建基于真实物理学的车辆、铰接机械和机器人模型提供了灵活的开发平台。Vortex 可模拟基于地面的车辆和机械，并使其具有真实的物理属性，包括刚体动力学、丰富的关节库、准确的碰撞检测及车辆动力学。Vortex 模块能够方便地创建齿轮、电机、悬架模型、水力学、轮、轨迹和其他组件，装配后能够组合成运动和行为准确的车辆和机械。此外，开发者能够对场景中的所有对象添加物理特征，真正实现交互式仿真效果。图 1.1.11 所示为 Vortex 效果图片。

Vortex 模块能够在真实感和速度中取得平衡，充分满足苛刻的工业要求。Vortex 模块具备通用工具包，可为多种模拟器开发提供灵活的开发平台，并且能够在实时仿真中进行配置，适用于操作训练、产品设计和测试。

效果呈现

图 1.1.11　Vortex 效果图片

1.1.4　Vega Prime 的应用领域

Vega Prime 的应用领域非常广泛，同时也正在进一步发展，现在主要的应用领域有：
- 制造业：虚拟设计、虚拟装配和教学演示；
- 军事：军事模拟、虚拟战场和电子对抗；
- 医学：虚拟手术、医学研究；
- 城市：房地产楼盘展示、城市规划和城市仿真；
- 地理信息：GIS；
- 其他：科研、矿藏开采、石油勘探及航空航天等。

1.2　Vega Prime 入门指引

1.2.1　Vega Prime 的版本问题

Vega Prime 发展到现在，技术非常成熟，功能非常强大，也产生了很多版本。现在常见的版本有 Vega Prime 2.0、Vega Prime 4.0、Vega Prime 5.0 和 Vega Prime 2013。

为什么强调 Vega Prime 的版本呢？主要是，要真正发挥 Vega Prime 的强大功能，在很大程度上都依赖于 C++编程控制，而不同版本的 Vega Prime 对应不同的 C++编译环境。在 Windows 环境下的 C++编译环境主要是 Visual Studio 下的 Visual C++，Vega Prime 与 Visual C++ 版本的对应关系如下：

Vega Prime 1.x：Visual C++ 6.0；

Vega Prime 2.0：Visual C++ 2003；

Vega Prime 2.1：Visual C++ 2003；

Vega Prime 2.2：有 Visual C++ 7.1（Visual C++ 2003.net）和 Visual C++ 8.0（Visual C++ 2005.net sp1 补丁）两个版本。

Vega Prime 3.x：有 Visual C++ 8（Visual C++ 2005.net sp1 补丁）和 Visual C++ 9（Visual C++ 2008.net sp1 补丁）两个版本。

Vega Prime 4.x：有 Visual C++ 8（Visual C++ 2005.net sp1 补丁）和 Visual C++ 9（Visual C++ 2008.net sp1 补丁）等多个版本。

Vega Prime 5.0：有 Visual C++ 9（Visual C++ 2008.net sp1 补丁）和 Visual C++ 10（Visual C++ 2010.net）等多个版本。

笔者主要使用的是 Vega Prime 2.0 和 VS2003。随着 Windows 系统的更新，笔者推荐开发者使用 Vega Prime 5.0 和 VS2008，在 Windows7/Windows8/Windows10 系统下的使用效果非常良好。

1.2.2 Vega Prime 入门指引

1. 明确目的

在学习 Vega Prime 之前，要想清楚使用 Vega Prime 进行开发的目的是什么？

Vega Prime 的长处不在于视觉效果特别强的视景演示，它更侧重于工业仿真、物流仿真、军事仿真等应用。当然，美工和烘焙模型也能够导入 Vega Prime 中。

Vega Prime 支持 Visual C++编译器、C++ builder 和 QT(跨平台 C++图形用户界面应用程序)，通过 COM（Component Object Model）技术也能支持浏览器和 ActiveX 跨平台应用。

但是，Vega Prime 是一款商业仿真软件，享受其强大功能的同时，需要付费使用，运行时需要 License，价格较昂贵。

2. 入门前提

因为 Vega Prime 是面向对象的类库，是基于 VSG 引擎的上层 C++类库，所以开发者需要具备一定的 C/C++语言知识和基本的面向对象编程知识（包括 Visual C++中 MFC 对话框编程）。

显卡的质量是 3D 场景渲染效果的物理基础，推荐开发者使用配有 3D 加速独立显卡的 PC 机，独立显卡为 Nvida 类型（俗称 N 卡）。

3. 安装的方法

Vega Prime 2.01 的安装方法如下：

http://www.52vr.com/bbs/forum.php?mod=viewthread&tid=11340

4. 初步认识和使用

打开 LynX Prime 图标，点击 Active Priview 按钮，就能够正常播放三维视景了（若软件安装成功）。

Vega Prime 软件安装目录 doc 中有 desktop tutor 的中文翻译教程。

5. 配置开发环境

目前根据版本要求，若无特别声明，Vega Prime 2.0，Vega Prime 2.1 和 Vega Prime 2.2 for Visual C++ 7 都可以使用 Visual C++ 2003.net 进行开发，Vega Prime 2.2 for 8.0 可以使用 Visual C++ 2005.net 进行开发。

Vega Prime 版本对应关系和开发配置如下：

http://www.52vr.com/bbs/forum.php?mod=viewthread&tid=1333

若学习了上述场景问题后,还是不会,可以打开 Vega Prime 自带的工程示例进行自学,工程示例一般放在安装目录"MultiGen-Paradigm\resources\samples"中。工程示例都是已配置好的文件,可以直接编译运行,也可以打开工程的属性,查看如何设置 Vega Prime 库和 Visual C++工程的绑定与库链接。

6. 编写 Vega Prime 程序

最基本的 Vega Prime 程序就是控制台程序,所以首先要学会编写控制台程序。

(1)控制台程序的工程示例的安装目录为 "C:\Program Files\MultiGen-Paradigm\resources\tutorials\ vegaprime\desktop_tutor"(假定软件安装在 C 盘),开发者可根据中文教程或者英文教程,试运行该程序。

(2)运行:打开安装目录中的其他程序示例,对 Vega Prime 控制台程序进行了解。

(3)下载论坛中的 Websample 程序包,了解程序包中的各类功能示例(包括 Opengl 扩展等)。

(4)下载 Vega Prime MFC 程序模板,使用这个模板即可对基于 MFC 框架的 Vega Prime 程序进行开发。

1.3 Vega Prime 应用开发

1.3.1 Vega Prime 的结构及资源

Vega Prime 安装完成后,很有必要认识一下它的文件结构及资源。与其他成功的商业软件一样,Vega Prime 提供了良好的学习资料。在 C:\Program Files\目录下找到 MultiGen-Paradigm 文件夹,其下的结构如图 1.3.1 所示。其中,docs 目录包含了大量的文档资料,而 resources 目录包含了大量的对象资源、丰富示例以及 1 个桌面教程。

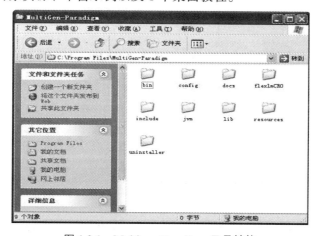

图 1.3.1 Multigen-Paradigm 目录结构

docs 目录下的 vegaprime 目录包含的文档资料主要包括 html 目录和 pdf 目录,如图 1.3.2 所示。

图 1.3.2 docs\vegaprime 目录结构

hmtl 目录包含了大量的超文本文件,打开其中的 index.html 文件,发现这些都是 Vega Prime 主要的混合列表、类层次结构、混合列表成员、命名空间、模块、命名空间成员以及类层次结构。pdf 目录如图 1.3.3 所示,这些是开始使用 Vega Prime 编程的第一手资料。pdf 目录包含 Vega Prime 的升级手册、参考指导、编程指导、选项指导、开始指导、桌面辅导等资料,这些是 Vega Prime 代码化编程的核心资料,如同微软的 MSDN。

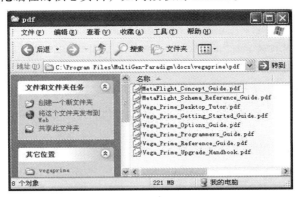

图 1.3.3 pdf 目录结构

Vega Prime 也提供了大量的示例供开发者学习借鉴。在图 1.3.1 所示的目录结构中,resources 目录就包含大量的学习示例,其结构如图 1.3.4 所示。

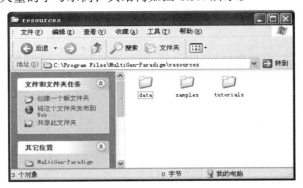

图 1.3.4 resources 目录结构

resources 目录包含了 data、samples 和 turorials 目录。其中的 data 目录包含大量的物体对象,为 Vega Prime 开发者提供了方便;samples 目录包含了大量的示例,可供 Vega Prime 开发者模仿学习;turorials 目录提供了一个辅导教程,一步一步指导 Vega Prime 应用开发者进行学习开发。

samples 目录包含了一个 vegaprime 目录，它提供的示例比较完整地展示了 Vega Prime 能够实现的功能，包括基本配置、声音操作、坐标控制、调试、环境控制、特效操作、输入控制、灯光效果、海洋控制、路径操作、影子控制等，可以看出 Vega Prime 功能之强大，完全能满足虚拟现实开发的各种要求。vegaprime 目录的结构如图 1.3.5 所示。

图 1.3.5　vegaprime 目录结构

在图 1.3.4 所示的 turorials 目录下能够找到一个 tornado 目录，其结构如图 1.3.6 所示。图 1.3.6 中的 application 目录包含了一个完整的 Vega Prime 应用程序的源代码，可以在 Visual C++ 2003 中编译通过（当然，计算机要按照前面的方法安装 Vega Prime 2.01）。

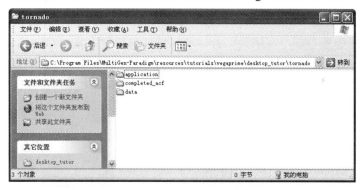

图 1.3.6　tornado 目录结构

完整了解 Vega Prime 的安装目录结构后，可以让开发者完全清楚自己所拥有的资源，这是进行 Vega Prime 程序开发的第一步。

1.3.2　Vega Prime 系统结构及应用组成

1. Vega Prime 系统结构

如图 1.3.7 所示，Vega Prime 的最底层是功能强大的图形库 OpenGL，紧接着的一层是 Presagis 公司自己的跨平台场景渲染引擎 VSG，VSG 上面是 Vega API 和 C++接口，而最上层是一个可视化的配置文件编辑器 LynX Prime,可以做到不写一行程序代码而实现虚拟现实仿真。这样，就把 OpenGL 的强大功能与 Vega Prime 方便、快捷、完美地结合起来，为实时场景仿真等应用程序开发提供了一个综合高效的平台。

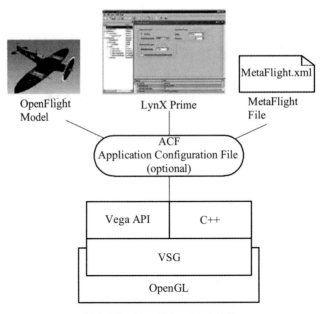

图 1.3.7　Vega Prime 系统结构

2. Vega Prime 应用组成

（1）应用程序。

应用程序控制场景、模型在场景中的移动和场景中其他大量的动态模型。实时应用程序包括汽车驾驶、动态模型的飞行、碰撞检测和特殊效果（如爆炸）。在 Vega Prime 外的开发平台(如 Visual C++ 7.10)创建应用程序，并使用 Vega Prime 的应用库，文件将以.ccp 格式存档，它就包含了 C++可以调用的 Vega Prime 库的功能，在编译完成后就形成了一个可执行的实时 3D 应用文件。

（2）应用配置文件。

应用配置文件(Application Configuration File,简称 ACF)包含了 Vega Prime 应用在初始化和运行时所需的一切信息。通过编译不同的 ACF 文件，Vega Prime 能够生成不同种类的应用。ACF 文件为现在流行的通用文件格式——扩展标识语言（XML）格式。可以使用 Vega Prime 的可视化编辑器 LynX Prime 来开发一个 ACF 文件，也可以使用任何文本编辑器来开发。当然，也可以使用 Vega Prime API 动态地改变应用中的模型运动。对于实时应用来说，ACF 文件不是必要的，只是一个方便快捷的选项。但它可以将改动信息进行译码，记录在.cpp 程序中，这样可以节省大量的时间，加快 Vega Prime 应用开发效率。现在，多数开发者已经习惯了使用 LynX Prime 来开发 ACF 文件，作为 Vega Prime 应用开发的第一个阶段。

（3）模型包。

以前，通常是通过计算机辅助设计系统或几何学来创建单个模型，但这些方法在实时应用时很难进行编码。现在，可以使用 MulitGen Creator 和 ModelBuilder 3D，以 OpenFlight 的格式来创建实时 3D 应用中所有独立的模型。可以使用 Creator Terrain Studio（CTS），以 MetaFlight 格式来生成大面积地形文件，并可以使用这两种格式在 Vega Prime 中增加模型文件。

3. Vega Prime 开发工具

Vega Prime 是一个开发实时三维驱动的工具包，其开发工具主要是 Vega Prime 模块库和 LynX Prime。

LynX Prime 是用来定义 Vega Prime 中的类及其参数的人机交互界面，定义好的内容可以保存到一个配置文件中。Vega Prime 最好与 LynX Prime 一起使用。尽管 Vega Prime 包含了创建一个应用所需的所有 API，但 LynX Prime 简化了开发过程，而且 LynX Prime 允许开发者无须编写代码即可创建一个应用。LynX Prime 是一个编辑器，用于增加不同种类的模型，为模型定义参数。这些参数都存储于应用配置文件（ACF）中的一个模型结构内，如观察者的位置、模型及它们在场景中的位置，在场景中的移动、光线、环境效果，以及目标硬件平台等。ACF 文件包含了 VP 在初始化和运行时所需的信息。还可以在 Active Preview（动画预览）中查看所定义的内容。Active Preview 可以允许使用交互式方法进行配置 ACF，Active Preview 会根据变化信息持续修改 ACF 内容。当出现变化时，Active Preview 将用新的数据更新 Vega Prime 仿真窗口内容。

开发者同样可以使用 C++语言，使用 Vega Prime 模块库编写程序控制场景及其对象的属性和状态，也可以根据应用中的特殊场景修改模型的参数。当一个模型建立完成后，可以修改它的位置、姿态等。Vega Prime 应用程序同样可以将 ACF 加载到一个图像数据流中。当一个 Vega Prime 应用程序编译完成后，它就成为一个可运行的 3D 实时应用。

本书后续的章节主要就是深入剖析如何利用 LynX Prime 和 Vega Prime 模块库开发虚拟现实仿真应用程序。

第 2 章　Vega Prime 主要功能模块

深入理解并掌握 Vega Prime 的主要功能模块是进行 Vega Prime 程序开发最关键的一步。Vega Prime 是一套完整地应用于开发交互式、可视化仿真的软件平台和工具集，它最基本的功能是驱动、控制、管理虚拟场景，并能够方便地实现大量特殊视觉和声音效果。Vega Prime 具有典型的面向对象特点，主要功能模块都以类的形式进行定义存在，同时也存在一定的继承关系，对基本功能进行扩充。具体而言，Vega Prime 的主要功能模块有应用模块(vpApp)、内核模块(vpKernel)、场景模块(vpScene)、对象模块(vpObject)和运动模块(vpMotion)等。各个功能模块为虚拟现实仿真提供了全方位的服务。

【本章重点】

- 六自由度；
- Vega Prime 中的基本概念；
- 应用模块(vpApp)；
- 内核模块(vpKernel)；
- 管道模块(vpPipeline)；
- 窗口模块(vpWindow)；
- 通道模块(vpChannel)；
- 观察者模块(vpObserver)；
- 场景模块(vpScene)；
- 对象模块(vpObject)；
- 运动模块(vpMotion)；
- 转换模块(vpTransform)；
- 碰撞检测模块(vpIsector)。

2.1　Vega Prime 基本概念

2.1.1　Vega Prime 中的六自由度

在现实中，任何一个物体都有一个坐标位置，比如 GPS(Global Position System)。在虚拟现实中，要定位一个物体对象，也同样如此。任何一个物体对象的位置由(X，Y，Z)三坐标决定，姿态由(H，P，H)决定。在 Vega Prime 的坐标系统中，用 X，Y，Z 来表示物体对象的位置，用坐标系中的朝向 H(Heading)、斜度 P(Pitch)和转角度 R(Roll)的值来表示物体对象的姿态。对于 X、Y、Z、H、P、H 六个值，就是 Vega Prime 中的六自由度，具体定义如图 2.1.1 所示，符合笛卡儿坐标系中的右手定则。在这个坐标系中，观察者站在负 Y 轴上，面向正 Y 轴，Vega Prime 的 X、Y、Z 遵循如下定义：

- +X 指向右；
- -X 指向左；
- +Y 指向前；
- -Y 指向后；
- +Z 指向上；
- -Z 指向下。

在这个坐标系中，观察者站在负 Y 轴上，面向正 Y 轴，Vega Prime 的 H、P、R 遵循如下定义：

朝向 Heading 指 Z 轴转向：

- +H 指向左旋转；
- -H 指向右旋转。

斜度 Pitch 指 X 轴转向：

- +P 指向上旋转；
- -P 指向下旋转。

转角度 Roll 指 Y 轴转向：

- +R 指运动转向右边旋转；
- -R 指运动转向左边旋转。

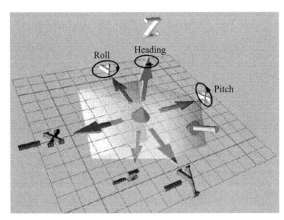

图 2.1.1　坐标系

2.1.2　Vega Prime 中的基本概念

利用 Vega Prime 进行虚拟现实开发，需要首先掌握一些基本概念。Vega Prime 充分利用自身面向对象的特点，这些基本概念主要以核心类的形式出现，如内核（vpKernel）、应用（vpApp）、管道（vpPipeline）、窗口（vpWindow）、场景（vpScene）、通道（vpChannel）、观察者（vpObserver）、对象（vpObject）、自由度（vsDOF）、几何体（vpGeometry）、碰撞（vpIsector）、碰撞服务（vpIsectorServiceInline）、灯光（vpLight）、路径（vpSearchPath）、转换（vpTransform）、运动模式（vpMotion）等。

对一个三维场景的视景感知，都是通过观察者的眼睛来观察。一个典型的 Vega Prime 应用与此类似，也是通过观察者（vpObserver）来视景感知。一个典型的 Vega Prime 应用，包含以下部分：内核（vpKernel），负责控制帧循环并负责管理各种服务；管道（vpPipeline），定义一个逻辑管道，负责窗体与硬件图形渲染管道之间的映射，同时负责剪切与绘制等多线程的控制；窗体（vpWindow），定义基本的窗体机制与消息处理，也提供了一条具体化帧缓冲与输入处理的途径；场景（vpScene），它是一个节点容器，也是所有场景图画的根节点；观察者（vpObserver），它是一个抽象的"摄像机"，用来定位、管理和渲染一系列的通道，让观察者视景感知"现实";通道（vpChannel），定义观察者对现实的观察，观察者观察所有的事物都是通过一定的"通道"来实现的，一个通道可以附加到一个或多个窗体；对象（vpObject），是场景中最基本的渲染库单元，可以是任何的几何体和材质的集合，是观察者控制与操作的主要目标；路径（vpSearchPath），用来控制输入对象的路径；转换（vpTransform），用来定义对象之间、对象与特效之间、对象与自由度之间等的关系；碰撞（vpIsector），用来负责处理对象之间的碰撞问题。

刚开始时，开发者可以粗糙认识这些对象。随着认识的深入，开发者需要深刻认识这些对象，才能方便快捷地进行实时三维虚拟现实开发。

2.2 Vega Prime 主要功能模块

Vega Prime 具有典型的面向对象特点，主要功能模块都以类的形式进行定义存在，同时也存在一定的继承关系，对基本功能进行扩充。Vega Prime 的主要功能模块代表不同的类型层次：Vega Prime ×××表示 Vega Prime 模块中的函数和类，这一层完全可以通过 LynX Prime 来操作完成；vs×××表示 VSG 这一层的函数和类，这一层是 Vega Prime 的核心、心脏；vr×××表示 render 层，也就是硬件接口层，可以具体理解为对 OpenGL 或 Direc×的封装层；vu×××表示一些内存管理、辅助、数学、工具类，它们对 Vega Prime 的其他功能进行了完善。

正确安装 Vega Prime 后，在桌面上双击 LynX Prime 图标，运行 LynX Prime，就可以得到如图 2.2.1 所示的 LynX Prime 图形操作界面，保存 ACF（Application Configuration File）文件后，点击工具栏的黄色三角图标就可以得到如图 2.2.2 所示的虚拟仿真场景。

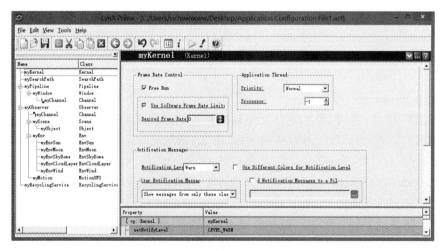

图 2.2.1　LynX Prime 图形操作界面

在图 2.2.1 的左半部分，可以看到 Vega Prime 项目相关的基本功能类和对象，主要有内核（Kernel）、路径搜索对象（SearchPath）、管道（Pipeline）、窗口（Window）、通道（Channel）、观察者（Observer）、场景（Scence）、对象（Object）、环境（Env）、运动模式（MotionUFO）、循环服务对象（RecyclingService）等，它们之间的结构关系如图 2.2.3 所示。一个典型的 Vega

Prime 应用包含内核、路径、管道、窗口、通道、观察者、场景、对象和循环服务模块。其中，管道负责管理硬件渲染与通道之间的映射，通道通过窗口呈现渲染结果；观察者利用通道，观察场景和环境；场景中可以有多个对象；环境中包含日、月、云、风等环境要素。

图 2.2.2　虚拟仿真场景

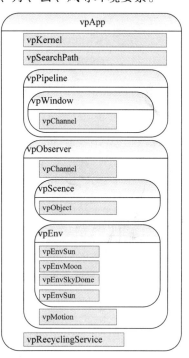

图 2.2.3　Vega Prime 的一个应用构成

2.2.1　应用 vpApp

应用(vpApp)定义了一个典型的 Vega Prime 应用框架，负责控制帧循环和管理各种服务。应用(vpApp)的成员函数采用内联函数方式，绝大多数函数都是对内核(vpKernel)模块的功能函数进行了包装，频繁用于管理 Vega Prime 应用，所以在 LynX Prime 中没有直接见到应用(vpApp)，在编写代码的时候也很少直接使用应用(vpApp)对象，很多时候都是通过内核(vpKernel)来实现其功能。具体内容见图 1.3.2 中 docs\vegaprime 目录结构下的 html 目录中的 classvpApp.html 文件和 classvpApp-members.html 文件。

应用(vpApp)提供了很多键盘输入控制功能，这些功能在控制台仿真窗口下可以直接使用。Esc 键：退出应用；s/S 键：显示所有通道的统计数据；Backspace 键：重置所有的观察者状态向量；Enter 键：抓取所有的观察者状态向量；c 键：使观察者位于场景中心；f 键：启用或停止所有通道的大雾效果；l 键：启用或停止所有通道的光照效果；p 键：打印所有观察者的绝对位置；t 键：启用或停止所有通道的纹理效果；T 键：启用或停止所有通道的透明效果；x 键：启用或停止所有观察者的位置策略。

在 LynX Prime 中看不到应用(vpApp)，但任何一个 ACF 文件已经是在一个应用管理之下，才能完成渲染仿真。在 LynX Prime 下，内核(vpKernel)完成了绝大多数应用(vpApp)的功能，可以通过表 2.2.1 所示的函数对应用(vpApp)的功能进行操作调用。

表 2.2.1　应用（vpApp）功能函数

函数原型	功能描述
define(const char *configFile)	通过文件定义
configure (void)	配置应用
isConfigured (void)	是否完成配置
unconfigure (void)	取消配置
run (void)	运行应用
beginFrame()	开始帧循环
breakFrameLoop()	暂停帧循环
endFrame()	结束帧循环
update(void)	更新数据
getFrameNumber()	获取帧号
getKernel ()	获取内核
centerObserver (vpObserver *observer)	中心化观察者

2.2.2　内核 vpKernel

内核(vpKernel)继承于服务管理(vsServiceMgr)，负责控制帧循环和管理各种服务。同时，内核(vpKernel)创建一个遍历更新(vsTraversalUpdate)实例并控制它的执行。这个遍历一旦被内核的更新方法(vpKernel::update)触发，就会访问所有用内核注册登记的场景对象，自然就可以更新场景中所有对象的参数。内核(vpKernel)提供了明确的 API 来操作它管理的场景(vpScene)。当观察者定位自己的通道时，内核(vpKernel)也会通过消息机制提供控制。内核(vpKernel)还要负责把相应的配置信息和帧循环控制消息发送给各个模块(vpModules)。另外，内核(vpKernel)还要负责发布帧号和模拟仿真时间。具体内容见图 1.3.2 中 docs\vegaprime 目录结构下的 html 目录中的 classvpKernel.html 文件和 classvpKernel-members.html 文件。

可以通过 LynX Prime 对内核(vpKernel)的功能进行可视化编辑使用，也可以通过表 2.2.2 所示的函数对内核(vpKernel)的功能进行操作调用。

表 2.2.2　内核（vpKernel）功能函数

函数原型	功能描述
instance()	创建内核实例
define(const char *configFile)	通过文件定义
configure()	配置
beginFrame()	开始帧循环
breakFrameLoop()	暂停帧循环
endFrame()	结束帧循环
getActualFrameRate()	获取确切的帧率
getBeginFrameTime()	获取帧开始时间
getDesiredFrameRate()	获取期望的帧率

续表

函数原型	功能描述
getFrameNumber()	获取帧号
setDesiredFrameRate(double rate)	设置期望的帧率
setFrameRateLimiter(FrameRateLimiter *ctrl)	设置帧率限制
setFrameRateLimiterEnable(bool bEnable)	使帧率限制设置有效
setPriority(vuThread::Priority pri)	设置线程优先级
setProcessor(int cpu)	设置处理器数目
update(void)	更新数据

2.2.3 管道 vpPipeline

管道(vpPipeline)定义了一条逻辑图形渲染管道，负责管理窗口（vpWindows）与硬件图形渲染管道之间的映射，同时提供了对绘制线程和剪切线程等多线程进行控制管理的机制。管道提供了常规的设置方法，如设置管道名称、设置多线程方式、设置管道 Id、设置剪切绘制线程数目、设置剪切绘制线程优先级、设置剪切绘制线程使用的处理器数目等。具体内容见图 1.3.2 中 docs\vegaprime 目录结构下的 html 目录中的 classvpPipeline.html 文件和 classvpPipelinemembers.html 文件。

可以通过 LynX Prime 对管道(vpPipeline)的功能进行可视化编辑使用，也可以采用其默认值，也可以通过表 2.2.3 所示的窗口功能函数对管道(vpPipeline)的功能进行操作调用。

表 2.2.3 管道（vpPipeline）功能函数

函数原型	功能描述
setName(const char *name, bool hash=true)	设置管道名称
setMultiThread(MultiThread mode)	设置管道多线程模式
setId(int pipeId)	设置管道 Id
setNumCullThreads(int n)	设置管道剪切线程数目
setCullThreadPriority(vuThread::Priority pri)	设置管道剪切线程优先级
setCullThreadProcessor(int cpu)	设置管道剪切线程处理器数目
setDrawThreadPriority(vuThread::Priority pri)	设置管道绘制线程优先级
setDrawThreadProcessor(int cpu)	设置管道绘制线程处理器数目
setDesiredPostDrawTime(double dTime)	设置管道期望的后绘制时间

2.2.4 窗口 vpWindow

窗口(vpWindow)定义了基本的窗口及其消息处理机制，提供了一条具体配置帧缓冲区和输入处理的途径。通常，窗口的创建应该在内核实例配置(vpKernel::instance()->configure())之前，它可以为窗口设置默认的键盘函数、鼠标函数、改变形状函数和关闭函数。否则，如果在内核实例配置之后创建窗口，用户需要自己明确的为窗口设置键盘函数、鼠标函数、改变形状函数以及关闭函数。这样，Vega Prime 为用户提供了一种灵活的机制：既可以采用默认输入控制，也可以个性化的自定义输入控制。同时，窗口也提供了常规的设置窗体的方法，如设置尺寸、设置全屏、设置边界等。具体内容见图 1.3.2 中 docs\vegaprime 目录结构下的 html 目录中的 classvpWindow.html 文件和 classvpWindow-members.html 文件。

可以通过 LynX Prime 对窗口(vpWindow)的功能进行可视化编辑使用，也可以通过表 2.2.4 所示的窗口功能函数对窗口(vpWindow)的功能进行操作调用。

表 2.2.4　窗口（vpWindow）功能函数

函数原型	功能描述
addChannel(vpChannel *chan, int loc=-1)	添加通道
getChannel(int loc)	获取通道
setFullScreenEnable(bool bFullScreen)	设置窗口全屏
setInputEnable(bool input)	设置允许输入
setKeyboardFunc(KeyboardFunc userFunc, void *data=NULL)	设置键盘处理函数
setMouseFunc(MouseFunc userFunc, void *data=NULL)	设置鼠标处理函数
setOrigin(int left, int bottom)	设置窗口位置
setParent(Window win)	设置窗口的父载体
setReshapeFunc(ReshapeFunc userFunc, void *data=NULL)	设置形状变化处理函数
setSwapInterval(int interval)	设置交换间隔
updateStress(double frameRate)	更新帧速率
KEY_1 ,KEY_a ,KEY_RIGHT	键盘键标示
MESSAGE_KEY_DOWN　　MESSAGE_MOUSE_LEFT_DOWN	鼠标键标示
getOrigin(int *left, int *bottom)	获取窗口原始位置

2.2.5　通道 vpChannel

观察者用双眼观察世界，形成了一个以眼睛为锥尖点、向四周扩展的椎体。通道(vpChannel)定义了观察世界的视点，控制视点内的绘制区域，允许通过设置具体参数来控制从视点出发形成的可视化椎体和剪切面。一个通道应该附加于一个窗体，而且只能附加于一个窗体。具体内容见图 1.3.2 中 docs\vegaprime 目录结构下的 html 目录中的 classvpChannel.html 文件和 classvpChannel-members.html 文件。

可以通过 LynX Prime 对通道(vpChannel)的功能进行可视化编辑使用，也可以通过表 2.2.5 所示的通道功能函数对通道(vpChannel)的功能进行配置使用。

表 2.2.5　通道（vpChannel）功能函数

函数原型	功能描述
setCullFunc(CullFunc f)	设置剪切函数
setCullMask(uint mask)	设置剪切掩码
setCullThreadPriority(vuThread::Priority)	设置剪切线程优先级
setCullThreadProcessor(int cpu)	设置剪切线程 CPU 数目
getViewMatrix()	获取视点矩阵
setDrawArea(double l, double r, double b, double t)	设置绘制区域
setDrawBuffer(DrawBuffer buffer)	设置绘画缓冲

续表

函数原型	功能描述
setDrawFunc(DrawFunc f)	设置绘画函数
setLODFilter(vsLODFilterBase *)	设置 LOD 过滤器
setLODTransitionRangeScale(double scale)	设置 LOD 转换范围
setLODVisibilityRangeScale(double scale)	设置 LOD 可视范围
setLogicalPipelineId(int logicalPipelineId)	设置逻辑管道
setNearFar(float nr, float fr)	设置远近
setFrustumAsymmetric(float l, float r, float b, float t)	设置不对称椎体
setLightPointComputeAsynchronousEnable(bool enable)	设置允许不对称光点
setLightPointThreadPriority(vuThread::Priority pri)	设置光点线程优先级
setLightPointThreadProcessor(int cpu)	设置光点线程 CPU 数目

2.2.6 场景 vpScene

场景(vpScene)是所有节点的容器，是一个场景图形的根节点，也是剪切遍历和更新遍历的起点。更新遍历需要内核更新方法(vpKernel::update())来触发。从观察者(vpObserver)角度出发，通过某个通道(vpChannel)，在某个窗口(vpWindow)中就可以见到某个场景(vpScene)，其他物体对象(vpObject)都是位于某个场景中呈现在观察者眼中。最常见的操作就是把其他物体作为"孩子"添加到场景中。场景(vpScene)具体内容见图 1.3.2 中 docs\vegaprime 目录结构下的 html 目录中的 classvpScene.html 文件和 classvpScene-members.html 文件。

可以通过 LynX Prime 对场景(vpScene)的功能进行可视化编辑使用，也可以通过表 2.2.6 所示的函数对场景(vpScene)的功能进行配置使用。

表 2.2.6 场景功能函数

函数原型	功能描述
addChild(vsNode *child, int loc=-1)	添加子节点
begin_child() end_child()	例举子节点
begin_parent() end_parent()	例举父节点
get_iterator_child(int loc)	获取子节点
get_iterator_parent(int loc)	获取父节点
getChild(int loc)	获取子节点
getNumChildren()	获取子节点个数
insert_child(vsNode *child, const_iterator_child it, uint mask=vsNode::DIRTY_UP_ALL)	插入子节点
push_back_child(vsNode *child, uint mask=vsNode::DIRTY_UP_ALL)	添加子节点到子节点列表后面
removeChild(vsNode *child) vpScene	移除子节点
replace_child(vsNode *child, const_iterator_child it, bool reparent, uint mask=vsNode::DIRTY_UP_ALL)	替换子节点
addSubscriber(Event event, Subscriber *subscriber)	添加订阅者

2.2.7 观察者 vpObserver

观察者(vpObserver)是一个具有位置特征的抽象"摄像机",用于定位、管理、渲染一系列通道。同时,观察者(vpObserver)负责发布参数对象"视点"。若果在一个应用中存在多个观察者(vpObserver)对象实例,最近一个被更新的观察者将负责发布参数对象"视点"。在场景中,呈现在面前的对象都是从观察者对象角度出发的,内核也是通过附加在窗口上的通道对场景中的对象进行绘制渲染的。不同观察者通过不同的通道,会见到场景中的不同视图,如左视图、右视图、俯视图等。观察者(vpObserver)具体内容见图 1.3.2 中 docs\vegaprime 目录结构下的 html 目录中的 classvpObserver.html 文件和 classvpObserver-members.html 文件。

可以通过 LynX Prime 对观察者(vpObserver)的功能进行可视化编辑使用,也可以通过表 2.2.7 所示的函数对观察者(vpObserver)的功能进行配置调用。

表 2.2.7 观察者(vpObserver)功能函数

函数原型	功能描述
addAttachment(Attachment *attr)	添加附属物
addChannel(vpChannel *chan, int loc=-1)	添加通道
begin_attachment() end_attachment()	例举附属物
begin_channel() end_channel()	例举通道
erase_attachment(Attachment *attr)	删除附属物
erase_channel(const_iterator_channel it)	删除通道
getTranslate(double *x, double *y, double *z)	获取观察者位置
getRotate(double *h, double *p, double *r)	获取观察者姿态
setLookAt(const vpPositionable *target)	设置目标观察物
setLookFrom(const vpPositionable *pos)	设置观察点
setPosition(double p0, double p1, double p2, const vpCoordConverter *from=NULL)	设置观察者位置
setRotate(double h, double p, double r, bool incr=false)	设置观察者姿态
setScene(const vpScene *scene)	设这场景
setStrategy(Strategy *strategy)	设置策略

2.2.8 对象 vpObject

对象(vpObject)是渲染的最基本数据库单元,可以是几何体与材质的任何集合。一个应用的整个渲染对象库可以是作为单个对象整体加载,也可以把每个模型作为分开的对象分别加载。这种加载选择,完全决定于应用程序如何使用这些对象,以及这些对象如何构造。要特别注意对象的引用计数,当对象被附加到场景中时自动加一,当对象被从场景中移除时自动减一;当引用计数为零时,对象将被删除。因此,如果一个对象在后来需要重新附加场景中,记住从场景中移除对象之前做对象应用。对象(vpObject)是虚拟现实中操作最频繁的目标,需要对此有深刻的认识。对象(vpObject)的具体内容见图 1.3.2 中 docs\vegaprime 目录结构下的 html 目录中的 classvpObject.html 文件和 classvpObject-members.html 文件。

可以通过 LynX Prime 对对象(vpObject)的功能进行可视化编辑使用，也可以通过表 2.2.8 所示的函数对对象(vpObject)的功能进行配置使用。

表 2.2.8　对象（vpObject）功能函数

函数原型	功能描述
addAttribute(Attribute *attr)	添加属性
addChild(vsNode *child, int loc=-1, uint mask=vsNode::DIRTY_UP_ALL)	添加子节点
begin_attribute()　　end_attribute()	例举属性
begin_child()　　end_child()	例举子节点
begin_geometry()　　end_geometry()	例举几何体
begin_named()　　end_named()	例举名字
begin_parent()　　end_parent()	例举父节点
begin_state()　　end_state()	例举状态
begin_texture()　　end_texture()	例举材质
copyFromSource()	复制
find_named(const char *name, int occurrence=0)	依据名字查找
setScale(double x, double y, double z, bool incr=false)	设置缩放比例
setTranslate(double x, double y, double z, bool incr=false)	设置位置
setRotate(double h, double p, double r, bool incr=false)	设置姿态

2.2.9　自由度 vsDOF

自由度(vsDOF)定义了一个节点，它为场景中的图形提供一种转换矩阵，这种转换矩阵典型应用于关节模型中。自由度(vsDOF)确切地封装了两个矩阵，第一个矩阵是相对于父节点的本地单元矩阵，第二个是相对于本地矩阵的转换矩阵。自由度(vsDOF)节点允许相对于场景里面本地坐标中任意一点进行位置与姿态转换。该节点提供了 API 来定义本地坐标和相对于本地坐标的转换，这个转换是自由度的集合与加载在这些自由度上的最大/最小值限制的集合。这些自由度包括：X、Y、Z 上的位移，H、P、R 上的旋转，沿 X、Y、Z 上的缩放因子。自由度(vsDOF)的具体内容见图 1.3.2 中 docs\vegaprime 目录结构下的 html 目录中的 classvsDOF.html 文件和 classvsDOF-members.html 文件。自由度函数如表 2.2.9 所示。

表 2.2.9　自由度（vsDOF）功能函数

函数原型	功能描述
begin_child()　　end_child()	例举子节点
begin_parent()　　end_parent()	例举父节点
push_back_child(vsNode *child, uint mask=vsNode::DIRTY_UP_ALL)	添加子节点到子节点列表后面
COMPONENT_ROTATE_H	朝向旋转限制
EVENT_ISECT	碰撞事件
update(vsTraversalUpdate *trav)	更新遍历
setRotateH(double h, bool incr=false)	设置朝向旋转
setRotateP(double p, bool incr=false)	设置倾斜度旋转

续表

函数原型	功能描述
setRotateR(double r, bool incr=false)	设置扭曲度旋转
setScaleX(double x, bool incr=false)	设置 X 轴上的缩放因子
setScaleY(double y, bool incr=false)	设置 Y 轴上的缩放因子
setScaleZ(double z, bool incr=false)	设置 Z 轴上的缩放因子
setTranslateX(double x, bool incr=false)	设置 X 轴上的位移
setTranslateY(double y, bool incr=false)	设置 Y 轴上的位移
setTranslateZ(double z, bool incr=false)	设置 Z 轴上的位移
setTranslate(double x, double y, double z, bool incr=false)	同时设置 X、Y、Z 轴上的位移

2.2.10 转换 vpTransform

转换(vpTransform)定义了一个节点,它为场景中的图形提供转换。转换(vpTransform)封装了一个矩阵,并且提供了操作这个矩阵的方法。转换(vpTransform)重载了节点的所有遍历函数,同时也相应地修改了遍历矩阵堆栈。转换(vpTransform)除了继承于 vsTransform 外,也继承于 vpPositionable,这样可以使转换(vpTransform)能够定位一个独立的坐标系统,且这个这个坐标系统不同于父场景中的坐标系统。转换(vpTransform)的具体内容见图 1.3.2 中 docs\vegaprime 目录结构下的 html 目录中的 classvpTransform.html 文件和 classvpTransform-members.html 文件。

可以通过 LynX Prime 对转换(vpTransform)的功能进行可视化编辑使用,也可以通过表 2.2.10 所示的函数对转换(vpTransform)的功能进行配置使用。

表 2.2.10 转换(vpTransform)功能函数

函数原型	功能描述
begin_child() end_child()	例举子节点
begin_parent() end_parent()	例举父节点
push_back_child(vsNode *child, uint mask=vsNode::DIRTY_UP_ALL)	添加子节点到子节点列表后面
update(vsTraversalUpdate *trav)	更新遍历
setRotateH(double h, bool incr=false)	设置朝向旋转
setRotateP(double p, bool incr=false)	设置倾斜度旋转
setRotateR(double r, bool incr=false)	设置扭曲度旋转
setScaleX(double x, bool incr=false)	设置 X 轴上的缩放因子
setScaleY(double y, bool incr=false)	设置 Y 轴上的缩放因子
setScaleZ(double z, bool incr=false)	设置 Z 轴上的缩放因子
setTranslateX(double x, bool incr=false)	设置 X 轴上的位移
setTranslateY(double y, bool incr=false)	设置 Y 轴上的位移
setTranslateZ(double z, bool incr=false)	设置 Z 轴上的位移
setTranslate(double x, double y, double z, bool incr=false)	同时设置 X、Y、Z 轴上的位移

2.2.11 运动模式 vpMotion

所有对象在场景中的运动都有一种方式，运动模式(vpMotion)就是定义了一种运动方式，这种方式本质上是一种位置策略，这种策略通过使用标准的输入设备，能够很好地执行定义好的位置驱动方法。作为其他运动模式的基础类，运动模式(vpMotion)是一个抽象的基础类，能够很好地为其他运动模式服务。其他运动模式包括驾驶模式(vpMotionDrive)、飞行模式(vpMotionFly)、游戏模式(vpMotionGame)、旋转模式(vpMotionSpin)、链条模式(vpMotionTether)、固定链条模式(vpMotionTetherFixed)、链条跟随模式(vpMotionTetherFollow)、链接旋转模式(vpMotionTetherSpin)、不明飞行物模式(vpMotionUFO)、行走模式(vpMotionWalk)等。所有的运动模式都拥有一个输入设备，通过这个输入设备可以驱动这种运动。输入设备可以是鼠标、键盘或者其他综合的输入设备。运动模式(vpMotion)的具体内容见图 1.3.2 中 docs\vegaprime 目录结构下的 html 目录中的 classvpMotion.html 文件和 classvpMotion-members.html 文件。

可以通过 LynX Prime 对运动模式(vpMotion)的功能进行可视化编辑使用，也可以通过表 2.2.11 所示的函数对运动模式(vpMotion)的功能进行设置使用。

表 2.2.11 模式模式（vpMotion）功能函数

函数原型	功能描述
getInput()	获取输入
getKeyboard()	获取键盘输入
getMouse()	获取鼠标
getName()	获取名字
getNextStrategy()	获取下一个策略
setInput(vpInput *input)	设置输入
setKeyboard(vpInputKeyboard *keyboard)	设置输入键盘
setMouse(vpInputMouse *mouse)	设置鼠标
setName(const char *name, bool hash=true)	设置名字
setNextStrategy(Strategy *nextStrategy)	设置下一个策略
getUserDataList(void) const vuBase [virtual]	获取用户数据列表

2.2.12 碰撞 vpIsector

检测场景中物体间的相交线段，是现在许多可视化仿真中一种必不可少的能力。如飞行仿真中，高度就是计算地面与飞机之间的垂直线段距离。碰撞(vpIsector)负责维护和管理用于碰撞检测的相交线段，并且提供了一个数据结构和一组方法来查询碰撞结果。为了检测哪一条线段被撞到，对场景中的图形不得不进行逐个节点的遍历，不同的节点要求不同的碰撞检测程序。碰撞(vpIsector)提供了 API 来配置和查询碰撞，另外，它是一个抽象基础类，可以设置检测或不检测，设置检测目标节点，定位碰撞，查询碰撞结果。但是，它不提供方法来配置碰撞线段，只提供访问内部指针的方法。具体内容见图 1.3.2 中 docs\vegaprime 目录结构下的 html 目录中的 classvpIsector.html 文件和 classvpIsector-members.html 文件。

可以通过 LynX Prime 对碰撞(vpIsector)的功能进行可视化编辑使用，也可以通过表 2.2.12 所示的函数对碰撞(vpIsector)的功能进行配置使用。

表 2.2.12　碰撞（vpIsector）功能函数

函数原型	功能描述
getEnable()	获取碰撞检测有效性
getHit()	获取碰撞
getHitGeometry()	获取碰撞几何体
getHitMatrix(vuMatrix< double > *matrix)	获取碰撞矩阵
getHitNode(const char *name=NULL, int occurrence=0)	获取碰撞节点
getHitNormal(vuVec3< float > *normal)	获取碰撞法线
getHitObject()	获取碰撞物体
getHitPoint(vuVec3< double > *point)	获取碰撞点
getTarget()	获取碰撞目标
setStrategy(Strategy *strategy)	设置碰撞策略
setEnable(bool option)	使碰撞检测有效
setIsectMask(unsigned int imask)	设置碰撞掩码
setTarget(const vsNode *target)	设置碰撞目标
setTranslate(double x, double y, double z, bool incr=false)	设置位置

第3章　LynX Prime 的使用

　　Vega Prime 包含所有创建虚拟现实应用程序的 API，另外还专门设计了图形操作工具 LynX Prime。LynX Prime 能够简化应用程序的开发过程，不用编写代码，只需通过可视化编辑界面进行配置，就能够开发出虚拟现实应用程序。LynX Prime 就是这样一个编辑器，通过它，开发者可以添加类的实例对象，并为这些实例对象定义参数，如观察者的位置、场景中的物体、场景中物体的移动、光照、环境效果、目标硬件平台等。这些参数都储存在一个应用程序的实例框架中，形成一个应用程序配置文件(Application Configuration File, ACF)。这个应用配置文件包含一个 Vega Prime 应用程序初始化和运行时所需要的信息。同时，LynX Prime 还提供了一个预览功能，让开发者对虚拟现实应用程序的开发达到所见即所得的效果。

【本章重点】

- LynX Prime 界面组成；
- LynX Prime 创建场景；
- LynX Prime 操作对象；
- LynX Prime 设置观察者；
- LynX Prime 建立转换；
- LynX Prime 创建运动模式；
- LynX Prime 应用碰撞检测；
- LynX Prime 建立特效；
- LynX Prime 设置窗口；
- LynX Prime 建立环境；
- LynX Prime 设置通道。

3.1　LynX Prime 的界面组成

3.1.1　LynX Prime 的启动与退出

1. LynX Prime 的启动

Windows 用户可以点击桌面上 LynX Prime 的快捷启动方式启动程序，这个快捷键应在安装程序时自动创建完成。当然，也可以按以下路径启动：开始→ 程序→ MultiGen-Paradigm→ Vega Prime 顺序点击，然后从第二级目录中启动 LynX Prime。

2. LynX Prime 的退出

如需退出 LynX Prime，可以选择 LynX Prime 的菜单：文件→退出，也可以同时按 Ctrl 和 Q 键，还可以单击窗口上方的关闭键。

3. LynX Prime 的保存

建议用户在 LynX Prime 操作过程中要经常进行保存。这样在出现突然停电或系统故障时，不会丢失数据。在文件菜单中提供了所有标准文件工具，用户必须学会使用这些工具。保存(Save)按钮在菜单栏中，建议用户在修改原文件前进行保存，快捷方式为 Ctrl+S 键。另外，点击另存为(Save As)按钮将对文件进行存档，默认文件扩展名为.acf。

3.1.2 LynX Prime 的界面构成

LynX Prime 的用户界面如图 3.1.1 所示。它包括 4 个部分：实例树形显示区（Instance Tree View）、用户操作区（GUI View）、应用程序区（API View）和工具条（Toolbar）与菜单区（Menus）。

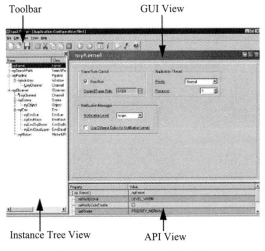

图 3.1.1　LynX Prime 用户界面

所有这些区域将显示同一选定对象的信息，但是这些信息是以不同格式进行安排的。用户可以在一个或多个工作区进行操作来定义 ACF。

1. 用户操作区 GUI View

用户操作区在用户界面中显示 ACF 模型及相关的参数，操作起来十分方便。可以从一个下拉菜单中选择参数，也可在空格处输入参数。

当打开 Lynx Prime，第一个显示的用户操作窗口叫作 myKernel，这是 Vega Prime 中 Kernel 类型中的一个实例，Kernel 即是应用程序的内核控制。

2. 实例树显示区 Instance Tree View

实例树形显示区显示当前正在操作的 ACF 文件和文件中包含的所有模型。实例树用一个等级结构显示了模型间的关系及与它们的上级模型和下级模型的关系。通过实例树可以直观地了解到应用中的模型之间的联系。

如果选定等级中一个模型，在所有的操作区中将同时显示这个模型的有关信息。如果一个模型在文件中的不同的地方被使用，在这个模型的文件名旁就标有蓝色的箭头。向下的箭头表示文件第一次被使用。向上的箭头表示这个模型的其他应用。

如图 3.1.1 所示，MyPipeline 的子目录 MyWindow 调用 MyChannel，这是 MyChannel 第一次被调用，所以在它旁边标有一个向下的箭头。MyChannel 后又被 myObserver 调用，所以这时在它旁边标有一个向上的箭头。

当选定一个图标时，图标的属性和当前的参数就会显示在用户操作区和 API 区。

3. 应用程序区 API View

应用程序区 API 显示选定的模型的所有可能的变量。在这里可以定义模型的值，就像在用户操作区一样。但是在 API 中，可以直接给变量赋值。当更改应用中的参数时，可参考应用程序区 API 中要使用的变量的默认值。

4. 工具条（Toolbar）与菜单区（Menus）

LynX Prime 工具条包括所有操作模型及属性的快捷按钮，如图 3.1.2 所示。这些功能同样可以在 LynX Prime 的菜单中找到。

图 3.1.2　LynX　Prime 工具条与菜单

（1）New File。
创建一个新文件——含有默认类型值的 ACF 文件。

（2）Open File。
打开文件浏览器，就可以选择一个 ACF 文件加载到 LynX Prime 中。所选定的 ACF 文件将替换目前正在显示的 ACF 文件。如果目前显示文件已经修改过并且还没有存盘，LynX Prime 将会在打开文件浏览器之前提醒保存所做修改。

（3）Save File。
保存当前的 ACF，如果这个文件之前没有保存，可以在显示的对话框中输入文件名和保存目录。

（4）Create Instance。
将显示一个对话框，其中包含可以增加的模型类型列表。也可以根据名称或范围在对话框中显示或分类这些项目。

（5）Cut Instance。
将当前选定的模型复制到剪切板，并从当前的 ACF 文件中移走。

（6） Copy Instance。

复制当前选定的模型到剪切板，但模型仍留在 ACF 文件中。

（7） Paste Instance。

将剪切板中所存的模型加到当前 ACF 文件中，只有剪切板中含有模型才能执行此操作。

（8） Delete Instance。

从当前的 ACF 中移走选定的模型，将会出现一个对话框提示进行删除或取消操作，删除操作不会在剪切板中留下模型的复制对象。

（9） Backward。

显示前一步。

（10） Forward。

显示下一个。只有操作过退后键才能操作这一步。

（11） Views。

在 LynX Prime 中 4 个不同的操作区之间转换。第一种：用户操作区在上面，同时 API 区在下面。第二种：API 区在上面，同时用户操作区在下面。第三种：只显示用户操作区。第四种：只显示 API 区。

（12） Active Preview。

按照 LynX Prime 中对 ACF 文件的设置，运行显示一个 Vega Prime 应用，这个应用按照 LynX Prime 设置的各个对象的具体参数，渲染出虚拟场景，使虚拟现实的开发也可以达到所见即所得的效果。

（13） i ACF Information。

弹出一个对话框，显示当前 ACF 文件的信息。这个对话框同样可以用于执行其他 ACF 文件的操作和生成一个在当前 ACF 文件中运行的应用。

（14） LP Documentation。

发布在线帮助阅览器，可以访问到每一个 API 和构成 API 各种方法。在线帮助提供 Vega PrimeAPI 的最新信息。

3.2 LynX Prime 的使用

本节通过开发一个简单的实时三维虚拟现实应用程序，让读者充分认识和掌握 LynX Prime，并能使用 LynX Prime 来开发 Vega Prime 应用程序。为了管理方便，首先在 C 盘根目录下建立一个文件夹 VegaPrimePractice，把所有完成的 ACF 文件保存在这个文件夹中。注意，在 LynX Prime 中所做的一切都会保存在 ACF 文件中，在 Vega Prime 应用程序的初始化以及运行时，都有可能使用 ACF 文件所保存的数据，为了更便捷有效的学习，请在最初学习过程中保持一致，特别是操作中关于名字的命名。

对于本节内容，读者可以反复练习，以求达到熟练的程度，尤其是对 Vega Prime 应用的各个功能对象、各种参数进行多种设置，反复运行查看。如果仅仅使用 Vega Prime 进行简单展示，精通这一节内容就足够了。如果需要深入掌握 Vega Prime 的开发，熟练掌握本节内容，既可以增加学习成就感,又可以为后继的学习打下良好的基础。通过 LynX Prime 的图形界面可以方便快捷地配置绝大多数固定对象的参数,而且还可以即时预览,满意后可以导出转换成 CPP 代码,能大大加快后继功能的开发。

3.2.1 创建场景 vpScene

地形创建是 Vega Prime 应用开发的基础,在 LynX Prime 中为场景类赋一个地形 OpenFlight 文件值,开发者就可以在 Active Preview 实时应用中浏览场景,并在其中运动。现在开始建造农场。

1. 添加场景

Step 1 在 C 盘中创建一个文件夹并命名为"VegaPrimePractice",把所有完成的 ACF 文件保存到这个目录中。

Step 2 单击"开始"→"程序"→"MultiGen-Paradigm"→"Vega Prime"→"LynX Prime",打开 LynXPrime;或者直接运行桌面快捷方式。

Step 3 单击 LynX Prime 菜单栏中的文件,选择另存为(Save as),弹出对话框,将对话框中的目录改为建立好的 C:\VegaPrimePractice,将 ACF 文件重命名为 VP3_2_1.acf。

Step 4 单击保存,将 Scene.acf 文件存于 C:\VegaPrimePractice 中,关闭对话框,LynX Prime 窗口标题条将显示新的 ACF 文件名。

Step 5 如图 3.2.1 所示,单击用户操作区顶部的实例键 ,选择 myObject,也可以在实例树中单击 myScene,在它下面选定 myObject。

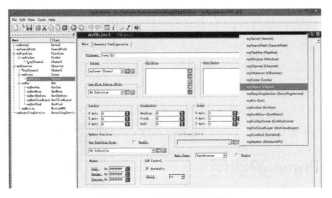

图 3.2.1 建立场景对象

myScene 实例是 myObject 的上一级实例,可单击实例键,从它的选择列表中选择 myObject。

注意,在文件名区中有 town.flt 文件,它是在默认 ACF 模板中 myScende 实例的默认子物体,必须用新的 OpenFlight 文件来替换它。

Step 6 点击 Filename 区旁的浏览键 ,在 C:\Program Files\Presagis\resources\tutorials\vegaprime\desktop_tutor\tornado\data\land 目录下选定 Prime_Junction.flt 文件。在对话框中点击 Open,Prime_Junction 将替换 town.flt 文件。

Step 7 如图 3.2.2 所示,在 LynX Prime 窗口下面的 API 区中,选择 myObject 的 value,

将其改为 terrain，在用户操作区和实例树区中，将 myObject 改为 terrain。

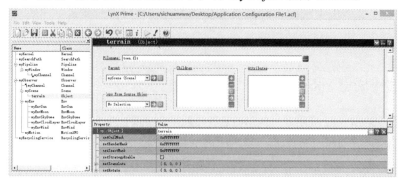

图 3.2.2 修改场景值

Step 8 在实例树区，点击 myScene，注意在所有的 3 个显示区内，都会显示 terrain 是 myScene 实例的子物体。

Step 9 保存。

2. 预览应用

利用 Active Preview，可以实时观测开发的应用。Active Preview 运行时，任何 ACF 参数的变化也会立即显示。选择 Tools 下的 Active Preview，弹出命令提示窗口，然后打开 Active Preview，开始运行 Scene.acf 文件，如图 3.2.3 所示，眼前是农场中的一块空地。如果 Active Preview 充满了整个屏幕，可以缩小它的尺寸。首先，按 Esc 键关闭 Active Preview 窗口，在实例树区内，点击 myWindow 实例打开 myWindow 用户操作区，这个实例控制 Active Preview 窗口，例如，将窗口的长和宽的值改为 400 和 300，这样便形成一个较小窗口。

另外，可以用鼠标和键盘来控制场景中的方向，若要在场景中向前进，持续按住鼠标左键。若要向后退，持续按住鼠标右键。若要停止（刹车），按下键盘中的 X 键。若要在场景中向上，将鼠标拉向自己。若要在场景中向下，将鼠标推离自己。鼠标固定在窗口的中央，可保持盘旋模式。总结如下：

- 在场景中向前进，按住鼠标左键。
- 在场景中向后退，按住鼠标右键。
- 若要停止（刹车），按下键盘中的 X 键。
- 将鼠标拉向自己，可在场景中向上。
- 将鼠标推离自己，可在场景中向下。
- 鼠标固定在窗口的中央，可保持盘旋模式。

图 3.2.3 预览场景

实际上，现在把鼠标放到 Active Preview 窗口上，稍不留神，就会出现"天旋地转"的感觉，现在还无法自如地控制自己的视野，最好的办法就是熟练使用鼠标控制，或者把鼠标放到窗口外，仔细观察。后续章节将做进一步开发，就能够做到自如地控制了。

3.2.2 操作对象 vpObject

继续建造农场，为其修建房屋，添加交通工具，修建粮仓。在场景中添加物体时，需要特别注意位置问题。否则，添加了物体，却无处寻找物体。在默认情况下，观察者都是位于 Y 轴的负轴上，正前面为 Y 轴正轴，左边为 X 轴负轴，右边为 X 轴正轴，垂直于地面向下为 Z 轴负轴，垂直于地面向上为 Z 轴正轴。

1. 添加房屋

场景中添加对象是很常见的操作，在本应用中，将首先添加一个房屋。这个房屋模型安装在 Vega Prime 目录中，具体路径为 c:\Program Files\Multigen-Paradigm\resources\tutorials\ vegaprime\ desktop_tutor\tornado\data\farmhouse 目录下选择 farmhouse.flt 文件。

Step 1 打开前面建立的 VP3_2_1.acf 文件，单击 LynX Prime 的菜单栏中的文件，选择另存为（Save as），弹出对话框，将对话框中的目录改为建立好的 C:\VegaPrimePractice，将 ACF 文件重命名为 VP3_2_2.acf。

Step 2 在实例树区，点击 myScene，它的用户操作区即显示出来，将房屋作为子物体加在这个用户操作区中的子区。

Step 3 在 Children 区点击创建实例键 ，为新实例选择 Object 类型。

Step 4 在对话框中选择 Object，点击 OK。

Step 5 在子区点击前进键 进入 myObject 用户操作区。

Step 6 在 API 区，选择 myObject 的 Value，将值改为 farmhouse。

Step 7 点击在文件名区旁的浏览键 ，在 c:\Program Files\Multigen-Paradigm\resources\tutorials\vegaprime\desktop_tutor\tornado\data\farmhouse 目录下选择 farmhouse.flt 文件。

Step 8 在地形中确定房屋的位置，在 Position 区中输入（X，Y，Z）的值为（2450，2460，0），其他保持不变。

Step 9 在工具条中点击 Active Preview 键 ，在场景中确定房屋的位置。用鼠标和键盘控制视角在场景中的位置，如图 3.2.4 所示。

Step 10 预览完毕后关闭 Active Preview。

Step 11 保存前面所有的操作。

现在，已经在沙地上修建了房屋。但只有屋子还远远不够，需要继续建造家园。

图 3.2.4 预览房屋

2. 增加汽车

有了前面的学习，在农场中添加一辆汽车非常容易，可以将汽车固定在靠近房屋的位置。后续章节将学习如何让汽车运动起来。

Step 1 打开前面建立的 VP3_2_2.acf 文件，单击 LynX Prime 的菜单栏中的文件，选择另存为（Save as），弹出对话框，将对话框中的目录改为建立好的 C:\VegaPrimePractice，将 ACF 文件重命名为 VP3_2_21.acf.acf。

Step 2 进入 myScene 用户操作区。

Step 3　在 Children 区点击创建实例键，选择物体类别为新实例。
Step 4　在创建实例对话框中选择 Object，单击 OK。
Step 5　在子区中 myObject 会突出显示，点击前进键　进入 myObject 用户操作区。
Step 6　在 API 区，选择 myObject 的 Value，将值改为 hummer。
Step 7　在用户操作区中，点击文件名称区旁的浏览键　，在 c:\Program Files\Multigen-Paradigm\resources\tutorials\vegaprime\desktop_tutor\tornado\data\ humv-dirty 目录下选择 humv-dirty.flt 文件。
Step 8　在位置区输入（2360，2490，0）。
Step 9　在方向区内输入汽车的头朝向，斜度，转弯度分别为：（-90，0，0），车的位置就定在了房屋旁边的沙地上。
Step 10　点击工具条动画预览键，在场景中房屋旁就有了一辆红色的汽车，如图 3.2.5 所示。
Step 11　关闭 Active Preview，保存。

现在，已经在沙地上修建了房屋，拥有了汽车。但还没解决饮食问题，继续建造家园。

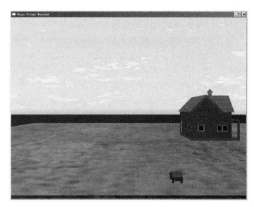

图 3.2.5　预览汽车场景

3. 添加谷仓

现在，添加一个大谷仓来储存粮食。
Step 1　打开前面建立的 VP3_2_21.acf 文件，单击 LynX Prime 菜单栏中的文件，选择另存为（Save as），弹出对话框，将对话框中的目录改为建立好的 C:\VegaPrimePractice，将 ACF 文件重命名为 VP3_2_22.acf。
Step 2　进入 myScene 用户操作区。
Step 3　在 Children 区点击创建实例键，选择 Object 类别为新实例。
Step 4　在创建实例对话框中选择 Object，单击 OK。
Step 5　在子区中 myObject 会突出显示，点击前进键　进入 myObject 用户操作区。
Step 6　在 API 区，选择 myObject 的 Value，将值改为 grainstorage。
Step 7　在用户操作区中，点击文件名称区旁的浏览键　，在 c:\Program Files\Multigen-Paradigm\resources\tutorials\vegaprime\desktop_tutor\tornado\data\grainstorage 目录下选择 grainstorage.flt 文件。
Step 8　在位置区输入（2450,2530,0），朝向、斜度、转角度都为 0。
Step 9　在方向区内输入谷仓的朝向、斜度、转角度分别为：（90，0，0），谷仓的位置就定在了房屋旁边的沙地上。
Step 10　运行 Active Preview，谷仓的位置与房屋保持了一定的距离,保存设置,如图 3.2.6 所示。

图 3.2.6　预览谷仓

4. 总　　结

创建了一些模型，作为场景的"孩子"，并把它们放入场景中。在场景中移动时，这些物体固定在原地，但这并不是希望的最后结果。每个物体都有它固有的特性，如汽车应该可以驾驶。后续章节将完善这些工作，让场景越来越逼真。

3.2.3　设置观察者 vpObserver

一个观察者就是仿真的观察点，虚拟仿真场景所有的物体都是从观察者角度渲染呈现的。在 Vega Prime 中，观察者的起始位置的默认值是地形的原点。原点的通常位置是在西南角或地形的中点。可以在 MultiGen Creator 或 ModelBuilder 3D 中的地形 OpenFlight 文件中找到原点位置。

在 Vega Prime 平面地面的坐标系统中，用 X，Y，Z 来表示观察者的位置，观察者的方向是用坐标系统中的朝向(Heading)、斜度(Pitch)和转角度(Roll)的 HPR 值来表示，符合笛卡尔坐标，符合右手准则。观察者的默认位置是在三维空间中的 XYZ 坐标原点，方向值默认也都为 0。

在 LynX Prime 操作界面中，读者可以选中左侧的 myObserver，右侧的窗口就会显示当前观察者的位置（X,Y,Z）和姿态（H,P,R）的值。读者可以尝试修改成不同的值，然后预览，查看效果。

接下来复习三维空间坐标的定义：

（1）XYZ 的定义：
- +X 指向右；
- -X 指向左；
- +Y 指向前；
- -Y 指向后；
- +Z 指向上；
- -Z 指向下。

（2）朝向(Heading)指 Z 轴转向：
- +H 指向左旋转；
- -H 指向右旋转。

（3）斜度(Pitch)指 X 轴转向：
- +P 指向上旋转；
- -P 指向下旋转。

（4）转角度(Roll)指 Y 轴转向：
- +R 指运动转向右边旋转；
- -R 指运动转向左边旋转。

现实生活中，都是通过双眼观察事物，而双眼会随着观察者自身的运动而运动。所以需要把观察者绑定在一个可以运动的物体上，让观察者运动起来，满足虚拟现实的要求。

3.2.4　建立转换 vpTransform

转换(vpTransform)是一个抽象的概念，是一个动态坐标系统，是一个不可见物体，是一种关系。在场景中设置的是物体、特殊效果或转换(vpTransform)的其他子系统，而转换(vpTransform)可以是这些可见对象互相关联的纽带。也就是说，转换(vpTransform)的值与父系统有关。

接下来将讨论如何将转换(vpTransform)作为出发点用于一个观察者,它将观察者设置于汽车的后面。观察者将随着场景中汽车位置的移动而移动，就如同观察者坐在汽车上观光一样。具体步骤如下：

Step 1 打开前面建立的 VP3_2_22.acf 文件，单击 LynX Prime 菜单栏中的文件，选择另存为（Save as），弹出对话框，将对话框中的目录改为建立好的 C:\VegaPrimePractice，将 ACF 文件重命名为 VP3_2_4.acf。

Step 2 在工具条中点击创建新实例键，打开创建实例的对话框。

Step 3 在实例列表中选择 transform，点击 create。在 LynX Prime 窗口中显示 myTransform 用户操作区。

Step 4 在 API 区，将名字改为"hummerTransform"，这个 Transform 将用于从汽车上设置观察点。

Step 5 在 hummerTransform 用户操作区，在 Parent 列表中选择 hummer，这样 Transform 的父系统就是 hummer。所有赋予 Transform 的值都与 hummer 物体相关联。

Step 6 将 Transform 的位置设为（0，-30，5），则 Transform 的位置就在汽车后的 30 个数据库单位（m），汽车上空 5 个数据库单位处。

Step 7 保存所做工作，目录为 C:\VegaPrimePractice，文件名为 VP3_2_4.acf。

把观察者的观察角度设置在汽车上，这样就相当于观察者在汽车上，随着汽车的运动而观察现实，具体步骤如下：

Step 1 打开前面建立的 VP3_2_4.acf 文件，单击 LynX Prime 的菜单栏中的文件，选择另存为（Save as），弹出对话框，将对话框中的目录改为建立好的 C:\VegaPrimePractice，将 ACF 文件重命名为 VP3_2_41.acf。

Step 2 在实例树区选择 myObserver，打开 myObserver 操作区，如图 3.2.7 所示。

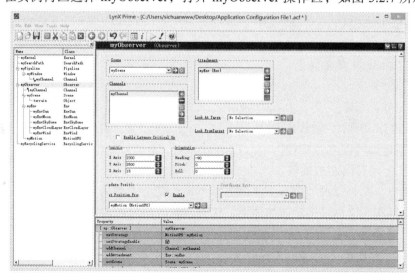

图 3.2.7 设置观察者

Step 3 注意在 Look At Target 中的设置为"无"。

Step 4 在 Look From Target 列表中选择 hummerTtransform。

Step 5 在更新位置（Update Position）区中清除 Enabled 的复选框，这样观察者的位置就随汽车的位改变而改变。（将在下节学习为汽车加载运动）

（6）打开 Active Preview（Ctrl+A），检查观察者的位置应在汽车后面，如图 3.2.8 所示。

（7）检查完毕后关闭 Active Preview。

（8）保存为 VP3_2_41.acf 文件（Ctrl+S）。

在浏览中用鼠标控制汽车的运行，希望观察者会随着汽车的运动而运动。但是，实际上会发现汽车根本不动。要让汽车真正运动起来，还需要添加运动模式。这就是在下节中要解决的运动模式问题。

图 3.2.8 观察者位于汽车后面

3.2.5 创建运动模式 vpMotion

运动模式是一个位置方法，它允许通过使用一些标准输入设备执行经过准确定义的定位方法，这些设备包括鼠标、键盘和操纵杆等。Vega Prime 中的 vpMotion 类是所有运动模式的基础。

在前几节中用来在场景中移动的默认的运动模式叫 MotionUFO。这种运动模式是一个无重力运动模式，它可以迅速移动，并且可以移动到任何地方。还有其他运动模式可以支持飞行、行走、驾驶等。

首先要将 UFO 运动模式添加到汽车上，然后学习如何更改运动模式。

1. 将运动模式添加到物体上

Step 1 打开前面建立的 VP3_2_41.acf 文件，单击 LynX Prime 菜单栏中的文件，选择另存为（Save as），弹出对话框，将对话框中的目录改为建立好的 C:\VegaPrimePractice，将 ACF 文件重命名为 VP3_2_5.acf。

Step 2 从 Instance Tree 区选择汽车。

Step 3 在 hummer 用户操作区的 Update Position 列表中选择 myMotion。

Step 4 注意选择 Enable 复选框，这样汽车就能从运动模式中接收并处理最新的信息。

Step 5 打开 Active Preview（Ctrl+A）。注意汽车正以 UFO 运动模式在地上运动，但这种运动方式对于普通汽车是不正常的。所以要更改，选择更好的模式以适应汽车的运动。

Step 6 关闭 Active Preview。

用鼠标控制汽车的运行时，会发现，汽车上天入地，无所不能，这不符合正常汽车的运动方式。出现这种情况的原因就是运动方式。MotionUFO 适合飞行穿越仿真，但是如果希望改为行驶仿真，就必须将运动模式改为 MotionDrive，在地面仿真汽车的运动应该应用 MotionDrive，可以仿真控制驾驶模式的速度和驾驭动作。

2. 更改运动模式

Step 1 打开前面建立的 VP3_2_5.acf 文件，单击 LynX Prime 菜单栏中的文件，选择另存为（Save as），弹出对话框，将对话框中的目录改为建立好的 C:\VegaPrimePractice，将 ACF 文件重命名为 VP3_2_51.acf。

Step 2 在 Instance Tree 中选择 myMotion，显示 myMotion 的用户操作区。

Step 3 在 Type 列表中选择 MotionDrive。

Step 4 在 Speed 区中将最高速度减小为 10.00，以便容易驾驭。

Step 5 打开 Active Preview，根据窗口下方提示，用鼠标在场景中行驶汽车。可以随意控制汽车的速度、方向等。

- 按鼠标左键为加速，朝各个方向拖拉鼠标，汽车的方向就随鼠标而动。
- 按鼠标右键为减速，连续按右键，汽车就慢慢减速。
- 停止运动（刹车），按鼠标中键。
- 后退，连续按鼠标右键。

Step 6 保存设置。

现在驾驶汽车，可以在地面上任意运动了，可以轻易地"穿墙而过"，这也不符合现实情况。那么继续往下走，正一步一步地接近真实。

3.2.6 应用碰撞检测 vpIsector

Isectors 是碰撞检测，一些碰撞检测有大量复杂的运算，可以支持用户用地线夹将运动模式固定在地面。还有一些相对简单的，包含几行代码的运算，用来区分目标。根据检测类型的不同，用户可以在 C++程序中编写适当的反应程序，如在墙体前停车，但现在不需要。

用户可以从以下内容选择运算：

- Tripod——三条聚集数据的直立线段，指引使用者到指到目标。Tripod 用于在水平地面上。
- Bump——六条线段，沿 X,Y,Z 轴正负方向聚集爆炸。
- LOS——单根视线线段，沿 Y 轴向辐射。
- HAT——单根线段，沿 Z 轴辐射，它计算地形上的高度。
- XYZPR——计算斜度和转向。
- ZPR——计算爆炸点 Z 轴方向的斜度和转向。
- Z——计算 Z 轴上的爆炸点。

接下来，将添加一个 Isectortripod 用地线夹将汽车固定在地形上，避免汽车钻到地下去。然后给汽车添加一个 bump 检测碰撞，检测汽车碰撞到地面上的物体。下面将使用 LynX Prime 添加这两个碰撞检测运算。

1. 添加一个 Tripod 检测

添加之前，必须创建一个实例用地夹线通过 Isector 来固定一个运动模式。

Step 1 打开前面建立的 VP3_2_51.acf 文件，单击 LynX Prime 菜单栏中的文件，选择另存为（Save as），弹出对话框，将对话框中的目录改为建立好的 C:\VegaPrimePractice，将 ACF 文件重命名为 VP3_2_6.acf。

Step 2 在 myMotion 的用户操作区中，点击创建新实例的键 （这个按键位于 Next Position Strategy 列表旁）。

Step 3 在创建新实例对话框中，选择 GroundClamp，点击 Ok。

Step 4 点前进键 显示 myGoundClamp 用户操作区。

Step 5 点击 Isector 列表旁边的创建新实例键 。

Step 6 在创建新实例的对话框中选择 IsectorTripod，Ground Clamp 实例将会使用这个功能来计算碰撞信息，最终运动方法将接触信息反馈给汽车，如图 3.2.9 所示。

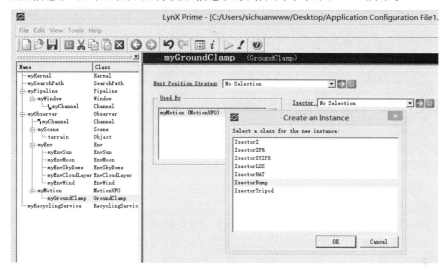

图 3.2.9 建立地面检测

Step 7 点击前进键显示 myIsector 用户操作区。

Step 8 在 API 区域将 myIsector 名称改为 tripodIsector。

Step 9 保存设置。

2. 设置 Tripod 检测

需要将地形作为 tripod 检测的目标，tripodIsector 将用地形来寻找接触，并将信息传递给所有相关的运动模式。

Step 1 在 tripodIsector 用户操作区，在 Target 列表中选择地形物体(terrain)作为目标。

Step 2 设置选择 Render Isectort 复选框用于 isector 显示。如果 Isector 找到地形目标，线条显示为绿色，否则线条显示为红色。

Step 3 运行 Active Preview，将能在屏幕上发现从汽车发出三条绿垂直线，如图 3.2.10 所示。

Step 4 设置 Tripod 检测的宽度为 5，长度为 6，这样可以检测更大的接触面。宽度和长度的参数表

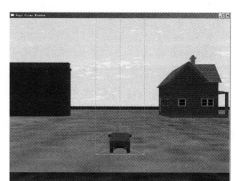

图 3.2.10 预览地面检测

示三根线彼此之间的距离。重新设置后,三根线之间的距离会拉大,将能看到地面上三根直线间的三角形联线。但这个三角形不处理任何碰撞信息,而是由这三根垂线处理所有的工作。

Step 5 现在,向前驾驭汽车。

Step 6 现在请大胆尝试,开始练习驾驭技术。

Step 7 将汽车驶离数据库区,当离开仿真世界的边缘时,汽车发出的三根线呈红色,因为这时他们无法再检测到与地面目标的接触。

Step 8 保存 VP3_2_6.acf,退出 Active Preview。

3. 添加 Bump 检测

如果汽车面前有房屋时,Bump 检测就会检验到。但是,如果在 LynX Prime 定义碰撞检测,那么它无法阻止汽车穿越房屋。必须在 API 中定义这个动作,这部分内容将不在本节学习范围之内。随后,可以在 LynX Prime 中添加汽车与房屋相撞时的效果。

Step 1 打开前面建立的 VP3_2_6.acf 文件,单击 LynX Prime 菜单栏中的文件,选择另存为(Save as),弹出对话框,将对话框中的目录改为建立好的 C:\VegaPrimePractice,将 ACF 文件重命名为 VP3_2_61.acf。

Step 2 在工具条中点击创建实例键。

Step 3 在对话框中选择 IsectorBump。

Step 4 点击 Create。

Step 5 将 myIsector 改名为 bumpIsector。

Step 6 将碰撞的宽、长和高的参数设置为 3,因为默认设置值太小。

Step 7 在 Target 列表中选择房屋(farmhouse)为目标。接触检测将根据边靠边标准检测与房屋的接触。

Step 8 打开 Isector Mask 旁的浏览按键,打开 Bitmask Editor,Bitmask 将会把指导地形目标(将在后面的步骤中建立这些地形目标)从一些检测中排除出去,在检测最新过程中节约时间。

Step 9 除了 Bit1 之外,清除所有的 Bit。Bitmask 将会显示为 0000001。

Step 10 点击 Ok,保存 Bitmask。

Step 11 在 Position Reference 列表中选择汽车(hummer)目标,碰撞检测自己就会与汽车建立联接。

Step 12 在用户操作区中选择 Render Isector 复选框,以便在 Active Preview 中查看线段。

Step 13 点击 Owning Isector Service 列表旁的创建实例键,并且选择 IsectorServiceInline。在线服务将会为每个结构提供最新的相关检测,这样就把接触数据储存起来为以后使用做准备。处理接触信息的流程与处理应用的流程一样。

不必给 Tripod 检测添加在线服务,因为三角检测不直接联接于运动模式,而运动模式会处理自己最新数据。但是,接触测试不与地夹线或运动模式联接。三角检测本身只确认它的位置与汽车直接相关。在线服务将确认每个结构的检测是最新的,这样才能储存数据。

4. 更改观察者位置

如果不喜欢在汽车后面驾驭汽车,可以将 Transform 添加在汽车内。

Step 1 在 hummerTransform 用户操作区,将 Transform 的位置改为(-9,-13.2,2.93)。
Step 2 运行 Active Preview 来检测新的设置,觉得驾驭变得容易了还是更难了?如果喜欢这个设置,就把它保存起来,如果不喜欢,可将设置改回原来的值(0,-30,5)。
Step 3 保存设置,退出 Active Preview。

3.2.7 建立特效 vpFx

特效能让用户的实时三维虚拟现实应用变得更加真实、丰富及夺目。用户可以定义烟火、爆炸、碎片等特效,也可以通过 Vega Prime 的粒子特效来模拟烟尘等效果。场景中的物体可以用来撞击或破坏,造成的结果可以加上特效。所有这些都可以通过 LynX Prime 来方便地进行定义。

Vega Prime 提供的特效包括以下几种:
● Blade(vpFxBlade)——旋转的螺旋桨,可以缩放和定位,适合用于直升机和飞机的螺旋桨效果。
● Missile Trail(vpFxMissileTrail)——用有烟的轨迹来代表飞机或导弹的飞行效果,可以随着时间而变淡。
● Particle System(vpFxParticleSystem)——多边形集聚体,用来模拟诸如焰火和烟雾等效果。
● Debris(vpFxDebris)——飞扬的碎片效果,经常用于爆炸。
● Explosion(vpFxExplosion)——地面或空中的爆炸效果。
● Fire(vpFxFire)——火焰效果。
● Smoke(vpFxSmoke)——翻腾的烟云效果。

Vega Prime 的特效模块提供了一个实时的特效库,可以通过 LynX Prime 来配置特效,也可以通过 Vega Prime 的 API 来配置特效。并且特效也可以进行形状、比例和颜色的变化。

从一个房屋的简单特效做起,当汽车撞击房屋时,碎片会四处散落。然后可以利用粒子效果创建一个龙卷风,并通过碰撞检测和场景中的物体进行交互。

1. 添加碰撞碎片特效

通过这个操作,学会给物体添加特效并和碰撞检测对应起来,同时要学会利用粒子系统来创建特效,并给粒子系统提供标靶。涉及以下概念:
● 给目标添加一至多个特效。
● 给碰撞检测添加特效。
● 利用粒子系统创建特效。
● 定义诸如颜色、粒子数量和尺寸、喷射间隔、方向、速率等属性。

在前面,已经给汽车和房子碰撞定义了一个 Bump 的碰撞检测。当汽车和房子相撞时,应该有一个撞后的效果。这里将给房子加上碎片效果,并且和汽车的碰撞关联起来。

Step 1 打开前面建立的 VP3_2_61.acf 文件,单击 LynX Prime 菜单栏中的文件,选择另存为(Save as),弹出对话框,将对话框中的目录改为建立好的 C:\VegaPrimePractice,将 ACF 文件重命名为 VP3_2_7.acf。

Step 2 点击 Create Instance 按钮。

Step 3 在 Create Instance 对话框里选中 vpFx，能看到特效的各种类。

Step 4 新建的数值采用默认值 1。

Step 5 选择 FxDebris 类。

Step 6 点击 Create 按钮。

Step 7 将 myFx 更名为 Debris。

Step 8 在 Debris 的界面里将 farmhouse（Object）作为 Parent。

Step 9 在 Trigger Method 列表里选中 Isector。

Step 10 在 Isector 列表里选择 bumpIsector。

Step 11 选择 Positioning 标签打开 Positioning 界面。

Step 12 将 Scale 值设为（20，20，20）。

Step 13 运行 Active Preview。将汽车撞向房屋，可以看到有碎片飞出。

Step 14 转到 bumpIsector 的界面，去掉 Render Isector 前面的钩，同样将 tripodIsector 界面里 Render Isector 前面的钩也去掉，这样看起来效果更好。

Step 15 关闭 Active Preview。

Step 16 将文件存为 VP3_2_7.acf。

用鼠标控制开动汽车与房屋发生碰撞，将会有碎片飞出的效果。其实，更真实的效果应该还包含汽车撞到房屋以后的反应，如减速停下或翻车等，而不是现在的穿墙而过。如果需要动态控制这些效果，需要调用 Vega Prime 的 API，进行编程控制。这是后续章节将要学习的内容。

2. 制作龙卷风效果

这里利用 LynX Prime 制作一个比较复杂的龙卷风效果，充分展示 Vega Prime 特效的巨大威力。

这个龙卷风的效果一共要使用三个粒子效果：一个为碎片层，一个是龙卷风的漏斗部分，一个为龙卷风的顶部。

Step 1 打开前面建立的 VP3_2_7.acf 文件，单击 LynX Prime 菜单栏中的文件，选择另存为（Save as），弹出对话框，将对话框中的目录改为建立好的 C:\VegaPrimePractice，将 ACF 文件重命名为 VP3_2_71.acf。

Step 2 点击 Create Instance 按钮。

Step 3 将 Instances to Create 设为 3。

Step 4 在 Class 列表里选择 FxParticleSystem 类。

Step 5 点击 Create，可以看到生成了三个新的类（myFx，myFx1，myFx2）。

Step 6 保存 ACF。

（1）配置顶部。

Step 1 将 myFx 更名为 wTop。

Step 2 在 wTop 面板里，打开 Main 所属界面。

Step 3 点击 Texture 旁边的 Browse 按钮，选择安装目录（如 C:\Program Files\MultiGen-

Paradigm\config\vegaprime\vpfx）下的 smoke.inta 文件，这个纹理是 tornado 的 3 个部分都要用到的。

Step 4　打开 Positioning 界面。

Step 5　将 Z 值设为 50，这样龙卷风的高度就是 50 m。

Step 6　打开 Particle Generation 界面。

Step 7　在 Maximum Number of Particles 栏里输入"1500"。

Step 8　在 Number of Particles to Release 栏里输入"30"。

Step 9　将 Release Interval 值设为 0.25 s，这是新粒子出现的时间间隔。

Step 10　将 Particle Life Cycle 值设为 10 s，这是新粒子出现前旧粒子所持续的时间。

Step 11　在 Source Domain Shape 栏里选择 Circle（默认值）。

Step 12　将 Size 值设为 250。

Step 13　打开 Particle Characteristics 界面。

Step 14　在 Size 栏里，将默认值改为 50，这是粒子的初始尺寸。

Step 15　点击 Scale Along Velocity 旁边的 Create Instance 按钮。

Step 16　输入如图 3.2.11 所示的数值，使球状体变得平一点。

Step 17　点击 Color 旁边的 Create Instance。

Step 18　点击新 Color 旁边的 Browse 按钮，打开 Color 窗口。为了造成龙卷风的浑浊效果，需要指定 3 个不同的灰度。这样在一个生命周期内，龙卷风首先是显示灰色，然后变亮，再变暗。

Step 19　将 Red，Green，Blue 值设为（193，193，193）。

Step 20　点击 OK。

Step 21　将 Color 旁边的 Create Instance 按钮点击两次。

Step 22　用如图 3.2.12 所示的值添加明暗两种颜色。

Time	Scale Along Velocity
0.0	1.0
1.0	5.0

图 3.2.11　设置粒子系统形状和时间

Time	Color
0.5	RGB values(198,198,198)
1.0	RGB values(163,163,163)

图 3.2.12　设置颜色值

Step 23　打开 Particle Movement 界面。

Step 24　在 Velocities 栏里添加如图 3.2.13 所示的值。

Step 25　在 Spherical Velocities 栏里添加如图 3.2.14 所示的值。

Time	Velocity Vector
0.0	(0.0,-50.0,0.0)
0.25	(100.0,50.0,0.0)
0.50	(-100.0,50.0,0.0)
0.75	(-100.0,-50.0,0.0)
1.0	(200.0,-50.0,0.0)

图 3.2.13　设置粒子运动时间和方向

Time	Velocity
0.0	7.0
0.50	8.0
0.75	-5.0
1.0	-3.0

图 3.2.14　设置粒子运动轨迹和时间

Step 26　在 Random Velocities 栏里添加如图 3.2.15 所示的两个值。

Step 27　以上步骤为龙卷风的顶部配置，先保存一下。

（2）定位。

Step 1　在 wTop 界面里打开 Main 窗口。

Step 2　创建一个新 parent 类，在 Class 列表里选择 Transform。

Step 3　打开 myTransform 界面。

Step 4　将 myTransform 更名为 wTopTransform。

Step 5　将 wTopTransform 的 parent 设为 myScene。

Step 6　输入（2450，2470，6）以定位 wTop。

Step 7　保存。

Step 8　运行 Active Preview，在正视图上就可以看见龙卷风的顶部移动，如图 3.2.16 所示。

Time	Velocity
0.0	2.0
1.0	5.0

图 3.2.15　设置粒子系统的随机运动　　　　图 3.2.16　观察龙卷风

Step 9　退出 Active Preview。

（3）配置漏斗。

Step 1　打开 myFx1 界面。

Step 2　将 myFx1 更名为 wFunnel。

Step 3　在 Main 面板里，将 Parent 设为 wTopTransform。

Step 4　点击 Texture 旁边的 Browse 按钮，选择安装目录下（如 C:\Program Files\MultiGen-Paradigm\config\vegaprime\vpfx）的 smoke.inta 文件。

Step 5　打开 Positioning 界面。

Step 6　将 Z 值设为 50，这样龙卷风的漏斗部分就和顶部一样高了。

Step 7　打开 Particle Generation 界面。

Step 8　将 Maximum Number of Particles 设为 25000。

Step 9　将 Number of Particles to Release 设为 75。

Step 10　将 Release Interval 设为 0.30 s。

Step 11　将 Particle Life Cycle 设为 12 s。

Step 12　将 Source Domain Shape 设为 Circle（默认值）。

Step 13　在 Size 输入 75。

Step 14　打开 Particle Characteristics 界面。

Step 15　在 Sizes 表里，将默认值改为 60，这是粒子的初始值。

Step 16　在 Sizes 表里添加如图 3.2.17 所示的值。

Step 17　在 Scale Along Velocity 表里输入如图 3.2.18 所示的值。

Time	Size
0.25	40.0
0.50	30.0
0.75	10.0
1.0	0.0

图 3.2.17 设置粒子系统的尺寸和时间

Time	Scale Along Velocity
0.0	1.0

图 3.2.18 设置粒子系统形状

Step 18 在 Color 表里添加如图 3.2.19 所示的值。

Step 19 打开 Particle Movement 界面。

Step 20 在 Velocities 表里输入如图 3.2.20 所示的值。

Time	Color
0.0	RGB values(172,172,172)
0.5	RGB values(195,195,195)
1.0	RGB values(170,170,170)

图 3.2.19 设置粒子系统的颜色和时间

Time	Velocity Vector
0.0	(0.0,−10.0,−40.0)
0.25	(25.0,10.0,−40.0)
0.50	(−25.0,10.0,−40.0)
0.75	(−25.0,−10.0,−40.0)
1.0	(50.0,−10.0,−40.0)

图 3.2.20 设置粒子系统的方向和时间

Step 21 在 Spherical Velocities 表里添加如图 3.2.21 所示的值。

Step 22 在 Random Velocities 表里添加如图 3.2.22 所示的值。

Time	Velocity
0.0	0.0
0.25	−10.0
0.50	−20.0
0.75	5.0
0.90	−200
1.0	0.0

图 3.2.21 设置粒子系统的形状和时间

Time	Velocity
0.0	1.0
1.0	5.0

图 3.2.22 设置粒子系统的随机值和时间

Step 23 至此漏斗部分已完成，保存设置。

Step 24 运行 Active Preview，龙卷风的漏斗部分几秒钟内就会从顶部降下并且往右移动一些。在原始视图浏览效果会更佳，如图 3.2.23 所示。

（4）配置碎片层。

Step 1 打开 myFx2 界面。

Step 2 将 myFx2 更名为 wTopDebris。

Step 3 在 Main 面板里，将 Parent 设为 tornado Transform。

Step 4 点击 Texture 旁边的 Browse 按钮，选择安装目录下（如 C:\Program Files\MultiGen-Paradigm\config\vegaprime\vpfx）的 smoke.inta 文件。

Step 5 打开 Positioning 界面。

图 3.2.23 预览龙卷风效果

Step 6 将 X 值设为 –50，这样碎片层就在龙卷风的偏左部。
Step 7 打开 Particle Generation 界面。
Step 8 将 Maximum Number of Particles 设为 1000。
Step 9 将 Number of Particles to Release 设为 10。
Step 10 将 Release Interval 设为 0.05 s。
Step 11 将 Particle Life Cycle 设为 5 s。
Step 12 将 Source Domain Shape 设为 Sphere。
Step 13 在 Size 输入 40。
Step 14 为 Velocity Distribution 选择 Sphere。
Step 15 打开 Particle Characteristics 界面。
Step 16 在 Sizes 表里，将默认值改为 15，这是粒子的初始尺寸。
Step 17 在 Scale Along Velocity 表里添加如图 3.2.24 所示的值，让球形变得稍微平一些。
Step 18 在 Color 表里添加如图 3.2.25 所示的值。

Time	Scale Along Velocity
0.0	1.0

图 3.2.24 设置粒子系统形状

Time	Color
0.0	RGB values(163,163,163)
0.5	RGB values(198,198,198)

图 3.2.25 设置粒子系统颜色

Step 19 打开 Particle Movement 界面。
Step 20 在 Velocities 表里添加如图 3.2.26 所示的值。
Step 21 在 Spherical Velocities 表里添加如图 3.2.27 所示的值。

Time	Velocity Vector
0.0	(0.0,0.0,0.0)
1.0	(1.25,1.25,0.0)

图 3.2.26 设置粒子系统位置和时间

Time	Velocity
0.0	2.0

图 3.2.27 设置粒子系统形状

Step 22 在 Random Velocities 表里添加如图 3.2.28 所示的值。
Step 23 至此龙卷风已经制作完成，保存 ACF 文件。
Step 24 运行 Active Preview，可以看到在漏斗部分下降之前，有一部分碎片层在地面运动。将视点驶离房子，然后转回头看，就能看到龙卷风的整体效果，如图 3.2.29 所示。

Time	Velocity
0.0	5.0

图 3.2.28 设置粒子系统随机值

图 3.2.29 预览完整的龙卷风效果

Step 25 保存设置，关闭 Active Preview。

（5）添加碰撞检测。

接上来添加一个碰撞检测来实现龙卷风在场景中的碰撞效果。通过碰撞检测能够判断出龙卷风是否碰撞到了场景中的物体，如粮仓、房子等。

为了排除龙卷风和地形的碰撞，需要关掉碰撞检测的 Bitmask 中的 Bit 1。

需要说明的是，以下步骤稍有简化，如果需要某些详细的配置过程，请参照前面的步骤。

Step 1 点击 Create Instance 按钮。

Step 2 创建一个 IsectorBump 类，命名为 tornadoBumpIsector。

Step 3 将 myIsectorService 添加给 tornadoBumpIsector（在 Owning Isector Service 域），myIsectorService 能让 tornadoBumpIsector 实现碰撞效果并接收碰撞数据。

Step 4 将 target 设为 Scene。在每一帧的渲染中，碰撞检测会检查场景中具有相同 lsector mask 的 bits 的物体。

Step 5 选择 tornadoTransform 为 tornadoBumpIsector 的 positon reference。这表示碰撞检测和龙卷风的位置关联起来。

Step 6 将 tornadoBumpIsector 的 Width，Length 和 Height 值设为 50，50 和 50。

Step 7 选上 Render Isector，这样就可以在 Active Preview 中看见碰撞检测的范围。

Step 8 将 lsector service 和 tornadoBumpIsector 关联起来。

Step 9 保存设置。

（6）给龙卷风添加目标。

接下来，将为龙卷风添加一个碰撞目标——一头奶牛，如图 3.2.30 所示。

Step 1 添加一个新 Object 类。

Step 2 将 myObject 更名为 cow。

Step 3 将安装目录下（如 C:\Program Files\MultiGen-Paradigm\resources\tutorials\ vegaprime\ desktop_tutor\tornado\data\cow）的 cow.flt 作为模型。

Step 4 将 cow 的 parent 设为 scene。

Step 5 将 cow 的 position 设为（2450，2625，0）。

Step 6 将 cow 的 X，Y，Z 的 scale 都设为 2。这样奶牛在碎片层中看起来就显眼一点，否则容易看成是碎片层的一部分。

Step 7 保存 ACF。

Step 8 运行 Active Preview，可以看到奶牛处在房子后面的原野上，如图 3.2.31 所示。

图 3.2.30 设置龙卷风目标

图 3.2.31 把奶牛放到场景中

运行 Active Preview，可以看到有趣的事情发生了，奶牛被直接卷入了龙卷风的碎片层中！

Step 9　退出 Active Preview。

（7）为粮仓制作特效。

接下来给粮仓加上特效。当龙卷风撞上粮仓时，会产生碎片飞扬和火焰冲天的效果。

需要说明的是，Active Preview 中看不到这些效果，因为这次的碰撞配置只在 runtime application 中生效。后续章节将对 application 进行编译运行。

Step 1　各创建一个 FxDebris、FxFire 和 FxExplosion 的类。

Step 2　打开 myFx 界面，将 myFx 更名为 grainStorageDebris。

Step 3　将 parent 设为 grainstorage。

Step 4　Triggering Method 中选择 Don't Trigger，必要时会使用一行代码来打开这个效果。

Step 5　打开 Position 界面。

Step 6　将 special effect 的 X，Y，Z 的 scale 设为 50。

Step 7　对 FxFire 和 FxExplosion 重复第 2 步至第 6 步，同时别忘了将这两个特效类更名为 grainStorageFire 和 grainStorageExplosion。

Step 8　保存 ACF。

到此为止，可以在 Active Preview 里看到龙卷风并不移动，只是原地打转。后续章节将编译一段代码，代码执行的时候可以让龙卷风按固定的路径移动，同时进行破坏！

3.2.8　设置窗口 vpWindow

Active Preview 窗口的默认名字为 Vega Prime Window，但是最好能给窗口一个独特的名字，这样在它运行时可以反映这个名字代表的程序。例如，如果观察者能在不同通道间转换，就将窗口名称改成现在正在使用的通道名称，如 View of Satellite from Shuttle 或 View of Target。还可以修改窗口的其他参数，如隐藏窗口的边框和窗口内的鼠标。

Step 1　打开前面建立的 VP3_2_7.acf 文件，单击 LynX Prime 菜单栏中的文件，选择另存为（Save as），弹出对话框，将对话框中的目录改为建立好的 C:\VegaPrimePractice，将 ACF 文件重命名为 VP3_2_8.acf。

Step 2　在实例树区选择 myWindow。

Step 3　在 Label 区输入 VegaPrimeTest。

Step 4　运行 Active Preview，就可以在顶边显示窗口的新名字，如图 3.2.32 所示。

Step 5　在 Options 区，取消 Border 的复选框，如果取消了边框，窗口的修饰和信息栏都会消失，可以随意打开或关闭边框。

Step 6　取消鼠标的复选框，那么在 Active Preview 窗口内就不再出现鼠标。

Step 7　设置窗口边框为打开。

Step 8　设置鼠标为无。

Step 9　保存设置，关闭 Active Preview。

图 3.2.32　预览窗口名称

现在汽车可以在地面上行驶了。在 Active Preview 中为观察者设定了位置,并将他们与场景中的物体联接起来,也为碰撞检测进行了定义,并且还添加一些特殊效果,将在其他章节继续学习这部分内容。接下来要改变环境,并给汽车加上车头灯,以适应在黑暗中行驶!

3.2.9 建立环境 vpEnv

LynX Prime 的环境效果包括阳光、月光、雾、风等。它还可以控制一天中时间的流逝速度,可以在几分钟内将黎明变成白天、黄昏和夜晚。下面将会讲解如何设置天时、天空和云层的效果以及添加光点。

1. 环境设置

应该学会为自己的应用设置环境效果,如依据天时而变化的阳光、月光、多云的天空。还可以给汽车加上光点,这样就可以在黑暗里行驶。环境设置主要包括控制天时、改变天空颜色、改变云层纹理、在物体上定位光源等。

如何定义环境呢?Vega Prime 中的环境是云、雾等大气现象的综合体。它包含了阳光和月光源。LynX Prime 中阳光和月光会依据天时进行自动调整。

LynX Prime 提供了以下几种效果来创建一个实时的环境:
- 太阳——由一个亮盘来代表太阳在天空中的位置。
- 月亮——由另一个亮盘来代表月亮在天空中的位置。
- 云层——云层的一层。
- 穹顶——刻划出天际线的效果。穹顶有一个无边的地平面,颜色可以自定义。

2. 改变天时

在实例树区选择 myEnv,默认天时是从午时 12 点开始,如图 3.2.33 所示。时间决定了光照的强度,一天中太阳和月亮交替出现。

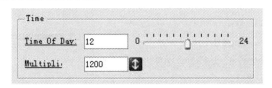

图 3.2.33 设置时间变化

Time Multiplier 是控制时间快慢的工具。默认值为 1,表示仿真时间和现实时间快慢相同;若值设为 0,表示时间静止;若值设为 60,表示仿真时间是现实时间速度的 60 倍,这时在半小时的现实时间内就可以看到几乎整天的仿真天时效果。

Step 1 打开前面建立的 VP3_2_8.acf 文件,单击 LynX Prime 菜单栏中的文件,选择另存为(Save as),弹出对话框,将对话框中的目录改为建立好的 C:\VegaPrimePractice,将 ACF 文件重命名为 VP3_2_9.acf。

Step 2 在实例树区选择 myWindow。

Step 3 转到 myEnv 的配置界面。

Step 4 将 Time Multiplier 的值设为 1200，这样一整天的天时渲染在 2 min 之内就可以完成。

Step 5 运行 Active Preview，这样就可以看到天色慢慢变黑的过程，同时还能看到太阳和月亮的升降效果。

Step 6 退出 Active Preview。

Step 7 将文件保存为 VP3_2_9.acf。

3. 改变天空的颜色

一场风暴就在眼前！可以改变天空的颜色和云层的纹理来制造出风暴的效果！

Step 1 在 myEnv 界面，点击 Sky Color 旁边的 Browse 按钮。

Step 2 将天空颜色选为红色。

Step 3 运行 Active Preview，就可以看到天空颜色和云层纹理混合后的效果。

Step 4 接下来将颜色换成灰色或蓝灰色，这样更接近于风暴的颜色，下一小节将会改变云层的纹理以接近风暴的效果。

Step 5 保存，退出 Active Preview。

4. 改变云层纹理

Step 1 在 myEnv 界面的 Environmental Effects 选项里选择 myEnvCloudLayer。

Step 2 点击旁边的转向按钮，打开 myEnvCloudLayer 界面。

Step 3 在 Texture 一栏里点击 Filename 旁边的 Browse 按钮。

Step 4 选择 E:\Program Files\Multigen-Paradigm\config\vegaprime\vpenv（安装目录）下的 clouds_storm.inta 文件。

Step 5 运行 Active Preview，就可以看到风暴来临前夕的天空效果。

Step 6 保存，退出 Active Preview。

5. 添加多云层

Step 1 在工具栏点击创建按钮。

Step 2 所创建类别的数量就保持默认值 1。

Step 3 在 Class 的列表中选择 myEnvcloudvolume。

Step 4 点击 Create，这样在 LP 中就出现一个 myEnvcloudvolume 类。

Step 5 在 EnvironmentParent 列表选择 myEnv。

Step 6 将 myEnvcloudvolume 更名为 "thundercloud"。

Step 7 在 Translate 列表里输入 thunderCloud 的位置，X:2460，Y:2450，Z:600，如图 3.2.34 所示。

Step 8 运行 Active preview，将会在不远处看见一团乌云。

Step 9 保存，退出 Active Preview。

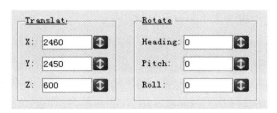

图 3.2.34　设置云层参数

6. 添加闪电

Step 1　在 thunderCloud 界面选择 effort 标签。

Step 2　在 lighting 列表里设置：Seveirty 为 1，IntraCloudSeverity 为 0.5，如图 3.2.35 所示。

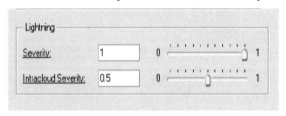

图 3.2.35　闪电参数设置

Step 3　运行 Active preview，将会看到闪电周期性的出现。

Step 4　保存，退出 Active Preview。

这样的天气里驾驶需要打开车前灯！下面要在汽车的前面添加光点来模拟车前灯！

7. 添加光点

为了模拟光照效果，需要加入多个光源。光源在仿真场景中用于照亮物体。可以通过 LynX Prime 来完成这些配置。光源分为：

- 定向光源——指向特定的方向。
- 位置光源——某一个位置上的漫反射光源。
- 点光源——某一位置上的定向光源。

这里要在汽车的车头上安装点光源。

（1）创建车头灯。

这里先创建一个车头灯，第二个车头灯可以通过复制来完成。

Step 1　打开前面建立的 VP3_2_9.acf 文件，单击 LynX Prime 菜单栏中的文件，选择另存为（Save as），弹出对话框，将对话框中的目录改为建立好的 C:\VegaPrimePractice，将 ACF 文件重命名为 VP3_2_91.acf。

Step 2　在工具栏点击创建按钮。

Step 3　所创建类别的数量就保持默认值 1。

Step 4　在 Class 的列表中选择 Light。

Step 5　点击 Create，这样在 LynX Prime 中就出现一个 myLight 类。

Step 6　将 myLight 更名为 leftHeadlight。

Step 7　在 Type 列表中选择 Spot。

Step 8 在 Parent 列表中选择 hummer（Object）。

Step 9 将光点的 Position 设为（-0.05, 2.24, 0.9），Pitch 值设为 -10。

Step 10 在 Attenuation 的 Constant 栏中输入 0.5，这表示光照强度在照明方向上随距离增加而减弱的程度。

Step 11 将 Spot Cone 的值设为：Inner = 20，Outer = 45，Falloff = 0.5。

Step 12 选上 Render 前面的小方框，光点将被一个小圆球代替。

Step 13 运行 Active Preview，当太阳落山时，就可以看到明显的车灯效果，如图 3.2.36 所示。

Step 14 保存，退出 Active Preview。

（2）创建第二个车头灯。

创建第二个车头灯的快捷方法便是复制第一个车头灯。

Step 1 在 Instance Tree 面板中点击 leftHeadlight。

Step 2 选择 Edit→Copy Instance。

Step 3 选择 Edit→Paste Instance，将会看见出现一个 leftHeadlight1。

Step 4 将 leftHeadlight 更名为"rightHeadlight"。

Step 5 将 rightHeadlight 的 X 坐标改为 0.05。

Step 6 将 Parent 改为 hummer（Object）。

Step 7 运行 Active Preview，查看车灯效果，如图 3.2.37 所示。

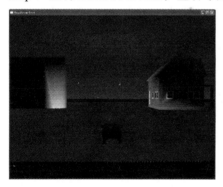 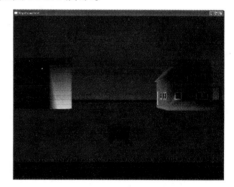

图 3.2.36 预览灯光效果　　　　　　图 3.2.37 预览车灯效果

Step 8 保存。

可以看出，在 LynX Prime 里设置天时和大气效果是很方便的。通过设置车头灯可以在黑暗中自由行驶。

3.2.10　设置通道 vpChannel

通道就是进入图形窗口的视角，一个窗口可能有几个通道。通道的位置由拖曳区来管理，拖曳区是根据相关窗口的规范值。

通道的默认值是与窗口大小一样，但是可以调节它的大小。同样可以调节拖曳区的值来排列或叠加通道。要用一些观察者来创建通道，然后调节它们，实现同时从多个视角观察场景。

1. 创造新的通道和观察者

Step 1 打开前面建立的 VP3_2_91.acf 文件，单击 LynX Prime 菜单栏中的文件，选择另存为（Save as），弹出对话框，将对话框中的目录改为建立好的 C:\VegaPrimePractice，将 ACF 文件重命名为 VP3_2_10.acf。

Step 2 从工具条中点击创建实例键，显示创建实例对话框。

Step 3 在 Class 列表中选择 Channel 和 Observer。

Step 4 点击 Create，这样可以同时创建通道和观察者。

Step 5 将 myChannel1 改名为"houseChannel"。

Step 6 将 myObserve1 改名为"houseObserver"。

2. 配置房屋通道

要配置房屋通道，需要先将房屋观察者和房屋通道联接起来，然后在 Active Preview 窗口中放置房屋的视角。

Step 1 在 houseChannel 用户操作区，在 Used By Observer 列表中选择 houseObserver。

Step 2 在 Used By Window 列表中选择 myWindow。

Step 3 在拖曳区输入以下值，在窗口的左上方设定房屋通道。

- 左边=0；
- 右边=0.5；
- 底边=0.5；
- 顶边=1。

Step 4 保存 ACF 文件。

3. 配置房屋观察者

Step 1 在 houseChannel 用户操作区中，选择 Used By Observer 列表中的 houseObserver，点击前进键，打开 houseObserver 用户操作区。

Step 2 在 Scene 列表中选择 myScene。

Step 3 将观察者的位置设为（2360，2490，2）。

Step 4 选择 farmhouse 作为注视目标，房屋就一直处于房屋观察者通道的中心。

Step 5 注意在 Get Position From 列表中选择 No Selection。

Step 6 在 Attachments 区，点击添加联接按键，创建一个环境，然后点浏览键，选择 myEnv 与房屋观察者联接，将在 houseChannel 中有光线显示与 myEnv 联接。

Step 7 运行 Active Preview，将会看到两个窗口，一个是原来的窗口，另一个左上角的窗口显示的是从一个固定位置看房屋的情景，如图 3.2.38 所示。

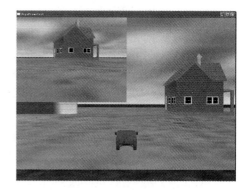

图 3.2.38　双通道观察场景

Step 8 退出 Active Preview 之前，注意保存设置。

在房屋前面增加一个通道，注意观察房屋前面的情况。

4. 从前方设定一个通道

首先，从前方设定一个视角。

Step 1 打开前面建立的 VP3_2_10.acf 文件，单击 LynX Prime 菜单栏中的文件，选择另存为（Save as），弹出对话框，将对话框中的目录改为建立好的 C:\VegaPrimePractice，将 ACF 文件重命名为 VP3_2_11.acf。

Step 2 点击工具条上的创建实例键，打开创建对话框。

Step 3 在 Instances to Create 区输入 2，创建 2 个通道。

Step 4 在 Class 列表中选择 Channel。

Step 5 单击 Create。

Step 6 转入 myChannel1 用户操作区。

Step 7 将 mychannel1 改名为 "portchChannel"。

Step 8 在拖曳区输入以下值，在 Active Preview 窗口的左下角设定前方通道位置：
- 左边=0；
- 右边=0.5；
- 底边=0；
- 顶边=0.5。

Step 9 在 Used By Window 列表中选择 myWindow。

Step 10 选择 Used By Observer 列表旁的创建实例键，这样就创建了一个新的观察者，并将 portchChannel 与新创建的观察者联接起来。

Step 11 保存前面的操作。

5. 配置前入口观察者

Step 1 在 Used By Observer 区域，点击前进键进入 myObserver1 用户操作区。

Step 2 将 myObserver1 改名为 "porchObserver"。

Step 3 在 Scene 列表中选定 myScene。

Step 4 在 Attachments 区，点击添加联接按键，创建一个环境，然后点浏览键，选择 myEnv 与房屋观察者联接，将在 porchChannel 中有光线显示与 myEnv 联接。

Step 5 点击 Look From Target 列表旁的增加实例键来增加一个新的 Transform，并将它与入口观察者建立联接。

Step 6 保存。

6. 配置入口 Transform

Step 1 在 porchObserver 用户操作区，在 Look From Targe 列表中 myTransform 旁有前进键，点击此键后显示 myTransform 用户操作区。

Step 2 将 myTransform 改名为 "porchTransform"。

Step 3 将父系统设置为 farmhouse。

Step 4 设置位置为（3.5，-4，5），设置朝向为180。这个朝向和位置位于房屋的前门向外看。

Step 5 运行 Active Preview，因为观察设置为入口 Transform，可以从入口 Transform 处看到景物，仍然可能看到房屋，也可从汽车后部看。

新设置的通道叠加在原始通道，通道按创建的先后进行排列。所以最新创建的通道在窗口的上方，将很快调整原始通道，如图 3.2.39 所示。

Step 6 保存 ACF 文件，退出 Active Preview。

图 3.2.39　预览三通道效果

7. 创建一个俯视通道

一个俯视通道可以像看地图一样看场景。

Step 1 打开前面建立的 VP3_2_11.acf 文件，单击 LynX Prime 菜单栏中的文件，选择另存为（Save as），弹出对话框，将对话框中的目录改为建立好的 C:\VegaPrimePractice，将 ACF 文件重命名为 VP3_2_12.acf。

Step 2 在实例树图中，选定 myChannel2 显示它的用户操作区。

Step 3 将 myChannel2 改名为"orthoChannel"。

Step 4 在 Projection 列表中选择 Orthographic，俯视投影不是一个透视俯视。

Step 5 在 Frustum 区中输入以下参数：

- 左=-1500；
- 右=1500；
- 下=-1500；
- 顶=1500。

这个截面参数确定了俯视图的周长大小。

Step 6 在 Draw 区输入以下参数：

- 左=0.5；
- 右=1；
- 底边=0.5；
- 顶边=1。

俯视图通道位于屏幕的右上角。

Step 7 将偏移量定为（0，0，600），将这些输入朝向、斜度和转向值。

- 朝向=0；
- 斜度=-90；
- 转向=0。

这个通道向下旋转 90°，截取的 Z 轴上的值为 600 m，这样屏幕将无法显示物体高于 600 m 的部分。

Step 8 在 Used By Window 列表选定 myWindow。

Step 9 点击 Used By Observer 列表旁的创建新实例键，为 OrthoChannel 创建一个新观察者。

Step 10 保存。

8. 配置 Orthographic 观察者

Step 1 在 orthoChannel 用户操作区中,点击刚创建的新观察者旁的前进键。

Step 2 将 myObserver1 改名为 "orthoObserver"。

Step 3 将 Scene 设置为 myScene。

Step 4 将 myEnv 加上光线。

Step 5 将观察者的位置设为(1500,1500,0)。

Step 6 运行 Active Preview,查看 4 个屏幕的内容,如图 3.2.40 所示。

图 3.2.40 预览多通道效果

正如前述提到的,需要调节原始窗口,后面将在 Active Preview 运行时来调节它。

第 4 章 运行 Vega Prime 应用

虽然 LynX Prime 能够简化应用程序的开发过程，不必编写代码，只需要通过可视化的编辑界面进行配置，就能开发出简单展示的虚拟现实应用程序。但是，要真正发挥 Vega Prime 的强大功能，就必须学会使用 API，在 C++环境中利用程序逻辑的灵活性，实现更为人性灵活、功能强大的虚拟现实应用程序。本章将逐步进入 API 的编程使用，让 Vega Prime 展现它本身所拥有的强大功能。首先，配置建立运行 Vega Prime 应用程序的环境，并在控制台下编译运行前面章节建立的应用程序，分为加载 ACF 文件方式和完全函数代码形式。其次，剖析应用程序的组成，清楚掌握应用程序的函数组成，为自由使用 Vega Prime 的 API 函数做好准备。最后，移植 Vega Prime 应用到 MFC 应用程序下，自由使用 API 来开发功能更加强大的 Vega Prime 虚拟现实应用程序。

开发者若想要进一步学习 Vega Prime 的编程功能，需要具有较好的 C++编程基础，尤其是具备在 VC 环境下的 C++面向对象编程能力。从某种程度上来说，开发者的 VC 编程能力决定了 Vega Prime 虚拟现实应用开发能力。

【本章重点】

- VC++编程基础；
- 进程与线程；
- 配置编译环境；
- 控制台程序加载 ACF 文件进行仿真；
- 建立 MFC 应用程序；
- Vega Prime 的最小应用程序；
- 导出 ACF 文件；
- 剖析 Vega Prime 应用组成。

4.1 VC++编程基础

4.1.1 安装 VC++

到现在为止，Vega Prime 较流行的有多个版本。本书所用的 Vega Prime 版本为 2.0，最后升级为 2.0.1。那么关心 Vega Prime 的版本，其原因在于不同的 Vega Prime 版本对应不同的

Visual C++开发环境：Vega Prime 1.1 对应 Visual C++ 6.0；Vega Prime 1.2 对应 Visual C++ 6.0；Vega Prime 2.0 对应 Visual C++ 7.0；Vega Prime 2.1 对应 Visual C++ 7.0 或 Visual C++ 8.0；Vega Prime 2.2 对应 Visual C++ 7.0 或 Visual C++ 8.0；Vega Prime 5.0 对应 Visual C++ 9.0。其中 Visual C++ 7.0 包含于 VS 2003，Visual C++ 8.0 包含于 VS 2005，Visual C++ 9.0 包含于 VS 2008。本书是在 WindowsXP 系统下，所用 Vega Prime 版本为 2.0，最终升级为 Vega Prime2.0.1，对应的确切开发环境为 Visual C++ 7.10。某些 VS 2003 版本包含的 Visual C++为 7.0 版本，请在安装 Visual C++环境时确认是 Visual C++ 7.10 版本。当然，Vega Prime 2.0 加上 Visual C++ 7.10 在 Win7、Win8 和 Win10 下也运行良好。

VS 2003 的安装界面如图 4.1.1 所示，其中，只选定安装 Visual C++。点击右下角的"立即安装"，自动完成安装。

图 4.1.1　安装 Visual C++

4.1.2　进程优先级

首先认识进程。操作系统将应用程序的每个正在执行的实例视为一个进程，每个进程有自己独立的系统资源，如内存和 CPU 处理时间。现在的 Windows 都支持多个进程同时运行。进程与线程之间有什么关系呢？关系就是，无论何时启动一个进程，在它内部就会创建一个线程，开始执行应用程序。简单来说，线程是进程内部的一个执行路径，可以定义为可执行代码的最小单位。另外，一个线程本身不能请求任何资源，它利用分配给进程的资源，而线程是这个进程的一部分。一个应用程序的每一个实例都会启动一个进程，而每一个进程至少跟一个线程相关联。这样，一个应用程序总会有一个相关的线程，这个线程称为这个应用程序的主线程。主线程包含应用程序的启动代码，以及 main 函数或 WinMain 函数的地址。当应用程序启动时，操作系统首先执行应用程序的主线程。

大多数进程用优先级 NORMAL_PRIORITY_CLASS 来开始其生命期。然而，进程一旦启动后，便可以通过调用::SetPriorityClass 函数来改变优先级。::SetPriorityClass 函数接受一个进程句柄（该句柄可以用::GetCurrentProcess 函数来获得），还接受表 4.1.1 中显示的一个指定项。若想把当前进程设置为实时优先级，则可使用以下代码：

::SetPriorityClass(::GetCurrentThread(),THREAD_PRIORITY_LOWEST);

表 4.1.1　进程相对优先级

优先级	说　　明
REALTIME_PRIORITY_CLASS	最高进程优先级
HIGH_PRIORITY_CLASS	进程高优先级
ABOVE_NORMAL_PRIORITY_CLASS	高于普通进程优先级
NORMAL_PRIORITY_CLASS	普通进程优先级
BELOW_NORMAL_PRIORITY_CLASS	低于普通进程优先级
IDLE_PRIORITY_CLASS	空闲进程优先级

大多数应用程序不需要改变其优先级。REALTIME_PRIORITY_CLASS 和 HIGH_PRIORITY_CLASS 会严重抑制系统的响应性，甚至会延迟关键性活动，如硬盘高速缓存的刷新(本应该以高优先级执行)。HIGH_PRIORITY_CLASS 主要用在系统应用中，这些应用程序大部分时间保持在隐藏状态，但是在某个输入事件发生时，则会弹出一个窗口。这些应用程序在阻塞着等待输入时的系统开销很小，但是一旦出现了输入，则比普通应用程序具有更高的优先级。REALTIME_PRIORITY_CLASS 主要是供实时数据获取程序使用，这些程序必须及时分享 CPU 时间，以便正常地发挥作用。IDLE_PRIORITY_CLASS 适用于屏幕保护程序、系统监视和其他后台操作的低优先级应用程序。

4.1.3　使用多线程

Vega Prime 应用程序都是多线程程序，至少包含绘制线程、剪切线程等。因此，很有必要对线程进行一定的了解。

线程可以分为两种：用户界面线程和工作者线程。

1. 用户界面线程

处理用户输入的线程称作用户界面线程。在多数情况下，一个应用程序只有一个用户界面线程。这个线程处理所有用户活动所产生的事件(如按键操作或鼠标单击)，用户界面线程还处理窗口消息。封装了线程所有功能的 MFC 类是 CWinThread 类。同样，在 MFC 类的层次结构中，CWinThread 类是 CWinApp 类的基类，所以 CWinApp 类从它的基类那里继承了所有线程处理功能。因此，用户界面线程的一个典型例子就是应用程序对象，它是 CWinApp 类或其派生类的一个实例。所以，当执行一个 MFC 应用程序时，就会启动一个用户界面线程。

2. 工作者线程

可以在应用程序中创建的另一种线程是工作者线程。工作者线程不处理用户输入，而是为应用程序执行后台处理工作。例如，当创建一个电子表格应用程序时，除用户界面线程外，还需要创建工作者线程来处理某些活动(如重算、绘制图表及打印任务等)。工作者线程没有消息处理程序，因此不能处理任何事件或窗口消息。一个应用程序可以有多个工作者线程来在后台处理工作。

要创建一个线程，需要调用 CWinThread 类的 AfxBengiThread 函数。AfxBengiThread 函数有两个重载版本：一个用于工作者线程，另一个用于用户界面线程。

用来创建工作者线程的 AfxBeginThread 函数的语法如图 4.1.2 所示。

```
CWinThread * AfxBeginThread（
    AFX_THREADPROC ThreadFunction,
    LPVOID Parameter,
    Int ThreadPriority=THREAD_PRIORITY_NORMAL,
    UINT StackSize=0,
    DWORD ControlFlags=0,
    LPSECURITY_ATTRIBUTES Security=NULL
）
```

图 4.1.2　创建线程

下面是 AfxBeginThread 函数的参数说明：

- ThreadFunction：需要在线程中执行的函数名称。当运行这个线程时，它将这个函数的地址提供给处理器。这个参数不能是 NULL,因为一个线程至少需要执行一个函数。ThreadFunction 必须是一个全局函数或一个类的静态函数成员。
- Parameter：需要给在线程中执行的函数提供的参数。
- ThreadPriority：指定想要为线程设置的优先级。表 4.1.2 所示为预定义的用来设置线程优先级的整数值。高优先级的线程先执行。

表 4.1.2　线程相对优先级

优先级	说　明
THREAD_PRIORITY_HIGHEST	指定一个线程为最高优先级
THREAD_PRIORITY_NORMAL	指定一个线程为普通优先级
THREAD_PRIORITY_ABOVE_NORMAL	指定一个线程高于普通优先级
THREAD_PRIORITY_BELOW_NORMAL	指定一个线程低于普通优先级
THREAD_PRIORITY_LOWEST	指定一个线程为最低优先级

- StackSize：指定分配给要创建的线程的堆栈内容的大小。当指定 StackSize 的值为 0 时，默认堆栈大小设置为与创建当前线程的线程堆栈大小相等。
- ControlFlags：指定一些附加状态，如线程创建时的状态。对于 ControlFlags,可以指定为 0 或 CREATE_SUSPENDED 值。当指定它的值为 0 时，线程在创建后立即启动；当指定为 CREATE_SUSPENDED 时，以挂起模式创建，这时要执行这个线程，需要调用 CWinThread::ResumeThread 函数。
- Security：指定一个线程的安全属性，它指向一个 SECURITY_ATTRIBUTESS_结构。这个结构包含用于一个对象的安全描述符。如果指定为 NULL,创建的线程的安全属性被设置为默认的安全属性。

线程的优先级从 0 到 31，数值越大，优先级越高。线程的优先级类型分为两类：

- 实时(real-time)：从 16 到 31，不会自动改变,如设备监控进程。
- 可变优先级(variable-priority)：从 1 到 15（级别 0 保留为系统使用），可由 OS 自动改变。线程都是相对于创建进程而言，分为 32 个级别，进程的优先级也分为 32 个级别。线程的优先级由母进程优先级和线程优先级共同决定。线程运行次序并不是按照创建他们时的顺序来运行的，CPU 处理线程的顺序是不确定的，如果需要确定，那么必须自己编成实现。使用 SetThreadPriority()方法设置优先级,设置线程的优先级函数如下：
　　BOOL SetThreadPriority(HANDLE hThread,int nPriority);

4.1.4 创建控制台程序

启动 VS 2003 程序，依次选择"文件"→"新建"，将会出现如图 4.1.3 所示的界面。

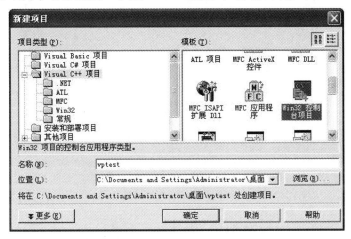

图 4.1.3　建立控制台程序

选中窗口左面的"Visual C++项目"，然后选中窗口右边的"Win32 控制台项目"，最后在"名称(N)"旁边的输入框中输入项目名称，这里输入"vptest"。最后双击"确定"按钮，将会出现图 4.1.4 所示的设置控制台程序的界面。

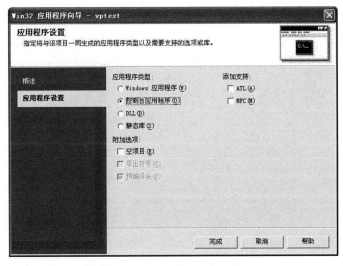

图 4.1.4　设置控制台程序

选中"控制台应用程序"单选按钮，然后点击"完成"按钮，将会出现如图 4.1.5 所示的开发界面。

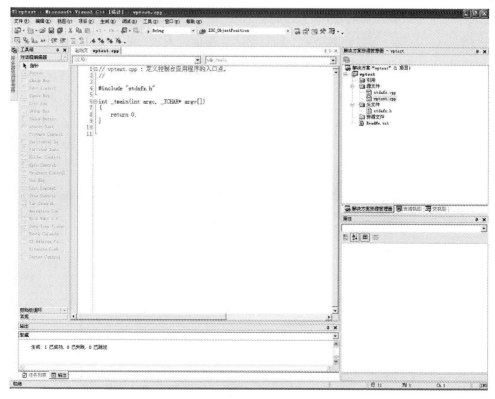

图 4.1.5 控制台程序开发界面

控制台程序开发界面包括：中间部分是代码区，右上角有"解决方案资源管理器""资源视图""类视图"，右下角是属性区等。

4.1.5 创建 MFC 对话框程序

启动 VS 2003 程序，依次选择"文件"→"新建"，将会出现如图 4.1.6 所示的界面。

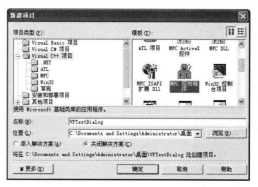

图 4.1.6 建立 MFC 对话框程序

选中窗口左面的"Visual C++项目"，然后选中窗口右边的"MFC 应用程序"，最后在"名称(N)"旁边的输入框中输入项目名称，这里输入"VPTestDialog"。最后双击"确定"按钮，将会出现图 4.1.7 所示的设置 MFC 对话框程序的界面。

第 4 章 运行 Vega Prime 应用

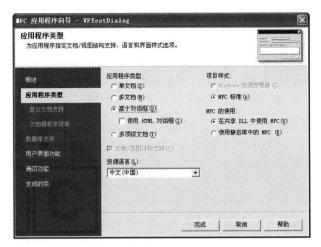

图 4.1.7 设置 MFC 对话框程序

选中左面"基于对话框"单选按钮和右面"MFC 标准"与"在共享 DLL 中使用 MFC"单选按钮，然后点击"完成"按钮，将会出现如图 4.1.8 所示的开发界面。

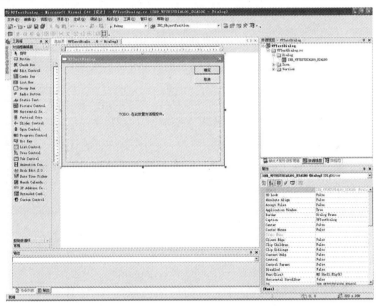

图 4.1.8 双通道观察场景

MFC 对话框程序开发界面包括：中间部分是代码区和可视化界面编辑区，右上角有"解决方案资源管理器""资源视图""类视图"，右下角是属性区等。

4.2 配置 Vega Prime 应用程序编译运行环境

编译运行 Vega Prime 应用程序，必须进行相应的设置，主要包含控制台程序和 MFC 对话框程序。

- 65 -

4.2.1 配置运行控制台仿真程序

对于控制台程序，配置相对简单。首先把 VP3_2_7.acf 文件复制到第 4.1.4 节所建立的控制台程序 vptest 所在的目录，然后打开项目，把图 4.1.5 所示的代码替换成如图 4.2.1 所示的代码。

```
#include "stdafx.h"
#include "vpApp.h"
int _tmain(int argc, _TCHAR* argv[])
{
vp::initialize(argc, argv);        // initialize vega prime
vpApp *app = new vpApp;            // create a vpApp instance
app->define("VP3_2_7.acf");        // load acf file
app->configure();                  // configure my app
app->run();                        // runtime loop
app->unref();                      // unref my app instance
vp::shutdown();                    // shutdown vega prime
    return 0;
}
```

图 4.2.1 Vega Prime 控制台应用程序

图 4.2.1 中的代码，是运行 Vega Prime 控制台程序最精简的代码：第 1 行是初始化 vega prime；第 2 行是创建一个 vpApp 应用实例；第 3 行是加载 acf 文件；第 4 行是配置应用；第 5 行是仿真循环；第 6 行是注销应用实例；第 7 行是关闭 vega prime。

点击图 4.2.2 所示的菜单：项目→vptest 属性，将出现如图 4.2.3 所示的附加包含目录的配置界面。选择 C/C++→常规，在右边"附加包含目录"里，分别填入：

$(MPI_LOCATE_VEGA_PRIME)\include\vsg

$(MPI_LOCATE_VEGA_PRIME)\include\vegaprime

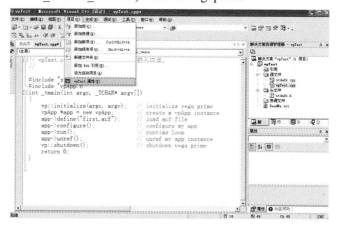

图 4.2.2 Vega Prime 最小控制台程序

如图 4.2.3 所示,选择"连接器"→"常规",将出现如图 4.2.4 所示的界面,在右边"附加库目录"里,填入:

$(MPI_LOCATE_VEGA_PRIME_LIB)

如图 4.2.4 所示,选择"C/C++"→"代码生成",将出现如图 4.2.5 所示的界面,在右边"运行时库"里,选择"多线程 DLL(/MD)"。

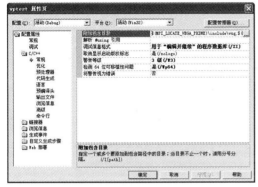

图 4.2.3 配置附加包含目录　　　　　　　　图 4.2.4 配置附加库目录

上述三项配置至关重要,确认完成后,就可以编译运行 Vega Prime 应用程序。编译运行 vptest,将出现如图 4.2.6 所示的界面。

图 4.2.5 配置运行时库　　　　　　　　　　图 4.2.6 编译运行 Vega Prime 应用程序

Vega Prime 应用程序全是以 main 函数开头,这样程序运行时会先有一个 DOS 窗口,这样在调试时非常方便,但在发布时就很不合时宜。可以这样做:如果是 debug 时,使用 DOS 窗口,是 Release 时直接使用 windows 的窗口,并不弹出 DOS 窗口。为了达到这个目的,需要在主文件的所有#include 文件之后加上以下代码即可:

#ifdef _DEBUG

#else if

#pragma comment(linker, "/subsystem:\"windows\" /entry:\"mainCRTStartup\"")

#endif

4.2.2 配置 MFC 对话框运行环境

打开第 4.1.5 节所建立的 MFC 对话框程序 VPTestDialog。首先，仿照控制台程序的配置来配置项目属性，选择菜单"项目"→"VPTestDialog 属性"，出现附加包含目录配置界面，选择"C/C++"→"常规"，在右边"附加包含目录"里，填入：

$(MPI_LOCATE_VEGA_PRIME)\include\vsg，

$(MPI_LOCATE_VEGA_PRIME)\include\vegaprime

接着选择"连接器"→"常规"，在右边"附加库目录"里，填入：

$(MPI_LOCATE_VEGA_ PRIME_LIB)

因为 MFC 对话框程序一定是多线程程序，所以不需要进行多线程配置。后续章节将重点处理基于 MFC 对话框程序的 Vega Prime 应用程序的代码。

4.3 导出 ACF 文件

4.3.1 导出 ACF 文件

ACF 文件本身采用了 XML 文件格式,记录了 Vega Prime 应用的固定配置。对于 Vega Prime 应用的 ACF 文件，不同的使用者会采用不同的处理方式。

就像在第 4.2.1 节控制台程序里应用的一样，直接使用 ACF 文件，可以进行实景仿真。系统自带的例子也采用着这种方式。但是，这里的 ACF 文件是独立存在的明码文件，一旦丢失或路径不正确，应用程序将无法正常运行。所以，更愿意采用第二种方式，把 ACF 文件完全导出成为纯粹的 CPP 代码，成为几个独立的函数，以方便使用。本书主要采用纯粹的 CPP 代码方式进行。

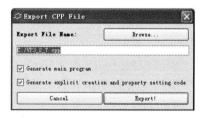

图 4.3.1　导出 ACF 文件

要配置运行纯粹的 CPP 代码 Vega Prime 应用程序,首先要学会从 LynX Prime 导出 CPP 代码。选择前面建立的 VP3_2_7_.acf 文件，因为这个文件功能比较齐全而又不过于复杂。启动 LynX Prime,打开前面建立的 VP3_2_7_.acf 文件,选择菜单"File"→"Export CPP File"，将会弹出图 4.3.1 所示的窗口。其中有两个选项，第二个选项是默认选中的，另一个是包含或不包含 main 函数的，选择包含 main 函数的这一项。导出完成后，将得到 VP3_2_7.cpp 文件。

4.3.2 Vega Prime 的最小应用程序

从上一节导出的 main 函数认识最小的 Vega Prime 应用程序。这个 main 是 Vega Prime 最基本的应用程序配置，如图 4.3.2 所示。

```
#include "stdafx.h"
#include "vpApp.h"
int _tmain(int argc, _TCHAR* argv[])
{
    vp::initialize(argc, argv);        //初始化 VP

    vpApp *app=new vpApp;              //创建一个 vpApp 类

    app->define("VP3_2_7.acf");        //装载 ACF 文件

    app->configure( );                 //配置应用

    app->run( );                       //应用运行和仿真循环

    app->unref( );                     //取消引用

    vp::shutdown( );                   //退出 VP
}
```

图 4.3.2　Vega Prime 最小应用

　　整个应用包含了配置 Vega Prime 应用的 API，包括用来初始化、定义、配置和退出仿真循环的 API 命令和类。

　　基于 STL（标准模板库）和 C++的 API 显得非常紧凑和灵活。Vega Prime 的 API 通过模板和继承性的使用使得仿真循环更加简洁而有效，实时控制包括装载 ACF、配置 ACF、应用运行和仿真循环、取消引用及最后退出 VP 仿真循环。

1. 初始化

vp::initialize 执行以下任务：
- 检查 license 是否正确；
- 初始化静态变量（static variables）和单例类（singleton classes）；
- 初始化内存分配（memory allocator）；
- 初始化渲染库（rendering library）；
- 初始化场景图（scene graph）；
- 初始化 ACF 剖析程序（ACF parser）；
- 初始化模块界面（module interface）；
- 初始化内核（kernel classes）。

需要注意的是，在自己定制的应用中，ACF 中的模块不需要初始化。

vpModule::initializeModule 是告诉用户如何初始化自己的应用中所添加的模块。初始化所添加模块和用户定制模块需要使用以下句法：

```
//
//初始化所有模块
vpModule::initializeModule( "vp" );
vpModule::initializeModule( "vpEnv" );
```

```
vpModule::initializeModule( "vpMotion" );
vpModule::initializeModule( "vpLADBM" );
vpModule::initializeModule( "vpFx" );
vpModule::initializeModule( "vpIR" );
```

2. vpApp 类

vpApp 类用来定义一个典型的 VP 应用的框架。它在 vpApp.h 中被定义了。所有子方法（member methods）都使用内联函数，使用者可以复制和修改 vpApp 类。

vpApp 的主体封装了 Vega Prime 应用中经常用到的 vpKernel 的功能，vpApp 类控制实时功能，包括定义 ACF、配置仿真类、仿真循环、更新和退出。

```
vpApp *app = new vpApp;
```

从 vpApp 类创建一个实例对象，这是虚拟仿真应用开始的最简单方法。如果对 C++很熟悉，现在就可以轻松运用 vpApp 了。

vpApp 类的所有成员都可以多次定制以满足应用的需求，用户可以通过继承 vpApp 类来创建自己的类。

3. 定义语句

定义语句可以替代 ACF 执行功能，参数就是用字符串来替代 ACF。与以往的 Vega 不同的是，在 Vega Prime 中，可以多次定制并且实时调用多个 ACF。这个功能代替了 Vega 的 vgScan 函数。

调入 ACF 文件，假定 argv[1]就是当前的 ACF 文件：

```
if (argv[1])
    app->define(argv[1]);
```

函数原型为：

```
define(const char* filename);
virtual int vpApp::define(const char *filename);
```

描述——装载和解析 ACF 文件。

自变量——filename 是所需 ACF 文件的名称。这个值可以用一行语句来表示，也可以通过复杂的语句来表示。

4. 配置

配置从 ACF 中分解而来，同时将不同的类关联起来。例如，它将系统中定义的 pipeline 添加给服务管理器，并且为每个类配置相关的联系。config 功能是相互的，通过 unconfig 可以将应用配置返回到 config 前的状态。config 方法经常被用户反复运用。

配置示例：

```
app->configure（）;
```

5. 仿真循环

仿真循环包括一个功能呼叫：

void vpApp::run（）

run（）执行主要的仿真循环，这个功能会持续呼叫 beginFrame（），接下来是 endFrame（）用来结束循环。当然还可以在循环过程中用 breakFrameLoop（）来结束循环，接着这项功能会呼叫 unconfigure（）。

这种方法也经常被用户反复运用于自己定义的应用中。

6. 仿真更新

更新 kernel 发生在主循环的中间：

vpKernel::update（）

这种方法在主循环的中间通过应用来调用，否则它会被 vpKernel::endFrame（）自动调用，这个过程发生在 pKernel::processNonLatencyCritical 信息传递给 kernel 之前，发生在仿真循环的 non-latency-critical 阶段。

关于帧的准确位置有几点说明。所有仿真对象都由应用程序来定位，要么是通过使用 vsTraversalUpdate 来自动更新，要么是通过代码来手动更新。因此，应用程序的仿真更新部分大概分为 3 部分：

- Pre-update：包括仿真循环的代码和由 vpKernel::EVENT_PRE_UPDATE_TRAVERSAL 事件执行的代码。
- VSG 更新：VSG 场景更新由 vpKernel::update 内部触发。
- Post-update：包括 vpKernel::EVENT_POST_UPDATE_TRAVERSAL 事件执行的代码和主仿真循环中 vpKernel::update 执行后的代码。

依赖于 vsTraversalLocate 或者直接获取信息（如 vsTransform）的位置查询是保持不变的，它们并不修改或更新场景。因此，通常来说位置查询可能产生不同的结果，这要看位置查询是在场景更新前还是更新后。如果在更新前执行，结果就会和预期的不同；如果在更新后执行，结果就会和预期的一致。因此，建议所有的位置查询都在仿真更新后执行。

7. 关闭

退出 Vega Prime。

vp::shutdown（）;

vp::shutdown 执行以下任务：

- 释放被 kernel classes 分配的内存；
- 结束各模块以释放它们在应用中所占用的内存；
- 终止多线程；
- 将 licenses 返回给 license server。

4.4 剖析 Vega Prime 应用程序组成

前面从 ACF 文件导出的 CPP 文件，没有编写一行代码，都是通过 LynX Prime 操作界面配置完成的。剖析 Vega Prime 应用，不仅仅是为了熟悉代码，更重要的是使用这些代码。本节剖析的代码，稍做修改，都可以作为开发其他 Vega Prime 应用的代码。这也为开发虚拟仿真程序提供了一个思路，Vega Prime 应用仿真的开发，基本功能都可以借助 LynX Prime 完成可视化图形配置编辑，然后再导出为 CPP 功能代码使用，可以大大地提高编码效率。

4.4.1 基本组成

导出的 CPP 文件，包含几个函数：void define(void)，void unconfigure(void)，void main(int argc, char *argv[])。其中只有一个 main 函数。首先，仔细分析头文件，主要的头文件如图 4.4.1 所示。

```
#include "vpModule.h"
#include "vpKernel.h"
#include "vpSearchPath.h"
#include "vpPipeline.h"
#include "vpWindow.h"
#include "vpChannel.h"
#include "vpObserver.h"
#include "vpScene.h"
#include "vpTransform.h"
#include "vpObject.h"
#include "vpIsectorBump.h"
#include "vpIsectorTripod.h"
#include "vpIsectorServiceInline.h"
#include "vpGroundClamp.h"
#include "vpRecyclingService.h"
#include "vpEnv.h"
#include "vpEnvSun.h"
#include "vpEnvMoon.h"
#include "vpEnvSkyDome.h"
#include "vpEnvCloudLayer.h"
#include "vpEnvWind.h"
#include "vpMotionDrive.h"
#include "vpFxDebris.h"
#include "vpApp.h"
```

图 4.4.1　Vega Prime 应用的头文件

从程序代码的头文件可以看出，Vega Prime 进行了很好的模块化设计，完全能够按照应用程序的需要进行加载配置。在这个应用中，依次为：模块头文件、内核头文件、路径搜索头文件、管道头文件、窗口头文件、通道头文件、观察者头文件、场景头文件、转换头文件、对象头文件、碰撞检测头文件、三角碰撞检测头文件、碰撞内联服务头文件、地面夹线头文件、循环服务头文件、环境头文件、太阳头文件、月亮头文件、天幕头文件、云层头文件、风头文件、驾驶模式头文件、粒子特效头文件、应用头文件。在使用某个功能类时，均需要加载相应的头文件。

基本组成如图 4.4.2 所示，就是所导出的 main 函数，与前面所描述的 Vega Prime 的最小应用相同。

```
void main(int argc, char *argv[])
{
    vp::initialize(argc, argv);      //初始化 Vega Prime

    vpApp *app = new vpApp;          //创建应用实例

    define();                        //定义应用

    app->configure();                //配置应用

    app->run();                      //执行主循环

    unconfigure();                   //取消配置

    app->unref();                    //删除应用实例

    vp::shutdown();                  //关闭 Vega Prime
}
```

图 4.4.2　Vega Prime 的最小应用

4.4.2　创建容器

需要创建一个静态列表容器实例 s_pInstancesToUnref。在这个应用中，创建了一个静态列表容器实例 s_pInstancesToUnref，具体代码如图 4.4.3 所示。当建立实例对象时，可以把对象放入容器；当取消配置时，可以删除对象。这样，便于管理实例对象。在后续章节中，建立完对象后都会把对象放进 s_pInstancesToUnref。

```
typedef vuVector< vuField< vuBase* > > InstancesList;
static InstancesList* s_pInstancesToUnref = NULL;
s_pInstancesToUnref = new InstancesList;
```

图 4.4.3　创建 Vega Prime 应用容器

4.4.3　初始化模块

在正式使用某些模块之前，需要对模块进行初始化。在这个应用中，分别对环境、运动模式、特效模块进行初始化，具体代码如图 4.4.4 所示。初始化时，直接调用模块类的静态成员 vpModule::initializeModule 对模块进行初始化，参数仅仅是某个模块的字符名称。

```
vpModule::initializeModule( "vpEnv" );
vpModule::initializeModule( "vpMotion" );
vpModule::initializeModule( "vpFx" );
```

图 4.4.4　初始化 Vega Prime 模块

4.4.4 建立内核

每个 Vega Prime 应用都必须有一个内核对象实例。在这个应用中，建立了内核实例 pKernel_myKernel，然后使用内核实例。设置了消息提示的级别，设置了内核的线程优先级，设置了处理器的数目，设置了帧循环率限制的使能，设置了期望的帧循环率，具体代码如图 4.4.5 所示。

```
vpKernel* pKernel_myKernel = vpKernel::instance();
pKernel_myKernel->setNotifyLevel( vuNotify::LEVEL_WARN );
pKernel_myKernel->setNotifyColorEnable( false );
pKernel_myKernel->setPriority( vuThread::PRIORITY_NORMAL );
pKernel_myKernel->setProcessor( -1 );
pKernel_myKernel->setFrameRateLimiterEnable( true );
pKernel_myKernel->setDesiredFrameRate( 0.000000 );
pKernel_myKernel->setNotifyClassTypeMode( vuNotify::CLASS_TYPE_MODE_INCLUSIVE );
```

图 4.4.5　创建 Vega Prime 应用内核实例

4.4.5 建立路径

每个 Vega Prime 应用都应该有一个路径搜索对象实例。在这个应用中，建立了路径搜索实例 pSearchPath_mySearchPath，其中的一个方法为 append，分别添加了地形、房屋、汽车和谷仓所在的目录。这样，在后面创建实例对象时就能找到相关文件，具体代码如图 4.4.6 所示。读者会发现，这里使用的都是绝对路径，很难满足实际应用需求。

```
vpSearchPath* pSearchPath_mySearchPath = vpSearchPath::instance();

pSearchPath_mySearchPath->append(
"C:/Program Files/MultiGen-Paradigm/resources/tutorials/vegaprime/desktop_tutor/tornado/data/land" );

pSearchPath_mySearchPath->append(
"C:/Program Files/MultiGen-Paradigm/resources/tutorials/vegaprime/desktop_tutor/tornado/data/farmhouse" );

pSearchPath_mySearchPath->append(
"C:/Program Files/MultiGen-Paradigm/resources/tutorials/vegaprime/desktop_tutor/tornado/data/humv-dirty" );

pSearchPath_mySearchPath->append(
"C:/Program Files/MultiGen-Paradigm/resources/tutorials/vegaprime/desktop_tutor/tornado/data/grainstorage");
```

图 4.4.6　创建 Vega Prime 应用路径搜索实例

4.4.6 建立管道

每个 Vega Prime 应用都应该有一个管道对象实例。在这个应用中，建立了管道实例 pPipeline_myPipeline，然后使用管道实例，设置了管道名字，设置了管道的线程优先级，设置了管道 Id 号，设置了剪切线程数目，设置了剪切线程优先级，设置了剪切线程处理器数目，设置了绘制线程优先级，设置了绘制线程处理器数目，具体代码如图 4.4.7 所示。最后，把管道放进在第 4.4.2 节创建的容器队列中：

  s_pInstancesToUnref->push_back(pPipeline_myPipeline)

```
vpPipeline* pPipeline_myPipeline = new vpPipeline();
    pPipeline_myPipeline->setName( "myPipeline" );
    pPipeline_myPipeline->setMultiThread( vsPipeline::MULTITHREAD_INLINE );
    pPipeline_myPipeline->setId( 0 );
    pPipeline_myPipeline->setNumCullThreads( 0 );
    pPipeline_myPipeline->setCullThreadPriority( vuThread::PRIORITY_NORMAL );
    pPipeline_myPipeline->setCullThreadProcessor( -1 );
    pPipeline_myPipeline->setDrawThreadPriority( vuThread::PRIORITY_NORMAL );
    pPipeline_myPipeline->setDrawThreadProcessor( -1 );
    pPipeline_myPipeline->setBeginFrameOnVsyncEnable( false );
    pPipeline_myPipeline->setDesiredPostDrawTime( -1.000000 );

    s_pInstancesToUnref->push_back( pPipeline_myPipeline );
```

图 4.4.7　创建 Vega Prime 应用管道实例

4.4.7　建立窗口

每个 Vega Prime 应用都应该有一个窗口对象实例。在这个应用中，建立了窗口实例 pWindow_myWindow，然后使用窗口实例，设置了窗口名字，设置了窗口的标签，设置了窗口起始点，设置了窗口尺寸，设置了窗口全屏选项，设置了窗口边界，设置了窗口输入选项，设置了窗口鼠标形状，设置了窗口缓存交换间隔，设置了窗口像素缓存模式，等等，具体代码如图 4.4.8 所示。最后，把窗口放进在第 4.4.2 节创建的容器队列中：

  s_pInstancesToUnref->push_back(pWindow_myWindow);

```
vpWindow* pWindow_myWindow = new vpWindow();
    pWindow_myWindow->setName( "myWindow" );
    pWindow_myWindow->setLabel( "Vega Prime Window" );
    pWindow_myWindow->setOrigin( 0 ,  0 );
    pWindow_myWindow->setSize( 1024 ,  768 );
    pWindow_myWindow->setFullScreenEnable( false );
    pWindow_myWindow->setBorderEnable( true );
    pWindow_myWindow->setInputEnable( true );
    pWindow_myWindow->setCursorEnable( true );
    pWindow_myWindow->setStereoEnable( false );
    pWindow_myWindow->setNumColorBits( 8 );
    pWindow_myWindow->setNumAlphaBits( 0 );
    pWindow_myWindow->setNumDepthBits( 24 );
    pWindow_myWindow->setNumStencilBits( 0 );
    pWindow_myWindow->setNumAccumColorBits( 0 );
    pWindow_myWindow->setNumAccumAlphaBits( 0 );
    pWindow_myWindow->setNumMultiSampleBits( 0 );
```

```
pWindow_myWindow->setSwapInterval( 1 );
pWindow_myWindow->setPixelBufferMode( vrDrawContext::PIXEL_BUFFER_MODE_OFF );
s_pInstancesToUnref->push_back( pWindow_myWindow );
```

图 4.4.8　创建 Vega Prime 应用窗口实例

4.4.8　建立通道

每个 Vega Prime 应用至少应该有一个通道对象实例。在这个应用中，建立了通道实例 pChannel_myChannel，然后使用通道实例，设置了通道名字，设置了通道的位置偏移，设置了通道的姿态偏移，设置了通道剪切掩码，设置了通道渲染掩码，设置了通道绘制区域，设置了通道远近，设置了通道剪切线程优先级，设置了通道剪切线程处理器数目，设置了通道图形模式(线框模式、透明模式、材质模式、光线模式、雾模式)，等等。具体代码如图 4.4.9 所示。最后，把通道放进在第 4.4.2 节创建的容器队列中：

```
s_pInstancesToUnref->push_back( pChannel_myChannel );
```

```
vpChannel* pChannel_myChannel = new vpChannel();
    pChannel_myChannel->setName( "myChannel" );
    pChannel_myChannel->setOffsetTranslate( 0.000000 , 0.000000 , 0.000000 );
    pChannel_myChannel->setOffsetRotate( 0.000000 , 0.000000 , 0.000000 );
    pChannel_myChannel->setCullMask( 0x0FFFFFFFF );
    pChannel_myChannel->setRenderMask( 0x0FFFFFFFF );
    pChannel_myChannel->setClearColor( 0.000000f , 0.500000f ,
              1.000000f , 0.000000f );
    pChannel_myChannel->setClearBuffers( 0x03 );
    pChannel_myChannel->setDrawArea( 0.000000 , 1.000000 , 0.000000 , 1.000000 );
    pChannel_myChannel->setFOVSymmetric( 45.000000f , -1.000000f );
    pChannel_myChannel->setNearFar( 1.000000f , 35000.000000f );
    pChannel_myChannel->setLODVisibilityRangeScale( 1.000000 );
    pChannel_myChannel->setLODTransitionRangeScale( 1.000000 );
    pChannel_myChannel->setCullThreadPriority( vuThread::PRIORITY_NORMAL );
    pChannel_myChannel->setCullThreadProcessor( -1 );

    pChannel_myChannel->setGraphicsModeEnable(
              vpChannel::GRAPHICS_MODE_WIREFRAME , false );
    pChannel_myChannel->setGraphicsModeEnable(
              vpChannel::GRAPHICS_MODE_TRANSPARENCY , true );
    pChannel_myChannel->setGraphicsModeEnable(
              vpChannel::GRAPHICS_MODE_TEXTURE , true );
    pChannel_myChannel->setGraphicsModeEnable(
              vpChannel::GRAPHICS_MODE_LIGHT , true );
    pChannel_myChannel->setGraphicsModeEnable(
              vpChannel::GRAPHICS_MODE_FOG , true );
    pChannel_myChannel->setLightPointThreadPriority( vuThread::PRIORITY_NORMAL );
    pChannel_myChannel->setLightPointThreadProcessor( -1 );
    pChannel_myChannel->setMultiSample( vpChannel::MULTISAMPLE_OFF );
```

```
pChannel_myChannel->setStatisticsPage( vpChannel::PAGE_OFF );
pChannel_myChannel->setCullBoundingBoxTestEnable( false );
pChannel_myChannel->setOpaqueSort(vpChannel::OPAQUE_SORT_TEXTURE, vpChannel:: OPAQUE_
                    SORT_MATERIAL );
pChannel_myChannel->setTransparentSort( vpChannel::TRANSPARENT_SORT_DEPTH );
pChannel_myChannel->setDrawBuffer( vpChannel::DRAW_BUFFER_DEFAULT );
pChannel_myChannel->setStressEnable( false );
pChannel_myChannel->setStressParameters( 1.000000f,  20.000000f,
                              0.750000f,  0.500000f,  2.000000f );

s_pInstancesToUnref->push_back( pChannel_myChannel );
```

图 4.4.9　创建 Vega Prime 应用通道实例

4.4.9　建立观察者

每个 Vega Prime 应用至少应该有一个观察者对象实例。在这个应用中，建立了观察者实例 pObserver_myObserver，然后使用观察者实例，设置了观察者名字，设置了观察者的策略选项，设置了观察者的位置，设置了观察者的姿态，设置了观察者潜在关键选项，具体代码如图 4.4.10 所示。最后，把观察者放进在第 4.4.2 节创建的容器队列中：

　　s_pInstancesToUnref->push_back(pObserver_myObserver);

```
vpObserver* pObserver_myObserver = new vpObserver();
pObserver_myObserver->setName( "myObserver" );
pObserver_myObserver->setStrategyEnable( false );
pObserver_myObserver->setTranslate( 2300.000000,  2500.000000,  15.000000 );
pObserver_myObserver->setRotate( -90.000000,  0.000000,  0.000000 );
pObserver_myObserver->setLatencyCriticalEnable( false );

s_pInstancesToUnref->push_back( pObserver_myObserver );
```

图 4.4.10　创建 Vega Prime 应用观察者实例

4.4.10　建立场景

每个 Vega Prime 应用至少应该有一个场景对象实例。在这个应用中，建立了场景实例 pScene_myScene，然后使用场景实例，设置了场景名字，具体代码如图 4.4.11 所示。最后，把场景放进在第 4.4.2 节创建的容器队列中：

　　s_pInstancesToUnref->push_back(pScene_myScene);

```
vpScene* pScene_myScene = new vpScene();
pScene_myScene->setName( "myScene" );

s_pInstancesToUnref->push_back( pScene_myScene );
```

图 4.4.11　创建 Vega Prime 应用场景实例

4.4.11 建立转换

Vega Prime 应用可以根据实际需求建立转换对象实例。在这个应用中，建立了转换实例 pTransform_hummerTransform，然后使用转换实例，设置了转换名字，设置了转换剪切掩码，设置了转换的渲染掩码，设置了转换的碰撞掩码，设置了转换的策略选项，设置了转换的位置，设置了转换的姿态，设置了转换的缩放比例，具体代码如图 4.4.12 所示。最后，把转换放进在第 4.4.2 节创建的容器队列中：

  s_pInstancesToUnref->push_back(pTransform_hummerTransform);

```
vpTransform* pTransform_hummerTransform = new vpTransform();
    pTransform_hummerTransform->setName( "hummerTransform" );
    pTransform_hummerTransform->setCullMask( 0x0FFFFFFFF );
    pTransform_hummerTransform->setRenderMask( 0x0FFFFFFFF );
    pTransform_hummerTransform->setIsectMask( 0x0FFFFFFFF );
    pTransform_hummerTransform->setStrategyEnable( true );
    pTransform_hummerTransform->setTranslate( 0.000000 , -30.000000 , 5.000000 );
    pTransform_hummerTransform->setRotate( 0.000000 , 0.000000 , 0.000000 );
    pTransform_hummerTransform->setScale( 1.000000 , 1.000000 , 1.000000 );
    pTransform_hummerTransform->setStaticEnable( false );

    s_pInstancesToUnref->push_back( pTransform_hummerTransform );
```

图 4.4.12 创建 Vega Prime 应用的转换

4.4.12 建立对象

每个 Vega Prime 应用可能包含较多的实例对象。在这个应用中，建立了地形、房屋、汽车、谷仓等实例，其中的代码绝大部分是相同的，主要不同的是：

- pObject->setName("xxxx");
- pObject->setTranslate(0.000000, 0.000000, 0.000000);
- pObject->setRotate(0.000000, 0.000000, 0.000000);
- pObject->setScale(1.000000, 1.000000, 1.000000);
- pObject->setFileName("xxxx.flt");

这 5 行代码也是创建对象的有所区别的代码，依次为：设置对象名称、设置对象位置、设置对象姿态、设置对象缩放比例、设置对象文件。

以建立房子实例为例，建立了对象实例 pObject_farmhouse，然后使用对象实例，设置了对象名字，设置了对象剪切掩码，设置了对象渲染掩码，设置了对象碰撞掩码，设置了对象策略选项，设置了对象位置，设置了对象姿态，设置了对象缩放比例，设置了对象静态选项，设置了对象文件，设置了对象大量的加载选项以及渲染选项，等等。具体代码如图 4.4.13 所示。最后，把对象放进在第 4.4.2 节创建的容器队列中：

  s_pInstancesToUnref->push_back(pObject_farmhouse);

```cpp
vpObject* pObject_farmhouse = new vpObject();
pObject_farmhouse->setName( "farmhouse" );
pObject_farmhouse->setCullMask( 0x0FFFFFFFF );
pObject_farmhouse->setRenderMask( 0x0FFFFFFFF );
pObject_farmhouse->setIsectMask( 0x0FFFFFFFF );
pObject_farmhouse->setStrategyEnable( true );
pObject_farmhouse->setTranslate( 2450.000000 ,   2460.000000 ,   0.000000 );
pObject_farmhouse->setRotate( 0.000000 ,   0.000000 ,   0.000000 );
pObject_farmhouse->setScale( 1.000000 ,   1.000000 ,   1.000000 );
pObject_farmhouse->setStaticEnable( false );
pObject_farmhouse->setFileName( "farmhouse.flt" );
pObject_farmhouse->setAutoPage( vpObject::AUTO_PAGE_SYNCHRONOUS );
pObject_farmhouse->setManualLODChild( -1 );
pObject_farmhouse->setLoaderOption( vsNodeLoader::Data::LOADER_OPTION_COMBINE
                                    _LIGHT_POINTS ,   true );
pObject_farmhouse->setLoaderOption( vsNodeLoader::Data::LOADER_OPTION_COMBINE
                                    _LODS ,   true );
pObject_farmhouse->setLoaderOption( vsNodeLoader::Data::LOADER_OPTION_IGNORE_DOF
                                    _CONSTRAINTS ,   false );
pObject_farmhouse->setLoaderOption( vsNodeLoader::Data::LOADER_OPTION_PRESERVE
                                    _EXTERNAL_REF_FLAGS ,   true );
pObject_farmhouse->setLoaderOption( vsNodeLoader::Data::LOADER_OPTION_PRESERVE
                                    _GENERIC_NAMES ,   false );
pObject_farmhouse->setLoaderOption( vsNodeLoader::Data::LOADER_OPTION_PRESERVE
                                    _GENERIC_NODES ,   false );
pObject_farmhouse->setLoaderOption( vsNodeLoader::Data::LOADER_OPTION
                                    _PRESERVE_QUADS ,   false );
pObject_farmhouse->setLoaderOption( vsNodeLoader::Data::LOADER_OPTION_ALL
                                    _GEOMETRIES_LIT ,   false );
pObject_farmhouse->setLoaderOption( vsNodeLoader::Data::LOADER_OPTION_USE
                                    _MATERIAL_DIFFUSE_COLOR ,   false );
pObject_farmhouse->setLoaderOption( vsNodeLoader::Data::LOADER_OPTION
                                    _MONOCHROME ,   false );
pObject_farmhouse->setLoaderOption( vsNodeLoader::Data::LOADER_OPTION
                                    _CREATE_ANIMATIONS ,   true );
pObject_farmhouse->setLoaderOption( vsNodeLoader::Data::LOADER_OPTION_SHARE
                                    _LIGHT_POINT_CONTROLS ,   true );
pObject_farmhouse->setLoaderOption( vsNodeLoader::Data::LOADER_OPTION_SHARE
                                    _LIGHT_POINT_ANIMATIONS ,   true );
pObject_farmhouse->setLoaderOption( vsNodeLoader::Data::LOADER_OPTION_SHARE
                                    _LIGHT_POINT_APPEARANCES ,   true );
pObject_farmhouse->setLoaderDetailMultiTextureStage( -1 );
pObject_farmhouse->setLoaderBlendTolerance( 0.050000f );
pObject_farmhouse->setLoaderUnits( vsNodeLoader::Data::LOADER_UNITS_METERS );
pObject_farmhouse->setBuilderOption( vsNodeLoader::Data::BUILDER_OPTION
                                    _OPTIMIZE_GEOMETRIES ,   true );
pObject_farmhouse->setBuilderOption( vsNodeLoader::Data::BUILDER_OPTION
```

```
                                  _COLLAPSE_BINDINGS,  true );
pObject_farmhouse->setBuilderOption( vsNodeLoader::Data::BUILDER_OPTION
                                  _COLLAPSE_TRIANGLE_STRIPS,  true );
pObject_farmhouse->setBuilderNormalMode( vsNodeLoader::Data::BUILDER
                                  _NORMAL_MODE_PRESERVE );
pObject_farmhouse->setBuilderColorTolerance( 0.001000f );
pObject_farmhouse->setBuilderNormalTolerance( 0.860000f );
pObject_farmhouse->setBuilderVertexTolerance( 0.000100f );
pObject_farmhouse->setGeometryOption( vsNodeLoader::Data::GEOMETRY
                                  _OPTION_GENERATE_DISPLAY_LISTS,  false );
pObject_farmhouse->setGeometryFormat( vrGeometryBase::FORMAT_VERTEX_ARRAY,  0x0FFF );
pObject_farmhouse->setPostLoadOption( vpGeometryPageable::POST_LOAD
                                  _OPTION_FLATTEN,  true );
pObject_farmhouse->setPostLoadOption( vpGeometryPageable::POST_LOAD
                                  _OPTION_CLEAN,  true );
pObject_farmhouse->setPostLoadOption( vpGeometryPageable::POST_LOAD_OPTION
                                  _MERGE_GEOMETRIES,  true );
pObject_farmhouse->setPostLoadOption( vpGeometryPageable::POST_LOAD_OPTION
                                  _COLLAPSE_BINDINGS,  true );
pObject_farmhouse->setPostLoadOption( vpGeometryPageable::POST_LOAD_OPTION
                                  _COLLAPSE_TRIANGLE_STRIPS,  true );
pObject_farmhouse->setPostLoadOption( vpGeometryPageable::POST_LOAD
                                  _OPTION_VALIDATE,  true );
pObject_farmhouse->setTextureSubloadEnable( false );
pObject_farmhouse->setTextureSubloadRender( vpGeometry::TEXTURE_SUBLOAD
                                  _RENDER_DEFERRED );

s_pInstancesToUnref->push_back( pObject_farmhouse );
```

图 4.4.13 创建 Vega Prime 应用对象实例

4.4.13 建立碰撞检测

Vega Prime 应用可以根据需求建立碰撞检测实例。在这个应用中，建立了碰撞检测实例 pIsectorBump_bumpIsector，然后使用碰撞检测实例，设置了碰撞检测名字，设置了碰撞检测选项，设置了碰撞检测渲染选项，设置了碰撞检测位置，设置了碰撞检测姿态，设置了碰撞检测模式，设置了碰撞检掩码，设置了碰撞检测策略选项，设置了碰撞检测空间，具体代码如图 4.4.14 所示。最后，把碰撞检测放进在第 4.4.2 节创建的容器队列中：

s_pInstancesToUnref->push_back(pIsectorBump_bumpIsector);

```
vpIsectorBump* pIsectorBump_bumpIsector = new vpIsectorBump();
    pIsectorBump_bumpIsector->setName( "bumpIsector" );
    pIsectorBump_bumpIsector->setEnable( true );
    pIsectorBump_bumpIsector->setRenderEnable( false );
    pIsectorBump_bumpIsector->setTranslate( 0.000000,  0.000000,  0.000000 );
    pIsectorBump_bumpIsector->setRotate( 0.000000,  0.000000,  0.000000 );
```

```
pIsectorBump_bumpIsector->setMode( 0x02 );
pIsectorBump_bumpIsector->setIsectMask( 0x0FFFFFFFF );
pIsectorBump_bumpIsector->setStrategyEnable( true );
pIsectorBump_bumpIsector->setDimensions( 3.000000f, 3.000000f, 3.000000f );

s_pInstancesToUnref->push_back( pIsectorBump_bumpIsector );
```

图 4.4.14　创建 Vega Prime 应用碰撞检测实例

4.4.14　建立碰撞服务

建立了碰撞检测以后，必须建立碰撞服务实例。在这个应用中，建立了碰撞检测实例服务实例 pIsectorServiceInline_myIsectorService，然后使用碰撞检测服务实例。只需要设置碰撞检测服务名字，具体代码如图 4.4.15 所示。最后，把碰撞检测服务实例放进在第 4.4.2 节创建的容器队列中：

　　　　s_pInstancesToUnref->push_back(pIsectorServiceInline_myIsectorService);

```
vpIsectorServiceInline* pIsectorServiceInline_myIsectorService = new vpIsectorServiceInline();
    pIsectorServiceInline_myIsectorService->setName( "myIsectorService" );

    s_pInstancesToUnref->push_back( pIsectorServiceInline_myIsectorService );
```

图 4.4.15　创建 Vega Prime 应用碰撞服务对象

4.4.15　建立循环服务

每个 Vega Prime 应用必须建立循环服务实例。在这个应用中，建立了循环服务实例 pRecyclingService_myRecyclingService，然后使用循环服务实例，设置了循环服务实例的线程方式，设置了循环服务实例线程优先级，设置了循环服务实例线程处理器数目，设置了循环服务实例的循环时间。循环服务实例与其他对象不同，不需要把循环服务实例放进在第 4.4.2 节创建的容器队列中。具体代码如图 4.4.16 所示。

```
vpRecyclingService* pRecyclingService_myRecyclingService = vpRecyclingService::instance();
pRecyclingService_myRecyclingService->setMultiThread( vpRecyclingService::MULTITHREAD_INLINE );
pRecyclingService_myRecyclingService->setThreadPriority( vuThread::PRIORITY_NORMAL );
pRecyclingService_myRecyclingService->setThreadProcessor( -1 );
pRecyclingService_myRecyclingService->setRecycleTime( -1.000000 );
```

图 4.4.16　创建 Vega Prime 应用循环服务实例

4.4.16　建立环境

每个 Vega Prime 应用可以建立环境对象。在这个应用中，建立了环境对象 pEnv_myEnv，然后使用环境对象，设置了环境的名字，设置了环境的日期，设置了环境的变化快慢，设置了环境的更新间隔，设置了环境的参考位置，设置了环境的时区差异，设置了环境的颜色，设置了环境的可见类型，设置了环境的日落可见范围，设置了环境的透明可见范围，设置了环境的可见颜色，设置了环境的本地可见选项，等等。具体代码如图 4.4.17 所示。最后，把环境对象放进在第 4.4.2 节创建的容器队列中：

```
s_pInstancesToUnref->push_back( pEnv_myEnv );
```

```
vpEnv* pEnv_myEnv = new vpEnv();
pEnv_myEnv->setName( "myEnv" );
pEnv_myEnv->setDate( 22, 6, 2002 );
pEnv_myEnv->setTimeOfDay( 12.000000f );
pEnv_myEnv->setTimeMultiplier( 1.000000f );
pEnv_myEnv->setEphemerisUpdateInterval( 0.000000f );
pEnv_myEnv->setReferencePosition( -96.790001f, 32.790001f );
pEnv_myEnv->setEphemerisTimeZoneOffset( -10000 );
pEnv_myEnv->setSkyColor( 0.513725f, 0.701961f, 0.941176f, 1.000000f );
pEnv_myEnv->setVisibilityType( vpEnvFx::VISIBILITY_TYPE_PIXEL_EXP2 );
pEnv_myEnv->setVisibilityRangeOnset( 0.000000f );
pEnv_myEnv->setVisibilityRangeOpaque( 80000.000000f );
pEnv_myEnv->setVisibilityColor( 1.000000f, 1.000000f, 1.000000f, 1.000000f );
pEnv_myEnv->setLocalViewerEnable( false );
pEnv_myEnv->setTwoSidedLightingEnable( false );
pEnv_myEnv->setAmbientLightingColor( 0.000000f, 0.000000f, 0.000000f, 1.000000f );
pEnv_myEnv->setLightColorScale( vpEnv::COLOR_AMBIENT, 1.000000f );
pEnv_myEnv->setLightColorScale( vpEnv::COLOR_DIFFUSE, 1.000000f );
pEnv_myEnv->setLightColorScale( vpEnv::COLOR_SPECULAR, 1.000000f );
pEnv_myEnv->setDominantCelestialLightingEnable( false );

s_pInstancesToUnref->push_back( pEnv_myEnv );
```

图 4.4.17　创建 Vega Prime 应用环境对象

4.4.17　建立太阳

每个 Vega Prime 应用建立环境对象以后，一般会接着建立环境对象中的太阳对象、月亮对象、云层对象等，这里以建立太阳对象为例进行说明。在这个应用中，建立了太阳对象 pEnvSun_myEnvSun，然后使用太阳对象，设置了太阳的名字，设置了太阳的使能选项，设置了太阳的黄昏设置，设置了太阳的几何体选项，设置了太阳的材质文件，设置了太阳的材质混合颜色，设置了太阳的材质混合模式，设置了太阳的颜色，设置了太阳的可见范围，设置了太阳的角度尺寸，设置了太阳的地平线角度，设置了太阳的地平线颜色，设置了太阳的地平线尺寸缩放比例，设置了太阳的地平线光线颜色比例，具体代码如图 4.4.18 所示。最后，把太阳对象放进在第 4.4.2 节创建的容器队列中：

```
s_pInstancesToUnref->push_back( pEnvSun_myEnvSun );
```

```
vpEnvSun* pEnvSun_myEnvSun = new vpEnvSun();
pEnvSun_myEnvSun->setName( "myEnvSun" );
pEnvSun_myEnvSun->setEnable( true );
pEnvSun_myEnvSun->setTwilightDip( -18.000000f );
```

```
pEnvSun_myEnvSun->setGeometryEnable( true );
pEnvSun_myEnvSun->setTextureFile( "sun.inta" );
pEnvSun_myEnvSun->setTextureBlendColor( 1.000000f,   1.000000f,   1.000000f,   1.000000f );
pEnvSun_myEnvSun->setTextureBlendMode( vpEnvSun::TEXTURE_BLEND_MODE_MODULATE );
pEnvSun_myEnvSun->setColor( 0.992156f,   1.000000f,   0.949019f,   1.000000f );
pEnvSun_myEnvSun->setVisibilityRange( 8000.000000f );
pEnvSun_myEnvSun->setAngularSize( 5.500000f );
pEnvSun_myEnvSun->setHorizonAngle( 40.000000f );
pEnvSun_myEnvSun->setHorizonColor( 1.000000f,   0.545098f,   0.239216f,   1.000000f );
pEnvSun_myEnvSun->setHorizonSizeScale( 2.000000f );
pEnvSun_myEnvSun->setHorizonLightColorScale( 0.250000f );

s_pInstancesToUnref->push_back( pEnvSun_myEnvSun );
```

图 4.4.18　创建 Vega Prime 应用太阳对象

4.4.18　建立运动模式

Vega Prime 应用通常包含一定的运动模式。在这个应用中，建立了驾驶运动模式对象 pMotionDrive_myMotion，然后使用驾驶模式对象，设置了驾驶模式的名字，设置了驾驶模式的速度，设置了驾驶模式的速度限制模式，设置了驾驶模式的速度变化量，设置了驾驶模式的朝向变化量，具体代码如图 4.4.19 所示。最后，把驾驶运动模式对象放进在第 4.4.2 节创建的容器队列中：

　　s_pInstancesToUnref->push_back(pMotionDrive_myMotion);

```
vpMotionDrive* pMotionDrive_myMotion = new vpMotionDrive();
    pMotionDrive_myMotion->setName( "myMotion" );
    pMotionDrive_myMotion->setSpeed( 0.000000 );
    pMotionDrive_myMotion->setSpeedLimits( -10.000000,   10.000000 );
    pMotionDrive_myMotion->setSpeedDelta( 1.000000,   1.000000 );
    pMotionDrive_myMotion->setHeadingDelta( 1.000000 );

    s_pInstancesToUnref->push_back( pMotionDrive_myMotion );
```

图 4.4.19　创建 Vega Prime 应用运动模式对象

4.4.19　建立特效

Vega Prime 应用通常包含一定的特效。在这个应用中，建立了碎片特效对象 pFxDebris_Debris，然后使用碎片特效对象，设置了碎片特效的名字，设置了碎片特效的剪切掩码，设置了碎片特效的渲染掩码，设置了碎片特的碰撞掩码，设置了碎片特效的策略选项，设置了碎片特效的位置，设置了碎片特效的姿态，设置了碎片特效的缩放比例，设置了碎片特效的静态选项，设置了碎片特效的重复选项，设置了碎片特效的颜色设置，设置了碎片特效的材质混合颜色，设置了碎片特效的材质模式，设置了碎片特效的材质文件，设置了碎片特效的持续时间，设置了碎片特效的消失持续时间，设置了碎片特效的透明深度补偿，设置了碎片特效的透明深度半径比例，设置了碎片特效的风源，具体代码如图 4.4.20 所示。最后，把特效对象放进在第 4.4.2 节创建的容器队列中：

  s_pInstancesToUnref->push_back(pFxDebris_Debris);

```
vpFxDebris* pFxDebris_Debris = new vpFxDebris();
pFxDebris_Debris->setName( "Debris" );
pFxDebris_Debris->setCullMask( 0x0FFFFFFFF );
pFxDebris_Debris->setRenderMask( 0x0FFFFFFFF );
pFxDebris_Debris->setIsectMask( 0x0FFFFFFFF );
pFxDebris_Debris->setStrategyEnable( true );
pFxDebris_Debris->setTranslate( 0.000000 , 0.000000 , 0.000000 );
pFxDebris_Debris->setRotate( 0.000000 , 0.000000 , 0.000000 );
pFxDebris_Debris->setScale( 20.000000 , 20.000000 , 20.000000 );
pFxDebris_Debris->setStaticEnable( false );
pFxDebris_Debris->setRepeatEnable( false );
pFxDebris_Debris->setOverallColor( 1.000000f , 1.000000f , 1.000000f , 1.000000f );
pFxDebris_Debris->setTextureBlendColor( 1.000000f , 1.000000f , 1.000000f , 1.000000f );
pFxDebris_Debris->setTextureMode( vpFx::TEXTURE_MODE_MODULATE );
pFxDebris_Debris->setTextureFile( "debris.inta" );
pFxDebris_Debris->setOverallDuration( 4.000000f );
pFxDebris_Debris->setFadeDuration( 0.000000f , 0.000000f );
pFxDebris_Debris->setTransparencyDepthOffset( 0.000000f );
pFxDebris_Debris->setTransparencyDepthRadiusScaler( 0.000000f );
pFxDebris_Debris->setWindSource( vpFxParticleSystem::WIND_SOURCE_TABLE );

s_pInstancesToUnref->push_back( pFxDebris_Debris );
```

图 4.4.20   Vega Prime 的最小应用

4.4.20 配置应用

在 Vega Prime 应用中，建立完各种对象以后，需要进行一定的配置。在这个应用中，进行以下配置：

1. 配置管道、窗口和通道

  pPipeline_myPipeline->addWindow(pWindow_myWindow);
  pWindow_myWindow->addChannel(pChannel_myChannel);

2. 配置观察者与策略、通道、环境、场景与观察点

  pObserver_myObserver->setStrategy(pMotionDrive_myMotion);
  pObserver_myObserver->addChannel(pChannel_myChannel);
  pObserver_myObserver->addAttachment(pEnv_myEnv);
  pObserver_myObserver->setScene(pScene_myScene);
  pObserver_myObserver->setLookFrom(pTransform_hummerTransform);

3. 配置场景

```
pScene_myScene->addChild( pObject_terrain );
pScene_myScene->addChild( pObject_farmhouse );
pScene_myScene->addChild( pObject_Hummer );
pScene_myScene->addChild( pObject_gainstore );
```

4. 配置特效

```
pObject_farmhouse->addChild( pFxDebris_Debris );
```

5. 配置运动模式与策略

```
pObject_Hummer->addChild( pTransform_hummerTransform );
pObject_Hummer->setStrategy( pMotionDrive_myMotion );
```

6. 配置碰撞检测目标与参考位置

```
pIsectorBump_bumpIsector->setTarget( pObject_farmhouse );
pIsectorBump_bumpIsector->setPositionReference( pObject_Hummer );
```

完整代码如图 4.4.21 所示。

```
pPipeline_myPipeline->addWindow( pWindow_myWindow );

pWindow_myWindow->addChannel( pChannel_myChannel );


pObserver_myObserver->setStrategy( pMotionDrive_myMotion );
pObserver_myObserver->addChannel( pChannel_myChannel );
pObserver_myObserver->addAttachment( pEnv_myEnv );
pObserver_myObserver->setScene( pScene_myScene );
pObserver_myObserver->setLookFrom( pTransform_hummerTransform );

pScene_myScene->addChild( pObject_terrain );
pScene_myScene->addChild( pObject_farmhouse );
pScene_myScene->addChild( pObject_Hummer );
pScene_myScene->addChild( pObject_gainstore );

pObject_farmhouse->addChild( pFxDebris_Debris );

pObject_Hummer->addChild( pTransform_hummerTransform );
pObject_Hummer->setStrategy( pMotionDrive_myMotion );


pIsectorBump_bumpIsector->setTarget( pObject_farmhouse );
pIsectorBump_bumpIsector->setPositionReference( pObject_Hummer );
```

```
pIsectorTripod_tripodIsector->setTarget( pObject_terrain );

pIsectorServiceInline_myIsectorService->addIsector( pIsectorBump_bumpIsector );

pGroundClamp_myGroundClamp->setIsector( pIsectorTripod_tripodIsector );


pEnv_myEnv->addEnvFx( pEnvSun_myEnvSun );
pEnv_myEnv->addEnvFx( pEnvMoon_myEnvMoon );
pEnv_myEnv->addEnvFx( pEnvSkyDome_myEnvSkyDome );
pEnv_myEnv->addEnvFx( pEnvCloudLayer_myEnvCloudLayer );
pEnv_myEnv->addEnvFx( pEnvWind_myEnvWind );


pMotionDrive_myMotion->setNextStrategy( pGroundClamp_myGroundClamp );

pFxDebris_Debris->setTriggerIsector( pIsectorBump_bumpIsector );
```

图 4.4.21　配置 Vega Prime 应用

第 5 章　建立基于 MFC 对话框的 Vega Prime 应用程序

窗口程序是当前 PC 机程序的主要方式。采用 MFC 对话框程序，充分展示了如何开发基于 MFC 对话框的 Vega Prime 应用程序。其中包括理解 Vega Prime 对话框应用程序、配置对话框编译环境、改造对话框界面、添加公共类和启动主线程。添加公共类是核心一步，把所有与 Vega Prime 虚拟仿真有关的业务逻辑进行封装，与用户界面完全没有关系。这样，按照 MVC 原则将 Vega Prime 虚拟仿真与界面分离，再加上适当的控制逻辑把二者进行关联。这样大大提高了 Vega Prime 虚拟仿真的代码重用性，加快了整个应用开发进程。本章将通过建立基于 MFC 对话框的 Vega Prime 应用程序，展示利用 Vega Prime 的 API 开发虚拟现实应用程序的步骤和过程。

【本章重点】

- 理解 MFC 对话框 Vega Prime 应用程序；
- 配置 MFC 对话框 Vega Prime 应用程序编译环境；
- MFC 对话框程序界面改造；
- 添加 Vega Prime 公共类 PublicMember；
- 启动 Vega Prime 应用主线程；
- 启动包含 ACF 文件的主线程。

5.1　配置 MFC 对话框 Vega Prime 应用程序的编译环境

5.1.1　MFC 对话框 Vega Prime 应用程序的理解

编写基于 MFC 对话框程序的 Vega Prime 程序主要是解决两个问题：
（1）在什么地方执行 Vega Prime 的主循环？
（2）在哪个窗口显示 Vega Prime 场景？

解决第一个问题可以采用的办法是：在 MFC 对话框程序里启动一个工作者线程，在线程的主函数里初始化 Vega Prime 和执行主循环。而第二个问题更简单了，也就是确定 Vega Prime 窗口的父窗口，Vega Prime 的场景在父窗口上显示，具体就是把父窗口的句柄传给函数 setParent()。什么是父窗口？在这里，父窗口就是想用来显示 Vega Prime 场景的窗口。例如，若想在 view 窗口显示 Vega Prime 场景，就把 view 窗口的句柄传给 setParent()；若想在一个对话框上显示 Vega Prime 场景，只需把对话框的句柄传给 setParent()；若想在对话框的某一部分上显示 Vega Prime 场景，只需在对话框添加一个合适控件，然后把该控件的句柄传给 setParent()。其实，基于 MFC 对话框的 Vega Prime 程序是很灵活的，灵活就在于传给 setParent() 的句柄。可以把任何一个窗口的句柄传给它，可以是 view 或 dialog 甚至一个按钮、文本框，更甚者，可以是另外一个进程的窗口。如果把桌面的句柄传给 setParent()，Vega Prime 的场景就直接在桌面上显示。

5.1.2 配置 MFC 对话框 Vega Prime 应用程序的编译环境

打开第 4.1.5 节建立的 MFC 对话框程序 VPTestDialog，如图 5.1.1 所示。

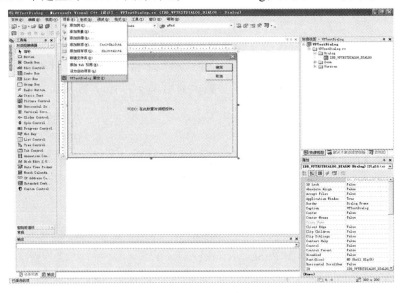

图 5.1.1 MFC 对话框程序

首先，仿照控制台程序的配置来配置项目属性，选择菜单"项目"→"VPTestDialog 属性"，出现配置界面，如图 5.1.2 所示。

图 5.1.2 配置附加包含目录

在图 5.1.2 中，在左侧选择"C/C++"→"常规"，在右边"附加包含目录"里，填入：

$(MPI_LOCATE_VEGA_PRIME)\include\vsg,

$(MPI_LOCATE_VEGA_PRIME)\include\vegaprime。

一定要填写正确，并且在两者之间用逗号隔开。

接着选择"连接器"→"常规"，如图 5.1.3 所示，在右边"附加库目录"里，填入：

$(MPI_LOCATE_VEGA_PRIME_LIB)。

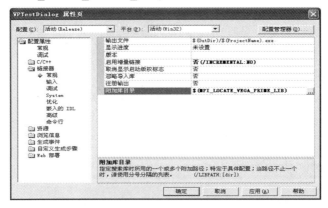

图 5.1.3　配置附加库目录

因为 MFC 对话框程序一定是多线程程序，所以不需要进行多线程配置，如图 5.1.4 所示。接下来将重点处理基于 MFC 对话框程序的 Vega Prime 应用程序的代码。

图 5.1.4　多线程选项

5.2　建立 MFC 对话框 Vega Prime 应用程序

MVC 全名是 Model View Controller，是模型(Model) – 视图(View) – 控制器(Controller)的缩写，是一种软件设计典范，用一种业务逻辑、数据、界面显示分离的方法组织代码，将业务逻辑聚集到一个部件里面，在改进和个性化定制界面及用户交互的同时，不需要重新编写业务逻辑。MVC 被独特地发展起来，用于映射传统的输入、处理和输出功能在一个逻辑的图形化用户界面的结构中。

在 Vega Prime 应用程序设计中，借鉴使用了 MVC 这种框架模式，主要目的就是把业务逻辑与界面分离，利用控制把二者相关联。把所有与 Vega Prime 虚拟仿真的内容进行封装，把 Vega Prime 业务逻辑与界面分离，再加上适当的控制逻辑把二者进行关联。这样大大提高了 Vega Prime 虚拟仿真的代码重用性，加快整个应用开发进程。

当然，本书主要介绍 Vega Prime 虚拟仿真，特别是在本章的设计中添加公共类，就是把所有与 Vega Prime 虚拟仿真的内容添加到一个类 PublicMember 里面，完全与界面分离，增加代码的重用性。在本章中，PublicMember 类的代码虽然较多，但都采用了静态成员方式，成员个数不多，就是为了读者更好的理解掌握 Vega Prime 虚拟仿真的整个过程。随着业务逻辑的增加，可以根据需要设计更多的成员函数或者分离出更多的业务类，就如在第 9 章所做的那样。

按照前面的方法配置完成后，对于建立的对话框程序，现在就是一个普通的对话框程序。在前面 4.2.1 导出的 VP3_2_7.cpp 文件也不能直接使用，需要进行改造。这些改造主要包括：界面改造，添加公共类和启动主线程。

5.2.1 MFC 对话框程序界面改造

为了便于 Vega Prime 应用程序的运行，对基于 MFC 的对话框程序进行界面改造。首先，删除界面上原来具有的静态文本框和"取消"按钮；然后，添加一个分组框，ID 为"IDC_grScene"，Caption 为"运行 Vega Prime"；最后，添加一个按钮，ID 为"IDC_btRun"，Caption"为"运行"。界面上的控件及其属性详细内容如表 5.2.1 所示。其中，ID 为"IDC_btRun"的运行按钮用来启动 Vega Prime 主线程；ID 为"IDC_grScene"的分组框用来作为场景窗口，改造后的效果如图 5.2.1 所示。

表 5.2.1 界面控件内容

类型	ID	Caption
按钮	IDC_btRun	运行
按钮	IDOK	关闭
分组框	IDC_grScene	运行 Vega Prime
对话框	IDD_VPTestDIALOG_DIALOG	VPTestDialog

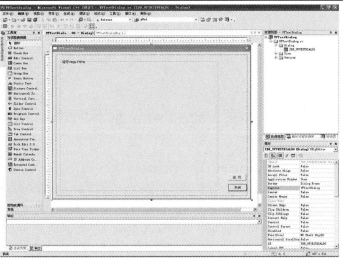

图 5.2.1 改造后的 MFC 对话框界面

5.2.2 添加 Vega Prime 应用公共类 PublicMember

为了便于管理以及理清思路，将有关 Vega Prime 的代码用类进行封装。对于这个类，尽可能地简化，对于其中的成员，全部采用静态成员。这样做，不仅仅是因为线程函数必须是类的静态成员，也是为了简化全局变量的管理。

在 MFC 对话框程序中添加新类非常简单，如图 5.2.2 所示。在图的右上部分，选中"类视图"标签，右键点击项目名称"VPTestDialog"，弹出右键菜单，依次选中"添加"→"添加类"，将会弹出如图 5.2.3 所示的窗口。

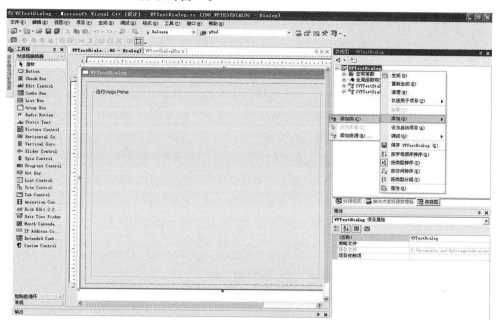

图 5.2.2　添加类

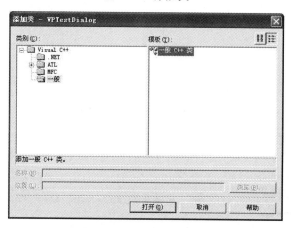

图 5.2.3　选择类别

在图 5.2.3 中，选择"一般 C++类"，然后点击"打开"按钮，将会出现图 5.2.4 所示的窗口。

在图 5.2.4 中，输入类名"PublicMember"，其.h 文件自动就为"PublicMember.h"，其.cpp 文件自动为"PublicMember.cpp"。另外，其访问为"public"。

Vega Prime 虚拟现实开发技术

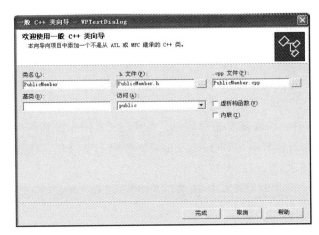

图 5.2.4 填写新类属性

添加完新类"PublicMember"后，首先要做的就是为新类添加头文件。对于在第 4.2.1 节得到的 VP3_2_7.cpp 文件，把其中的所有头文件复制到"PublicMember.h"的前面，如图 5.2.5 所示。

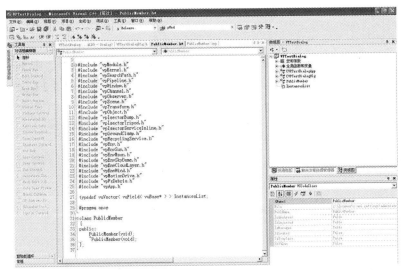

图 5.2.5 添加头文件

PublicMember 类对 VP3_2_7.cpp 文件主要进行了以下改造：
- 把 define 函数改造为 PublicMember::CTS_Define 函数；
- 把 main 函数改造为 PublicMember::CTS_RunBasicThread 函数；
- 增加了场景窗口句柄静态变量 PublicMember::CTS_RunningWindow ；
- 增加了控制 VP 线程运行的静态变量 PublicMember::CTS_continueRunVP；
- 增加了 VP 线程变量静态变量 PublicMember::CTS_vegaThread；
- 把对象容器的局部变量 s_pInstancesToUnref 改造为静态变量 PublicMember:: CTS_s_pInstancesToUnref;
- 把路径对象的局部变量 pSearchPath_mySearchPath 改造为静态变量 PublicMember::CTS_pSearchPath_mySearchPath；

● 把场景对象的局部变量 pScene_myScene 改造为静态变量 PublicMember::CTS_pScene_myScene。

（1）对于对象容器，先有：

typedef vuVector< vuField< vuBase* > > InstancesList;

然后有：

定义：static InstancesList* CTS_s_pInstancesToUnref；

初始化：PublicMember::CTS_s_pInstancesToUnref = new InstancesList；

（2）对于路径对象和场景对象有：

定义：static vpSearchPath * CTS_pSearchPath_mySearchPath；

定义：static vpScene* CTS_pScene_myScene；

初始化：PublicMember:: CTS_pSearchPath_mySearchPath = vpSearchPath::instance()；

初始化：PublicMember::CTS_pScene_myScene = new vpScene()；

（3）对于 PublicMember::CTS_Define 函数，只是把原来 define 函数中的 3 个局部变量替换为了相应的静态变量，即：

s_pInstancesToUnref 替换为：PublicMember::CTS_s_pInstancesToUnref

pSearchPath_mySearchPath 替换为：PublicMember:: CTS_pSearchPath_mySearchPath

pScene_myScene 替换为：PublicMember::CTS_ pScene_myScene

（4）把 main 函数改造为 PublicMember::CTS_RunBasicThread 函数，则进行了比较彻底的改造，以此来作为 Vega Prime 应用程序的主线程函数。详细代码如图 5.2.6 所示。后续章节还会继续添加修改适当的代码，以此来满足需要。

```
//VP 运行主线程。
UINT PublicMember::CTS_RunBasicThread(LPVOID)
{
    //初始化
    vp::initialize(__argc,__argv);//每一个变量前是两根下划线
    //定义场景
    PublicMember::CTS_Define();
    //配置场景
    vpKernel::instance()->configure();
    //设置窗体
    vpWindow * vpWin= * vpWindow::begin();
    vpWin->setParent(PublicMember::CTS_RunningWindow);
    vpWin->setBorderEnable(false);
    vpWin->setFullScreenEnable(true);
    vpWin->open();
    ::SetFocus(vpWin->getWindow());
    //帧循环
    while(vpKernel::instance()->beginFrame()!=0)
    {
        vpKernel::instance()->endFrame();
        if(!PublicMember::CTS_continueRunVP)
        {
```

```
            vpKernel::instance()->unconfigure();
            vp::shutdown();
            return 0;
        }
    }
    return 0;
}
```

图 5.2.6　Vega Prime 应用程序主线程函数

对于 VP3_2_7.cpp 文件的改造都包含在 PublicMember 类中，详细内容如图 5.2.7 和图 5.2.8 所示。

```
#include "vpModule.h"
#include "vpKernel.h"
#include "vpSearchPath.h"
#include "vpPipeline.h"
#include "vpWindow.h"
#include "vpChannel.h"
#include "vpObserver.h"
#include "vpScene.h"
#include "vpTransform.h"
#include "vpObject.h"
#include "vpIsectorBump.h"
#include "vpIsectorTripod.h"
#include "vpIsectorServiceInline.h"
#include "vpGroundClamp.h"
#include "vpRecyclingService.h"
#include "vpEnv.h"
#include "vpEnvSun.h"
#include "vpEnvMoon.h"
#include "vpEnvSkyDome.h"
#include "vpEnvCloudLayer.h"
#include "vpEnvWind.h"
#include "vpMotionDrive.h"
#include "vpFxDebris.h"
#include "vpApp.h"

typedef vuVector< vuField< vuBase* > > InstancesList;

#pragma once

class PublicMember
{
public:
/*
***********************************************************
公共成员函数
包含主要的公共函数.
```

```
*/
    PublicMember(void);
    ~PublicMember(void);
    //运动场景配置
    static void CTS_Define( void );
    //VP 主线程
    static UINT CTS_RunBasicThread(LPVOID) ;

    /*
    **********************************************************
    公共成员变量
    包含主要的公共数据成员.
    **********************************************************
    */
    //场景窗口句柄
    static HWND CTS_RunningWindow ;

    //控制 VP 线程运行的变量
    static bool   CTS_continueRunVP;

    //VP 线程变量
    static CWinThread *   CTS_vegaThread;

    //VP 对象容器
    static InstancesList* CTS_s_pInstancesToUnref ;

    //VP 对象文件路径
    static vpSearchPath * CTS_pSearchPath_mySearchPath;

    //VP 场景对象
    static vpScene* CTS_pScene_myScene;
};
```

图 5.2.7　PublicMember.h 文件

```
#include "StdAfx.h"
#include ".\publicmember.h"
 /*
 **********************************************************
 公共成员变量
 包含主要的公共成员变量.
 **********************************************************
 */
 //场景窗口句柄
 HWND   PublicMember::CTS_RunningWindow =NULL;
```

```cpp
//控制VP线程运行的变量
bool    PublicMember::CTS_continueRunVP=NULL;

//VP线程变量
CWinThread *   PublicMember::CTS_vegaThread=NULL;

//VP对象容器
InstancesList* PublicMember::CTS_s_pInstancesToUnref=NULL ;

//VP对象文件路径
vpSearchPath * PublicMember:: pSearchPath_mySearchPath=NULL;

//VP场景对象
vpScene* PublicMember::CTS_pScene_myScene=NULL;

/*
************************************************************
    公共成员函数
    包含主要的公共函数.
************************************************************
*/
PublicMember::PublicMember(void)
{
}
PublicMember::~PublicMember(void)
{
}

//运动场景配置函数
void PublicMember::CTS_Define( void )
{
PublicMember::CTS_s_pInstancesToUnref = new InstancesList;
//
//初始化各模块
vpModule::initializeModule( "vpEnv" );
vpModule::initializeModule( "vpMotion" );
vpModule::initializeModule( "vpFx" );

//建立各种实例----------------------------------------------------
//建立内核实例
vpKernel* pKernel_myKernel = vpKernel::instance();
pKernel_myKernel->setNotifyLevel( vuNotify::LEVEL_WARN );
pKernel_myKernel->setNotifyColorEnable( false );
pKernel_myKernel->setPriority( vuThread::PRIORITY_NORMAL );
pKernel_myKernel->setProcessor( -1 );
pKernel_myKernel->setFrameRateLimiterEnable( true );
```

```cpp
pKernel_myKernel->setDesiredFrameRate( 0.000000 );
pKernel_myKernel->setNotifyClassTypeMode( vuNotify::CLASS_TYPE_MODE
                                    _INCLUSIVE );
//建立路径搜索实例
PublicMember:: CTS_pSearchPath_mySearchPath = vpSearchPath::instance();
PublicMember:: CTS_pSearchPath_mySearchPath ->append
                            ( "C:/Program Files/MultiGen-Paradigm/resources
                            /tutorials/vegaprime/desktop_tutor/tornado/data/land" );
PublicMember:: CTS_pSearchPath_mySearchPath ->append
                            ( "C:/Program Files/MultiGen-Paradigm/resources
                            /tutorials/vegaprime/desktop_tutor/tornado/data/farmhouse" );
PublicMember:: CTS_pSearchPath_mySearchPath ->append
                            ( "C:/Program Files/MultiGen-Paradigm/resources
                            /tutorials/vegaprime/desktop_tutor/tornado/data/humv-dirty" );
PublicMember:: CTS_pSearchPath_mySearchPath ->append
                            ( "C:/Program Files/MultiGen-Paradigm/resources
                            /tutorials/vegaprime/desktop_tutor/tornado/data/grainstorage" );
//建立管道
vpPipeline* pPipeline_myPipeline = new vpPipeline();
pPipeline_myPipeline->setName( "myPipeline" );
pPipeline_myPipeline->setMultiThread( vsPipeline::MULTITHREAD_INLINE );
pPipeline_myPipeline->setId( 0 );
pPipeline_myPipeline->setNumCullThreads( 0 );
pPipeline_myPipeline->setCullThreadPriority( vuThread::PRIORITY_NORMAL );
pPipeline_myPipeline->setCullThreadProcessor( -1 );
pPipeline_myPipeline->setDrawThreadPriority( vuThread::PRIORITY_NORMAL );
pPipeline_myPipeline->setDrawThreadProcessor( -1 );
pPipeline_myPipeline->setBeginFrameOnVsyncEnable( false );
pPipeline_myPipeline->setDesiredPostDrawTime( -1.000000 );

PublicMember::CTS_s_pInstancesToUnref->push_back( pPipeline_myPipeline );

//建立窗体
vpWindow* pWindow_myWindow = new vpWindow();
pWindow_myWindow->setName( "myWindow" );
pWindow_myWindow->setLabel( "Vega Prime Window" );
pWindow_myWindow->setOrigin( 0 ,   0 );
pWindow_myWindow->setSize( 1024 ,   768 );
pWindow_myWindow->setFullScreenEnable( false );
pWindow_myWindow->setBorderEnable( true );
pWindow_myWindow->setInputEnable( true );
pWindow_myWindow->setCursorEnable( true );
pWindow_myWindow->setStereoEnable( false );
pWindow_myWindow->setNumColorBits( 8 );
pWindow_myWindow->setNumAlphaBits( 0 );
pWindow_myWindow->setNumDepthBits( 24 );
pWindow_myWindow->setNumStencilBits( 0 );
```

```cpp
pWindow_myWindow->setNumAccumColorBits( 0 );
pWindow_myWindow->setNumAccumAlphaBits( 0 );
pWindow_myWindow->setNumMultiSampleBits( 0 );
pWindow_myWindow->setSwapInterval( 1 );
pWindow_myWindow->setPixelBufferMode( vrDrawContext::PIXEL_BUFFER_MODE_OFF );

PublicMember::CTS_s_pInstancesToUnref->push_back( pWindow_myWindow );

//建立通道
vpChannel* pChannel_myChannel = new vpChannel();
pChannel_myChannel->setName( "myChannel" );
pChannel_myChannel->setOffsetTranslate( 0.000000 , 0.000000 , 0.000000 );
pChannel_myChannel->setOffsetRotate( 0.000000 , 0.000000 , 0.000000 );
pChannel_myChannel->setCullMask( 0x0FFFFFFFF );
pChannel_myChannel->setRenderMask( 0x0FFFFFFFF );
pChannel_myChannel->setClearColor( 0.000000f , 0.500000f , 1.000000f , 0.000000f );
pChannel_myChannel->setClearBuffers( 0x03 );
pChannel_myChannel->setDrawArea( 0.000000 , 1.000000 , 0.000000 , 1.000000 );
pChannel_myChannel->setFOVSymmetric( 45.000000f , -1.000000f );
pChannel_myChannel->setNearFar( 1.000000f , 35000.000000f );
pChannel_myChannel->setLODVisibilityRangeScale( 1.000000 );
pChannel_myChannel->setLODTransitionRangeScale( 1.000000 );
pChannel_myChannel->setCullThreadPriority( vuThread::PRIORITY_NORMAL );
pChannel_myChannel->setCullThreadProcessor( -1 );
pChannel_myChannel->setGraphicsModeEnable( vpChannel::GRAPHICS_MODE
                                _WIREFRAME , false );
pChannel_myChannel->setGraphicsModeEnable( vpChannel::GRAPHICS_MODE
                                _TRANSPARENCY , true );
pChannel_myChannel->setGraphicsModeEnable( vpChannel::GRAPHICS_MODE
                                _TEXTURE , true );
pChannel_myChannel->setGraphicsModeEnable( vpChannel::GRAPHICS_MODE
                                _LIGHT , true );
pChannel_myChannel->setGraphicsModeEnable( vpChannel::GRAPHICS_MODE
                                _FOG , true );
pChannel_myChannel->setLightPointThreadPriority( vuThread::PRIORITY_NORMAL );
pChannel_myChannel->setLightPointThreadProcessor( -1 );
pChannel_myChannel->setMultiSample( vpChannel::MULTISAMPLE_OFF );
pChannel_myChannel->setStatisticsPage( vpChannel::PAGE_OFF );
pChannel_myChannel->setCullBoundingBoxTestEnable( false );
pChannel_myChannel->setOpaqueSort( vpChannel::OPAQUE_SORT_TEXTURE
                                , vpChannel::OPAQUE_SORT_MATERIAL );
pChannel_myChannel->setTransparentSort( vpChannel::TRANSPARENT_SORT_DEPTH );
pChannel_myChannel->setDrawBuffer( vpChannel::DRAW_BUFFER_DEFAULT );
pChannel_myChannel->setStressEnable( false );
pChannel_myChannel->setStressParameters( 1.000000f, 20.000000f , 0.750000f , 0.500000f , 2.000000f );

PublicMember::CTS_s_pInstancesToUnref->push_back( pChannel_myChannel );
```

第 5 章　建立基于 MFC 对话框的 Vega Prime 应用程序

```cpp
//建立观察者
vpObserver* pObserver_myObserver = new vpObserver();
pObserver_myObserver->setName( "myObserver" );
pObserver_myObserver->setStrategyEnable( false );
pObserver_myObserver->setTranslate( 2300.000000 ,   2500.000000 ,   15.000000 );
pObserver_myObserver->setRotate( -90.000000 ,   0.000000 ,   0.000000 );
pObserver_myObserver->setLatencyCriticalEnable( false );

PublicMember::CTS_s_pInstancesToUnref->push_back( pObserver_myObserver );

//建立场景
PublicMember::CTS_pScene_myScene = new vpScene();
PublicMember::CTS_pScene_myScene->setName( "myScene" );

PublicMember::CTS_s_pInstancesToUnref->push_back( PublicMember::CTS_pScene_myScene );

//建立转换
vpTransform* pTransform_hummerTransform = new vpTransform();
pTransform_hummerTransform->setName( "hummerTransform" );
pTransform_hummerTransform->setCullMask( 0x0FFFFFFFF );
pTransform_hummerTransform->setRenderMask( 0x0FFFFFFFF );
pTransform_hummerTransform->setIsectMask( 0x0FFFFFFFF );
pTransform_hummerTransform->setStrategyEnable( true );
pTransform_hummerTransform->setTranslate( 0.000000 ,   -30.000000 ,   5.000000 );
pTransform_hummerTransform->setRotate( 0.000000 ,   0.000000 ,   0.000000 );
pTransform_hummerTransform->setScale( 1.000000 ,   1.000000 ,   1.000000 );
pTransform_hummerTransform->setStaticEnable( false );

PublicMember::CTS_s_pInstancesToUnref->push_back( pTransform_hummerTransform );

//建立地形
vpObject* pObject_terrain = new vpObject();
pObject_terrain->setName( "terrain" );
pObject_terrain->setCullMask( 0x0FFFFFFFF );
pObject_terrain->setRenderMask( 0x0FFFFFFFF );
pObject_terrain->setIsectMask( 0x0FFFFFFFF );
pObject_terrain->setStrategyEnable( false );
pObject_terrain->setTranslate( 0.000000 ,   0.000000 ,   0.000000 );
pObject_terrain->setRotate( 0.000000 ,   0.000000 ,   0.000000 );
pObject_terrain->setScale( 1.000000 ,   1.000000 ,   1.000000 );
pObject_terrain->setStaticEnable( false );
pObject_terrain->setFileName( "Prime_Junction.flt" );
pObject_terrain->setAutoPage( vpObject::AUTO_PAGE_SYNCHRONOUS );
pObject_terrain->setManualLODChild( -1 );
pObject_terrain->setLoaderOption( vsNodeLoader::Data::LOADER_OPTION_COMBINE
                                                  _LIGHT_POINTS,   true );
```

```
pObject_terrain->setLoaderOption( vsNodeLoader::Data::LOADER_OPTION_COMBINE
                                  _LODS , true );
pObject_terrain->setLoaderOption( vsNodeLoader::Data::LOADER_OPTION_IGNORE
                                  _DOF_CONSTRAINTS , false );
pObject_terrain->setLoaderOption( vsNodeLoader::Data::LOADER_OPTION_PRESERVE
                                  _EXTERNAL_REF_FLAGS , true );
pObject_terrain->setLoaderOption( vsNodeLoader::Data::LOADER_OPTION_PRESERVE
                                  _GENERIC_NAMES , false );
pObject_terrain->setLoaderOption( vsNodeLoader::Data::LOADER_OPTION_PRESERVE
                                  _GENERIC_NODES , false );
pObject_terrain->setLoaderOption( vsNodeLoader::Data::LOADER_OPTION_PRESERVE
                                  _QUADS , false );
pObject_terrain->setLoaderOption( vsNodeLoader::Data::LOADER_OPTION_ALL
                                  _GEOMETRIES_LIT , false );
pObject_terrain->setLoaderOption( vsNodeLoader::Data::LOADER_OPTION_USE
                                  _MATERIAL_DIFFUSE_COLOR , false );
pObject_terrain->setLoaderOption( vsNodeLoader::Data::LOADER_OPTION
                                  _MONOCHROME , false );
pObject_terrain->setLoaderOption( vsNodeLoader::Data::LOADER_OPTION_CREATE
                                  _ANIMATIONS , true );
pObject_terrain->setLoaderOption( vsNodeLoader::Data::LOADER_OPTION_SHARE
                                  _LIGHT_POINT_CONTROLS , true );
pObject_terrain->setLoaderOption( vsNodeLoader::Data::LOADER_OPTION_SHARE
                                  _LIGHT_POINT_ANIMATIONS , true );
pObject_terrain->setLoaderOption( vsNodeLoader::Data::LOADER_OPTION_SHARE
                                  _LIGHT_POINT_APPEARANCES , true );
pObject_terrain->setLoaderDetailMultiTextureStage( -1 );
pObject_terrain->setLoaderBlendTolerance( 0.050000f );
pObject_terrain->setLoaderUnits( vsNodeLoader::Data::LOADER_UNITS_METERS );
pObject_terrain->setBuilderOption( vsNodeLoader::Data::BUILDER_OPTION_OPTIMIZE
                                   _GEOMETRIES , true );
pObject_terrain->setBuilderOption( vsNodeLoader::Data::BUILDER_OPTION_COLLAPSE
                                   _BINDINGS , true );
pObject_terrain->setBuilderOption( vsNodeLoader::Data::BUILDER_OPTION_COLLAPSE
                                   _TRIANGLE_STRIPS , true );
pObject_terrain->setBuilderNormalMode( vsNodeLoader::Data::BUILDER_NORMAL_MODE
                                       _PRESERVE );
pObject_terrain->setBuilderColorTolerance( 0.001000f );
pObject_terrain->setBuilderNormalTolerance( 0.860000f );
pObject_terrain->setBuilderVertexTolerance( 0.000100f );
pObject_terrain->setGeometryOption( vsNodeLoader::Data::GEOMETRY_OPTION
                                    _GENERATE_DISPLAY_LISTS , true );
pObject_terrain->setGeometryFormat( vrGeometryBase::FORMAT_VERTEX
                                    _ARRAY , 0x0FFF );
pObject_terrain->setPostLoadOption( vpGeometryPageable::POST_LOAD_OPTION
                                    _FLATTEN , true );
pObject_terrain->setPostLoadOption( vpGeometryPageable::POST_LOAD_OPTION
```

```cpp
                                                _CLEAN , true );
pObject_terrain->setPostLoadOption( vpGeometryPageable::POST_LOAD_OPTION
                                                _MERGE_GEOMETRIES , true );
pObject_terrain->setPostLoadOption( vpGeometryPageable::POST_LOAD_OPTION
                                                _COLLAPSE_BINDINGS , true );
pObject_terrain->setPostLoadOption( vpGeometryPageable::POST_LOAD_OPTION
                                                _COLLAPSE_TRIANGLE_STRIPS , true );
pObject_terrain->setPostLoadOption( vpGeometryPageable::POST_LOAD_OPTION
                                                _VALIDATE , true );
pObject_terrain->setTextureSubloadEnable( false );
pObject_terrain->setTextureSubloadRender( vpGeometry::TEXTURE_SUBLOAD
                                                _RENDER_DEFERRED );

PublicMember::CTS_s_pInstancesToUnref->push_back( pObject_terrain );

//建立房屋
vpObject* pObject_farmhouse = new vpObject();
pObject_farmhouse->setName( "farmhouse" );
pObject_farmhouse->setCullMask( 0x0FFFFFFF );
pObject_farmhouse->setRenderMask( 0x0FFFFFFF );
pObject_farmhouse->setIsectMask( 0x0FFFFFFF );
pObject_farmhouse->setStrategyEnable( true );
pObject_farmhouse->setTranslate( 2450.000000 , 2460.000000 , 0.000000 );
pObject_farmhouse->setRotate( 0.000000 , 0.000000 , 0.000000 );
pObject_farmhouse->setScale( 1.000000 , 1.000000 , 1.000000 );
pObject_farmhouse->setStaticEnable( false );
pObject_farmhouse->setFileName( "farmhouse.flt" );
pObject_farmhouse->setAutoPage( vpObject::AUTO_PAGE_SYNCHRONOUS );
pObject_farmhouse->setManualLODChild( -1 );
pObject_farmhouse->setLoaderOption( vsNodeLoader::Data::LOADER_OPTION_COMBINE
                                                _LIGHT_POINTS , true );
pObject_farmhouse->setLoaderOption( vsNodeLoader::Data::LOADER_OPTION_COMBINE
                                                _LODS , true );
pObject_farmhouse->setLoaderOption( vsNodeLoader::Data::LOADER_OPTION_IGNORE
                                                _DOF_CONSTRAINTS , false );
pObject_farmhouse->setLoaderOption( vsNodeLoader::Data::LOADER_OPTION_PRESERVE
                                                _EXTERNAL_REF_FLAGS , true );
pObject_farmhouse->setLoaderOption( vsNodeLoader::Data::LOADER_OPTION_PRESERVE
                                                _GENERIC_NAMES , false );
pObject_farmhouse->setLoaderOption( vsNodeLoader::Data::LOADER_OPTION_PRESERVE
                                                _GENERIC_NODES , false );
pObject_farmhouse->setLoaderOption( vsNodeLoader::Data::LOADER_OPTION_PRESERVE
                                                _QUADS , false );
pObject_farmhouse->setLoaderOption( vsNodeLoader::Data::LOADER_OPTION_ALL
                                                _GEOMETRIES_LIT , false );
pObject_farmhouse->setLoaderOption( vsNodeLoader::Data::LOADER_OPTION_USE
                                                _MATERIAL_DIFFUSE_COLOR , false );
```

```cpp
pObject_farmhouse->setLoaderOption( vsNodeLoader::Data::LOADER_OPTION
                                   _MONOCHROME , false );
pObject_farmhouse->setLoaderOption( vsNodeLoader::Data::LOADER_OPTION
                                   _CREATE_ANIMATIONS , true );
pObject_farmhouse->setLoaderOption( vsNodeLoader::Data::LOADER_OPTION_SHARE
                                   _LIGHT_POINT_CONTROLS , true );
pObject_farmhouse->setLoaderOption( vsNodeLoader::Data::LOADER_OPTION_SHARE
                                   _LIGHT_POINT_ANIMATIONS , true );
pObject_farmhouse->setLoaderOption( vsNodeLoader::Data::LOADER_OPTION_SHARE
                                   _LIGHT_POINT_APPEARANCES , true );
pObject_farmhouse->setLoaderDetailMultiTextureStage( -1 );
pObject_farmhouse->setLoaderBlendTolerance( 0.050000f );
pObject_farmhouse->setLoaderUnits( vsNodeLoader::Data::LOADER_UNITS_METERS );
pObject_farmhouse->setBuilderOption( vsNodeLoader::Data::BUILDER_OPTION_OPTIMIZE
                                    _GEOMETRIES , true );
pObject_farmhouse->setBuilderOption( vsNodeLoader::Data::BUILDER_OPTION_COLLAPSE
                                    _BINDINGS , true );
pObject_farmhouse->setBuilderOption( vsNodeLoader::Data::BUILDER_OPTION_COLLAPSE
                                    _TRIANGLE_STRIPS , true );
pObject_farmhouse->setBuilderNormalMode( vsNodeLoader::Data::BUILDER_NORMAL
                                        _MODE_PRESERVE );
pObject_farmhouse->setBuilderColorTolerance( 0.001000f );
pObject_farmhouse->setBuilderNormalTolerance( 0.860000f );
pObject_farmhouse->setBuilderVertexTolerance( 0.000100f );
pObject_farmhouse->setGeometryOption( vsNodeLoader::Data::GEOMETRY_OPTION
                                     _GENERATE_DISPLAY_LISTS , false );
pObject_farmhouse->setGeometryFormat( vrGeometryBase::FORMAT_VERTEX
                                     _ARRAY , 0x0FFF );
pObject_farmhouse->setPostLoadOption( vpGeometryPageable::POST_LOAD
                                     _OPTION_FLATTEN , true );
pObject_farmhouse->setPostLoadOption( vpGeometryPageable::POST_LOAD_OPTION
                                     _CLEAN , true );
pObject_farmhouse->setPostLoadOption( vpGeometryPageable::POST_LOAD_OPTION
                                     _MERGE_GEOMETRIES , true );
pObject_farmhouse->setPostLoadOption( vpGeometryPageable::POST_LOAD_OPTION
                                     _COLLAPSE_BINDINGS , true );
pObject_farmhouse->setPostLoadOption( vpGeometryPageable::POST_LOAD_OPTION
                                     _COLLAPSE_TRIANGLE_STRIPS , true );
pObject_farmhouse->setPostLoadOption( vpGeometryPageable::POST_LOAD_OPTION
                                     _VALIDATE , true );
pObject_farmhouse->setTextureSubloadEnable( false );
pObject_farmhouse->setTextureSubloadRender( vpGeometry::TEXTURE_SUBLOAD
                                           _RENDER_DEFERRED );

PublicMember::CTS_s_pInstancesToUnref->push_back( pObject_farmhouse );
//建立汽车
vpObject* pObject_Hummer = new vpObject();
```

```cpp
pObject_Hummer->setName( "Hummer" );
pObject_Hummer->setCullMask( 0x0FFFFFFFF );
pObject_Hummer->setRenderMask( 0x0FFFFFFFF );
pObject_Hummer->setIsectMask( 0x0FFFFFFFF );
pObject_Hummer->setStrategyEnable( true );
pObject_Hummer->setTranslate( 2360.000000 , 2490.000000 , 0.000000 );
pObject_Hummer->setRotate( -90.000000 , 0.000000 , 0.000000 );
pObject_Hummer->setScale( 1.000000 , 1.000000 , 1.000000 );
pObject_Hummer->setStaticEnable( false );
pObject_Hummer->setFileName( "humv-dirty.flt" );

/*    其中的代码与配置 farmhouse 相同      */

PublicMember::CTS_s_pInstancesToUnref->push_back( pObject_Hummer );

//建立粮仓
vpObject* pObject_gainstore = new vpObject();
pObject_gainstore->setName( "gainstore" );
pObject_gainstore->setCullMask( 0x0FFFFFFFF );
pObject_gainstore->setRenderMask( 0x0FFFFFFFF );
pObject_gainstore->setIsectMask( 0x0FFFFFFFF );
pObject_gainstore->setStrategyEnable( true );
pObject_gainstore->setTranslate( 2450.000000 , 2530.000000 , 0.000000 );
pObject_gainstore->setRotate( 90.000000 , 0.000000 , 0.000000 );
pObject_gainstore->setScale( 1.000000 , 1.000000 , 1.000000 );
pObject_gainstore->setStaticEnable( false );
pObject_gainstore->setFileName( "grainStorage.flt" );

/*    其中的代码与配置 farmhouse 相同      */

PublicMember::CTS_s_pInstancesToUnref->push_back( pObject_gainstore );

//建立碰撞
vpIsectorBump* pIsectorBump_bumpIsector = new vpIsectorBump();
pIsectorBump_bumpIsector->setName( "bumpIsector" );
pIsectorBump_bumpIsector->setEnable( true );
pIsectorBump_bumpIsector->setRenderEnable( false );
pIsectorBump_bumpIsector->setTranslate( 0.000000 , 0.000000 , 0.000000 );
pIsectorBump_bumpIsector->setRotate( 0.000000 , 0.000000 , 0.000000 );
pIsectorBump_bumpIsector->setMode( 0x02 );
pIsectorBump_bumpIsector->setIsectMask( 0x0FFFFFFFF );
pIsectorBump_bumpIsector->setStrategyEnable( true );
pIsectorBump_bumpIsector->setDimensions( 3.000000f , 3.000000f , 3.000000f );

PublicMember::CTS_s_pInstancesToUnref->push_back( pIsectorBump_bumpIsector );
//建立三角碰撞
vpIsectorTripod* pIsectorTripod_tripodIsector = new vpIsectorTripod();
```

```cpp
pIsectorTripod_tripodIsector->setName( "tripodIsector" );
pIsectorTripod_tripodIsector->setEnable( true );
pIsectorTripod_tripodIsector->setRenderEnable( false );
pIsectorTripod_tripodIsector->setTranslate( 0.000000 , 0.000000 , 0.000000 );
pIsectorTripod_tripodIsector->setRotate( 0.000000 , 0.000000 , 0.000000 );
pIsectorTripod_tripodIsector->setMode( 0x02A );
pIsectorTripod_tripodIsector->setIsectMask( 0x0FFFFFFFF );
pIsectorTripod_tripodIsector->setStrategyEnable( true );
pIsectorTripod_tripodIsector->setSegmentZExtent( -5000.000000f , 5000.000000f );
pIsectorTripod_tripodIsector->setDimensions( 5.000000f , 6.000000f );

PublicMember::CTS_s_pInstancesToUnref->push_back( pIsectorTripod_tripodIsector );

//建立碰撞服务
vpIsectorServiceInline* pIsectorServiceInline_myIsectorService = new vpIsectorServiceInline();
pIsectorServiceInline_myIsectorService->setName( "myIsectorService" );

PublicMember::CTS_s_pInstancesToUnref->push_back(
pIsectorServiceInline_myIsectorService );

//建立地面碰撞
vpGroundClamp* pGroundClamp_myGroundClamp = new vpGroundClamp();
pGroundClamp_myGroundClamp->setName( "myGroundClamp" );
pGroundClamp_myGroundClamp->setOffsetFromGround( 0.000000f );

PublicMember::CTS_s_pInstancesToUnref->push_back( pGroundClamp_myGroundClamp );

//建立循环服务
vpRecyclingService* pRecyclingService_myRecyclingService = vpRecyclingService::instance();
pRecyclingService_myRecyclingService->setMultiThread(
vpRecyclingService::MULTITHREAD_INLINE );
pRecyclingService_myRecyclingService->setThreadPriority(
vuThread::PRIORITY_NORMAL );
pRecyclingService_myRecyclingService->setThreadProcessor( -1 );
pRecyclingService_myRecyclingService->setRecycleTime( -1.000000 );

//建立环境
vpEnv* pEnv_myEnv = new vpEnv();
pEnv_myEnv->setName( "myEnv" );
pEnv_myEnv->setDate( 22 , 6 , 2002 );
pEnv_myEnv->setTimeOfDay( 12.000000f );
pEnv_myEnv->setTimeMultiplier( 1.000000f );
pEnv_myEnv->setEphemerisUpdateInterval( 0.000000f );
pEnv_myEnv->setReferencePosition( -96.790001f , 32.790001f );
pEnv_myEnv->setEphemerisTimeZoneOffset( -10000 );
pEnv_myEnv->setSkyColor( 0.513725f , 0.701961f , 0.941176f , 1.000000f );
```

```cpp
pEnv_myEnv->setVisibilityType( vpEnvFx::VISIBILITY_TYPE_PIXEL_EXP2 );
pEnv_myEnv->setVisibilityRangeOnset( 0.000000f );
pEnv_myEnv->setVisibilityRangeOpaque( 80000.000000f );
pEnv_myEnv->setVisibilityColor( 1.000000f, 1.000000f, 1.000000f, 1.000000f );
pEnv_myEnv->setLocalViewerEnable( false );
pEnv_myEnv->setTwoSidedLightingEnable( false );
pEnv_myEnv->setAmbientLightingColor( 0.000000f, 0.000000f, 0.000000f, 1.000000f );
pEnv_myEnv->setLightColorScale( vpEnv::COLOR_AMBIENT, 1.000000f );
pEnv_myEnv->setLightColorScale( vpEnv::COLOR_DIFFUSE, 1.000000f );
pEnv_myEnv->setLightColorScale( vpEnv::COLOR_SPECULAR, 1.000000f );
pEnv_myEnv->setDominantCelestialLightingEnable( false );

PublicMember::CTS_s_pInstancesToUnref->push_back( pEnv_myEnv );

//建立太阳
vpEnvSun* pEnvSun_myEnvSun = new vpEnvSun();
pEnvSun_myEnvSun->setName( "myEnvSun" );
pEnvSun_myEnvSun->setEnable( true );
pEnvSun_myEnvSun->setTwilightDip( -18.000000f );
pEnvSun_myEnvSun->setGeometryEnable( true );
pEnvSun_myEnvSun->setTextureFile( "sun.inta" );
pEnvSun_myEnvSun->setTextureBlendColor( 1.000000f, 1.000000f, 1.000000f, 1.000000f );
pEnvSun_myEnvSun->setTextureBlendMode( vpEnvSun::TEXTURE_BLEND
                                        _MODE_MODULATE );
pEnvSun_myEnvSun->setColor( 0.992156f, 1.000000f, 0.949019f, 1.000000f );
pEnvSun_myEnvSun->setVisibilityRange( 8000.000000f );
pEnvSun_myEnvSun->setAngularSize( 5.500000f );
pEnvSun_myEnvSun->setHorizonAngle( 40.000000f );
pEnvSun_myEnvSun->setHorizonColor( 1.000000f, 0.545098f, 0.239216f, 1.000000f );
pEnvSun_myEnvSun->setHorizonSizeScale( 2.000000f );
pEnvSun_myEnvSun->setHorizonLightColorScale( 0.250000f );

PublicMember::CTS_s_pInstancesToUnref->push_back( pEnvSun_myEnvSun );

//建立月亮
vpEnvMoon* pEnvMoon_myEnvMoon = new vpEnvMoon();
pEnvMoon_myEnvMoon->setName( "myEnvMoon" );
pEnvMoon_myEnvMoon->setEnable( true );
pEnvMoon_myEnvMoon->setBrightness( 0.250000f );
pEnvMoon_myEnvMoon->setGeometryEnable( true );
pEnvMoon_myEnvMoon->setTextureFile( "moon.inta" );
pEnvMoon_myEnvMoon->setTextureBlendColor( 1.000000f, 1.000000f, 1.000000f, 1.000000f );
pEnvMoon_myEnvMoon->setTextureBlendMode( vpEnvMoon::TEXTURE_BLEND
                                          _MODE_MODULATE );
pEnvMoon_myEnvMoon->setColor( 0.811765f, 0.886275f, 0.937255f, 1.000000f );
pEnvMoon_myEnvMoon->setVisibilityRange( 15000.000000f );
pEnvMoon_myEnvMoon->setAngularSize( 1.500000f );
```

```cpp
pEnvMoon_myEnvMoon->setHorizonAngle( 30.000000f );
pEnvMoon_myEnvMoon->setHorizonColor( 0.811765f, 0.886275f, 0.937255f, 1.000000f );
pEnvMoon_myEnvMoon->setHorizonSizeScale( 1.700000f );
pEnvMoon_myEnvMoon->setHorizonLightColorScale( 0.000000f );

PublicMember::CTS_s_pInstancesToUnref->push_back( pEnvMoon_myEnvMoon );

//建立天幕
vpEnvSkyDome* pEnvSkyDome_myEnvSkyDome = new vpEnvSkyDome();
pEnvSkyDome_myEnvSkyDome->setName( "myEnvSkyDome" );
pEnvSkyDome_myEnvSkyDome->setEnable( true );
pEnvSkyDome_myEnvSkyDome->setGroundColor( 0.211765f, 0.286275f, 0.149020f, 1.000000f );
pEnvSkyDome_myEnvSkyDome->setGroundEnable( true );
pEnvSkyDome_myEnvSkyDome->setVisibilityRange( 0,    3500.000000f );
pEnvSkyDome_myEnvSkyDome->setVisibilityRange( 1,   10500.000000f );
pEnvSkyDome_myEnvSkyDome->setVisibilityRange( 2,   26250.000000f );
pEnvSkyDome_myEnvSkyDome->setVisibilityRange( 3,   31500.000000f );
pEnvSkyDome_myEnvSkyDome->setVisibilityRange( 4,   35000.000000f );
pEnvSkyDome_myEnvSkyDome->setVisibilityRange( 5,   35000.000000f );
pEnvSkyDome_myEnvSkyDome->setVisibilityRange( 6,   35000.000000f );

PublicMember::CTS_s_pInstancesToUnref->push_back( pEnvSkyDome_myEnvSkyDome );

//建立云层
vpEnvCloudLayer* pEnvCloudLayer_myEnvCloudLayer = new vpEnvCloudLayer();
pEnvCloudLayer_myEnvCloudLayer->setName( "myEnvCloudLayer" );
pEnvCloudLayer_myEnvCloudLayer->setEnable( true );
pEnvCloudLayer_myEnvCloudLayer->setColor( 0.956863f, 0.976471f, 0.984314f, 1.000000f );
pEnvCloudLayer_myEnvCloudLayer->setElevation( 3000.000000f,   5000.000000f );
pEnvCloudLayer_myEnvCloudLayer->setTransitionRange( 500.000000f,   500.000000f );
pEnvCloudLayer_myEnvCloudLayer->setScudEnable( false );
pEnvCloudLayer_myEnvCloudLayer->setTextureFile( "cloud_scattered.inta" );
pEnvCloudLayer_myEnvCloudLayer->setTextureBlendColor( 1.000000f, 1.000000f, 1.000000f,
                                                      1.000000f );
pEnvCloudLayer_myEnvCloudLayer->setTextureBlendMode(
    vpEnvCloudLayer::TEXTURE_BLEND_MODE_MODULATE );
pEnvCloudLayer_myEnvCloudLayer->setTextureTiling( 3.000000f,   3.000000f );
pEnvCloudLayer_myEnvCloudLayer->setMinVisibilityScale( 0.000000f );
pEnvCloudLayer_myEnvCloudLayer->setHorizonColorScale( 1.000000f );

PublicMember::CTS_s_pInstancesToUnref->push_back( pEnvCloudLayer_myEnvCloudLayer );

//建立风
vpEnvWind* pEnvWind_myEnvWind = new vpEnvWind();
pEnvWind_myEnvWind->setName( "myEnvWind" );
pEnvWind_myEnvWind->setEnable( true );
pEnvWind_myEnvWind->setSpeed( 0.000000f );
```

```cpp
pEnvWind_myEnvWind->setDirection( 0.000000f,   1.000000f,   0.000000f );

PublicMember::CTS_s_pInstancesToUnref->push_back( pEnvWind_myEnvWind );

//建立驾驶模式
vpMotionDrive* pMotionDrive_myMotion = new vpMotionDrive();
pMotionDrive_myMotion->setName( "myMotion" );
pMotionDrive_myMotion->setSpeed( 0.000000 );
pMotionDrive_myMotion->setSpeedLimits( -10.000000,   10.000000 );
pMotionDrive_myMotion->setSpeedDelta( 1.000000,   1.000000 );
pMotionDrive_myMotion->setHeadingDelta( 1.000000 );

PublicMember::CTS_s_pInstancesToUnref->push_back( pMotionDrive_myMotion );

//建立碎片特效
vpFxDebris* pFxDebris_Debris = new vpFxDebris();
pFxDebris_Debris->setName( "Debris" );
pFxDebris_Debris->setCullMask( 0x0FFFFFFFF );
pFxDebris_Debris->setRenderMask( 0x0FFFFFFFF );
pFxDebris_Debris->setIsectMask( 0x0FFFFFFFF );
pFxDebris_Debris->setStrategyEnable( true );
pFxDebris_Debris->setTranslate( 0.000000,   0.000000,   0.000000 );
pFxDebris_Debris->setRotate( 0.000000,   0.000000,   0.000000 );
pFxDebris_Debris->setScale( 20.000000,   20.000000,   20.000000 );
pFxDebris_Debris->setStaticEnable( false );
pFxDebris_Debris->setRepeatEnable( false );
pFxDebris_Debris->setOverallColor( 1.000000f,   1.000000f,   1.000000f,   1.000000f );
pFxDebris_Debris->setTextureBlendColor( 1.000000f,   1.000000f,   1.000000f,   1.000000f );
pFxDebris_Debris->setTextureMode( vpFx::TEXTURE_MODE_MODULATE );
pFxDebris_Debris->setTextureFile( "debris.inta" );
pFxDebris_Debris->setOverallDuration( 4.000000f );
pFxDebris_Debris->setFadeDuration( 0.000000f,   0.000000f );
pFxDebris_Debris->setTransparencyDepthOffset( 0.000000f );
pFxDebris_Debris->setTransparencyDepthRadiusScaler( 0.000000f );
pFxDebris_Debris->setWindSource( vpFxParticleSystem::WIND_SOURCE_TABLE );

PublicMember::CTS_s_pInstancesToUnref->push_back( pFxDebris_Debris );

//
// 配置各种对象及实例
//
pPipeline_myPipeline->addWindow( pWindow_myWindow );

pWindow_myWindow->addChannel( pChannel_myChannel );


pObserver_myObserver->setStrategy( pMotionDrive_myMotion );
```

```
pObserver_myObserver->addChannel( pChannel_myChannel );
pObserver_myObserver->addAttachment( pEnv_myEnv );
pObserver_myObserver->setScene( PublicMember::CTS_pScene_myScene );
pObserver_myObserver->setLookFrom( pTransform_hummerTransform );

PublicMember::CTS_pScene_myScene->addChild( pObject_terrain );
PublicMember::CTS_pScene_myScene->addChild( pObject_farmhouse );
PublicMember::CTS_pScene_myScene->addChild( pObject_Hummer );
PublicMember::CTS_pScene_myScene->addChild( pObject_gainstore );

pObject_farmhouse->addChild( pFxDebris_Debris );

pObject_Hummer->addChild( pTransform_hummerTransform );
pObject_Hummer->setStrategy( pMotionDrive_myMotion );

pIsectorBump_bumpIsector->setTarget( pObject_farmhouse );
pIsectorBump_bumpIsector->setPositionReference( pObject_Hummer );

pIsectorTripod_tripodIsector->setTarget( pObject_terrain );

pIsectorServiceInline_myIsectorService->addIsector( pIsectorBump_bumpIsector );

pGroundClamp_myGroundClamp->setIsector( pIsectorTripod_tripodIsector );

pEnv_myEnv->addEnvFx( pEnvSun_myEnvSun );
pEnv_myEnv->addEnvFx( pEnvMoon_myEnvMoon );
pEnv_myEnv->addEnvFx( pEnvSkyDome_myEnvSkyDome );
pEnv_myEnv->addEnvFx( pEnvCloudLayer_myEnvCloudLayer );
pEnv_myEnv->addEnvFx( pEnvWind_myEnvWind );

pMotionDrive_myMotion->setNextStrategy( pGroundClamp_myGroundClamp );

pFxDebris_Debris->setTriggerIsector( pIsectorBump_bumpIsector );
}

//VP 运行主线程。
UINT PublicMember::CTS_RunBasicThread(LPVOID)
{
//初始化
vp::initialize(__argc,__argv);    //每个参数前都是两根下划线

//定义场景
PublicMember::CTS_Define();

//绘制场景
vpKernel::instance()->configure();
```

```
//设置窗体
vpWindow * vpWin= * vpWindow::begin();
vpWin->setParent(PublicMember::CTS_RunningWindow);
vpWin->setBorderEnable(false);
vpWin->setFullScreenEnable(true);
vpWin->open();
::SetFocus(vpWin->getWindow());

//帧循环
while(vpKernel::instance()->beginFrame()!=0)
{
    vpKernel::instance()->endFrame();

    if(!PublicMember::CTS_continueRunVP)
    {
        vpKernel::instance()->unconfigure();
        vp::shutdown();
        return 0;
    }
}
return 0;
}
```

图 5.2.8　PublicMember.cpp 文件

对于改造后的代码的使用，将在后续章节进行详细介绍。

5.2.3　启动 Vega Prime 应用主线程

第 5.2.2 节对 VP3_2_7.cpp 中的 main 函数进行了改造，改造为 PublicMember 的静态成员 PublicMember::CTS_RunBasicThread。这其中还包含 3 个控制变量：

- 场景窗口句柄："static HWND CTS_RunningWindow;"这个变量就是某个控件的句柄，在这个应用中，把 Vega Prime 仿真窗口显示在一个分组控件上。
- 控制 VP 线程运行的变量："static bool　CTS_continueRunVP;"，Vega Prime 应用程序配置完成后，就进入帧循环，要退出程序，首先应该退出帧循环，就用这个布尔变量来控制是否继续进行帧循环。
- VP 线程变量："static CWinThread *　CTS_vegaThread;"，用这个变量控制主线程。

在图 5.2.9 中，PublicMember::CTS_RunBasicThread 函数首先调用静态函数 vp::initialize (__argc,__argv)对 Vega Prime 进行初始化，其两个参数前面都是两根下划线；然后调用 PublicMember::CTS_Define()函数进行场景定义；接着调用 vpKernel::instance()-> configure()进行配置；接下来就是设置窗口，其中要特别注意的是 vpWin->setParent (PublicMember::CTS_RunningWindow)，就把 Vega Prime 的窗口设置在了指定控件上，当然，在启动线程之前要给 PublicMember::CTS_RunningWindow 赋值；最后就是进入帧循环，在每次帧循环后，可以在

这里完成某些工作,现在,只是简单判断是否要停止帧循环,如果要停止,首先用 vpKernel::instance()->unconfigure()取消配置,然后调用 vp::shutdown(),关闭 Vega Prime。

```
//VP 运行主线程。
UINT PublicMember::CTS_RunBasicThread(LPVOID)
{
    //初始化
    vp::initialize(__argc,__argv); //每一个变量前是两根下划线
    //定义场景
    PublicMember::CTS_Define();
    //配置场景
    vpKernel::instance()->configure();
    //设置窗体
    vpWindow * vpWin= * vpWindow::begin();
    vpWin->setParent(PublicMember::CTS_RunningWindow);
    vpWin->setBorderEnable(false);
    vpWin->setFullScreenEnable(true);
    vpWin->open();
    ::SetFocus(vpWin->getWindow());
    //帧循环
    while(vpKernel::instance()->beginFrame()!=0)
    {
        vpKernel::instance()->endFrame();
        if(!PublicMember::CTS_continueRunVP)
        {
            vpKernel::instance()->unconfigure();
            vp::shutdown();
            return 0;
        }
    }
    return 0;
}
```

图 5.2.9 Vega Prime 应用程序主线程函数

在第 5.2.1 节中改造了对话框程序,在"运行"按钮下添加以下代码:

　　CWnd *pWnd=GetDlgItem(IDC_grScene);
　　PublicMember::CTS_RunningWindow=pWnd->GetSafeHwnd();
　　PublicMember::CTS_continueRunVP=true;
　　PublicMember::CTS_vegaThread=AfxBeginThread(PublicMember::CTS_RunBasicThread,this);

前两句代码就是获取 ID 为"IDC_grScene"的分组框的句柄,并赋值给 PublicMember::CTS_RunningWindow,作为 Vega Prime 应用程序运行的场所;接着为控制变量赋值,要允许帧循环;最后启动线程主函数。

在"关闭"按钮下添加以下代码:

　　PublicMember::CTS_continueRunVP=false;
　　Sleep(1000);

OnOK();

第一行代码就是停止 Vega Prime 应用程序的帧循环；接着等待 1 秒，使主线程正确地停止返回。

这两个按钮的代码完成，就可以运行 Vega Prime 应用程序了。首先选择 VC 的菜单"调试"，然后选择其下的"直接运行（不调试）"。这样就可以编译生成第一个可运行的 Vega Prime 应用程序了，点击"运行"按钮，将会出现如图 5.2.10 所示的界面。Vega Prime 应用程序运行在了 MFC 对话框程序里面。与 LynX Prime 里运行 ACF 文件一样，在运行界面里可以用鼠标操作汽车：按鼠标左键，汽车加速前进；按鼠标右键，汽车后退。实例代码为 VPTestDialogBasic。

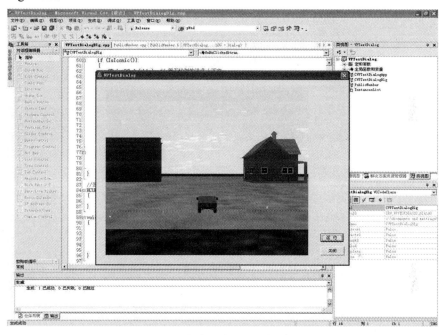

图 5.2.10　改造后的 MFC 对话框界面

5.2.4　启动包含 ACF 文件的主线程

在本章前面所有的改造过程中，PublicMember::CTS_Define(void)函数包含了非常多的代码。在实际工程中，有可能进行适当的功能分割，分解成多个函数。也有另外一种办法，把对应的 ACF 文件复制到项目所在目录，直接使用 ACF 文件，与使用纯粹 CPP 代码最大的不同就是主线程中不再使用 CTS_Define(void)函数，而是直接加载对应的 ACF 文件：

　　//定义场景

　　vpKernel::instance()->define("VP3_2_7.acf");

包含 ACF 文件的 Vega Prime 应用程序主线程函数代码如图 5.2.11 所示。

```
//VP 运行主线程。
UINT PublicMember::CTS_RunBasicThread(LPVOID)
{
    //初始化
```

```
vp::initialize(__argc,__argv); //每一个变量前是两根下划线

//定义场景
 vpKernel::instance()->define("VP3_2_7.acf");

//配置场景
vpKernel::instance()->configure();
//设置窗体
vpWindow * vpWin= * vpWindow::begin();
vpWin->setParent(PublicMember::CTS_RunningWindow);
vpWin->setBorderEnable(false);
vpWin->setFullScreenEnable(true);
vpWin->open();
::SetFocus(vpWin->getWindow());
 //帧循环
while(vpKernel::instance()->beginFrame()!=0)
{
    vpKernel::instance()->endFrame();
    if(!PublicMember::CTS_continueRunVP)
    {
       vpKernel::instance()->unconfigure();
       vp::shutdown();
       return 0;
    }
}
return 0;
}
```

图 5.2.11　包含 ACF 文件的 Vega Prime 应用程序主线程函数

其实，使用 define 函数和使用 acf 文件的主线程，实现的功能完全相同。define 函数方式，只是把 acf 文件转换成了纯粹的 CPP 代码，不用再担心 acf 文件的路径或丢失问题，但 define 函数包含了较多的代码。使用 acf 文件方式则相反，完全省略了 define 函数包含的代码，但需要包含一个 acf 文件，必须考虑文件路径或丢失等问题。

第 6 章　Vega Prime 编程对象的实例使用

前面的章节主要是对 Vega Prime 对象的配置使用，通常都通过 LynX Prime 来完成，通过 ACF 文件就可以导出相关的 CPP 代码，再进行适当的改造使用，这样可以大大地提高开发效率。当然，要能充分发挥 Vega Prime 的功能，必须对 Vega Prime 的编程对象进行反复认识，以达到灵活应用的程度。本章主要对 Vega Prime 的编程对象的 C++代码进行了进一步讲解，以期达到自由使用的目的。本章主要内容包括内核对象、管道对象、窗口对象、物体文件路径、运动模式、转换对象、观察者对象、键盘函数、碰撞检测、特效控制、灯光效果、DOF 操作等。本章主要采用实例讲解的方式，充分展示了利用 C++代码操控 Vega Prime 的主要对象，实现了虚拟仿真应用程序的强大功能。

【本章重点】

- 建立内核对象；
- 建立管道对象；
- 建立窗口对象；
- 物体文件路径；
- 动态加载物体；
- 设置运动模式；
- 建立转换对象；
- 建立观察者对象；
- 设置键盘函数；
- 物体缩放控制；
- 特效控制；
- 幻影效果；
- DOF 操作；
- 获取 DOF 坐标；
- Switch 操作；
- 配置多通道；
- 物体平面影子效果；
- 物体颜色控制；
- 雨雪天气控制；
- 场景能见度控制；
- 全屏控制；
- 加快物体加载速度。

6.1 建立内核实例

引入内核类的头文件为：#include "vpKernel.h"。

每个 Vega Prime 应用都必须有一个内核对象实例，内核对象的类视图如图 6.1.1 所示。在这个应用中，建立了内核实例 pKernel_myKernel，然后使用内核实例，设置了消息提示的级别，设置了内核的线程优先级，设置了处理器的数目，设置了帧循环率限制的使能，设置了期望的帧循环率。在这里没有把内核对象 pKernel_myKernel 压入容器 PublicMember::CTS_s_pInstancesToUnref，在后面建立的对象，却千万不能忘记把建立的对象压入容器中。建立内核实例的代码如图 6.1.2 所示。

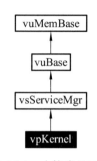

图 6.1.1 内核类继承图

```
//建立内核实例
vpKernel* pKernel_myKernel = vpKernel::instance();
pKernel_myKernel->setNotifyLevel( vuNotify::LEVEL_WARN );
pKernel_myKernel->setNotifyColorEnable( false );
pKernel_myKernel->setPriority( vuThread::PRIORITY_NORMAL );
pKernel_myKernel->setProcessor( -1 );
pKernel_myKernel->setFrameRateLimiterEnable( true );
pKernel_myKernel->setDesiredFrameRate( 0.000000 );
pKernel_myKernel->setNotifyClassTypeMode( vuNotify::CLASS_TYPE_MODE_INCLUSIVE );
```

图 6.1.2 建立内核实例

6.2 建立管道对象

引入管道类的头文件为：#include "vpPipleline.h"。

每个 Vega Prime 应用都应该有一个管道对象实例，管道类视图如图 6.2.1 所示。在这个应用中，建立了管道实例 pPipeline_myPipeline，然后使用管道实例，设置了管道名字，设置了管道的线程优先级，设置了管道 Id 号，设置了剪切线程数目，设置了剪切线程优先级，设置了剪切线程处理器数目，设置了绘制线程优先级，设置了绘制线程处理器数目。最后，把管道放进容器中：PublicMember::CTS_s_pInstancesToUnref->push_back(pPipeline_myPipeline)，具体代码如图 6.2.2 所示。

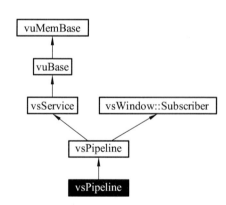

图 6.2.1 管道类继承图

```
//建立管道
vpPipeline* pPipeline_myPipeline = new vpPipeline();
pPipeline_myPipeline->setName( "myPipeline" );
pPipeline_myPipeline->setMultiThread( vsPipeline::MULTITHREAD_INLINE );
```

```
pPipeline_myPipeline->setId( 0 );
pPipeline_myPipeline->setNumCullThreads( 0 );
pPipeline_myPipeline->setCullThreadPriority( vuThread::PRIORITY_NORMAL );
pPipeline_myPipeline->setCullThreadProcessor( -1 );
pPipeline_myPipeline->setDrawThreadPriority( vuThread::PRIORITY_NORMAL );
pPipeline_myPipeline->setDrawThreadProcessor( -1 );
pPipeline_myPipeline->setBeginFrameOnVsyncEnable( false );
pPipeline_myPipeline->setDesiredPostDrawTime( -1.000000 );

PublicMember::CTS_s_pInstancesToUnref->push_back( pPipeline_myPipeline );
```

图 6.2.2　建立管道对象

在 Vega Prime 应用程序中，内核对象负责处理调用 Vega Prime 核心模块，处理结果通过管道的剪切与绘制呈现在窗体中，而观察者通过不同的通道会见到不同的场景。具体的配置代码如下：

```
pPipeline_myPipeline->addWindow( pWindow_myWindow );
pWindow_myWindow->addChannel( pChannel_myChannel );
pObserver_myObserver->addChannel( pChannel_myChannel );
```

第 1 行代码为管道添加窗体，第 2 行代码为窗体添加通道，第 3 行代码为观察者添加通道。

6.3　建立窗口

引入窗口类的头文件为：#include "vpWindow.h"。

每个 Vega Prime 应用都应该有一个窗口对象实例，窗口类视图如图 6.3.1 所示。在这个应用中，建立了窗口实例 pWindow_myWindow，然后使用窗口实例，设置了窗口名字，设置了窗口的标签，设置了窗口起始点，设置了窗口尺寸，设置了窗口全屏选项，设置了窗口边界，设置了窗口输入选项，设置了窗口鼠标形状，设置了窗口缓存交换间隔，设置了窗口像素缓存模式，等等。最后，把窗口放进容器队列中：
　　PublicMember::CTS_s_pInstancesToUnref->push_back(pWindow_myWindow);

具体代码如图 6.3.2 所示。

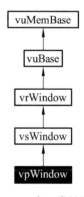

图 6.3.1　窗口类继承图

```
//建立窗体
vpWindow* pWindow_myWindow = new vpWindow();
pWindow_myWindow->setName( "myWindow" );
pWindow_myWindow->setLabel( "Vega Prime Window" );
pWindow_myWindow->setOrigin( 0 ,    0 );
pWindow_myWindow->setSize( 1024 ,    768 );
pWindow_myWindow->setFullScreenEnable( false );
```

```
pWindow_myWindow->setBorderEnable( true );
pWindow_myWindow->setInputEnable( true );
pWindow_myWindow->setCursorEnable( true );
pWindow_myWindow->setStereoEnable( false );
pWindow_myWindow->setNumColorBits( 8 );
pWindow_myWindow->setNumAlphaBits( 0 );
pWindow_myWindow->setNumDepthBits( 24 );
pWindow_myWindow->setNumStencilBits( 0 );
pWindow_myWindow->setNumAccumColorBits( 0 );
pWindow_myWindow->setNumAccumAlphaBits( 0 );
pWindow_myWindow->setNumMultiSampleBits( 0 );
pWindow_myWindow->setSwapInterval( 1 );
pWindow_myWindow->setPixelBufferMode( vrDrawContext::PIXEL_BUFFER_MODE_OFF );
PublicMember::CTS_s_pInstancesToUnref->push_back( pWindow_myWindow );
```

图 6.3.2　建立窗口

6.4　建立场景

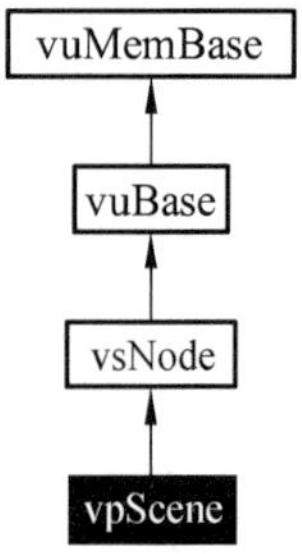

引入场景类的头文件为：#include <vpScene.h>。

场景类视图如图 6.4.1 所示。场景作为一个经常使用的对象，定义一个静态指针指向场景，以方便使用，定义如下：

static vpScene * PublicMember::CTS_pScene_myScene;

建立场景的代码非常简洁，其代码如下：

//建立场景

PublicMember::CTS_pScene_myScene = new vpScene();

PublicMember::CTS_pScene_myScene->setName("myScene");

PublicMember::CTS_s_pInstancesToUnref->push_back(PublicMember::CTS_pScene_myScene);

图 6.4.1　场景类继承图

在进行了这样的场景定义之后，就非常方便地把物体添加进场景了。

6.5　改造路径搜索对象

引入路径搜索类的头文件为：#include "vpSearchPath.h"。

路径搜索类视图如图 6.5.1 所示。在场景中加载的对象主要是物体，而物体很有可能是以文件形式储存在某个位置。如何来管理对象的路径呢？Vega Prime 提供了 vpSearchPath 对象来完成这个功能。在这个应用中，对象都储存于 Vega

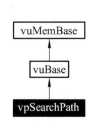

图 6.5.1　路径搜索类继承图

Prime 安装目录下的某个子目录中，所以在配置物体路径时采用了以下代码：

PublicMember::CTS_pSearchPath = vpSearchPath::instance();
PublicMember::CTS_pSearchPath->append("C:/Program Files/MultiGen-Paradigm/resources/tutorials/vegaprime/desktop_tutor/tornado/data/land");
PublicMember::CTS_pSearchPath->append("C:/Program Files/MultiGen-Paradigm/resources/tutorials/vegaprime/desktop_tutor/tornado/data/farmhouse");
PublicMember::CTS_pSearchPath->append("C:/Program Files/MultiGen-Paradigm/resources/tutorials/vegaprime/desktop_tutor/tornado/data/humv-dirty");
PublicMember::CTS_pSearchPath->append("C:/Program Files/MultiGen-Paradigm/resources/tutorials/vegaprime/desktop_tutor/tornado/data/grainstorage");

从代码中可以看出，使用了绝对路径，这样就限制了对物体对象的管理和加载，更不便于物体和程序的统一发布管理，须进行必要的处理。把"C:/Program Files/MultiGen-Paradigm/resources/tutorials/vegaprime/desktop_tutor/tornado/data"目录复制到应用程序 VPTestDialog 目录下，这样，只要获取当前程序所在的目录，再加上 data 目录，就能得到物体的路径。

首先，为 PublicMember 类添加一个静态数据成员：

    static CString CTS_RunPath; //存储程序运行路径

然后，利用下面的代码获取程序所在的路径：

    char buf[MAX_PATH];
    ::GetCurrentDirectory(MAX_PATH,buf);
    PublicMember::CTS_RunPath.Format("%s",buf);

做完这些工作，就可以把原来的路径搜索代码替换为以下代码：

PublicMember::CTS_pSearchPath = vpSearchPath::instance();
PublicMember::CTS_pSearchPath->append(PublicMember::CTS_RunPath +"/data/land");
PublicMember::CTS_pSearchPath->append(PublicMember::CTS_RunPath+"/data/farmhouse");
PublicMember::CTS_pSearchPath->append(PublicMember::CTS_RunPath+"/data/humv-dirty");
PublicMember::CTS_pSearchPath->append(PublicMember::CTS_RunPath+"/data/grainstorage");

在进行了这样的路径配置之后，对物体的管理和应用就与应用程序的管理和应用统一了起来，非常灵活方便，特别是在动态加载物体对象时非常方便。当然，这样配置就要求 data 目录与可执行文件必须在同一个目录中。例如，如果要在 VPTestDialog 目录中找到 Release 目录，直接运行 VPTestDialog.exe 文件，就要求把 data 目录复制到 Release 目录下。

当然，只要真正明白了 Vega Prime 目录对象的作用，就可以灵活地配置和管理目录。

6.6 加载物体函数设计

引入对象类的头文件为：#include "vpObject.h"。

物体对象类视图如图 6.6.1 所示。将物体加载进场景通常是使用最频繁的操作，实现加载物体函数化设计能给开发者带来很大的方便。仔细分析物体加载的代码，就会发现其中的关键所在。其中，每个物体必需的是物体对象指针、文件路径、文件名、物体名称、位置姿态参数、容器和加入场景。进一步分析还可以发现：物体对象指针可以使用函数返回值进行赋值；文件路径可以使用前面所讲的路径搜索对象提前完成；容器也可以使用改造时定义的容器 PublicMember::CTS_s_pInstancesToUnref；位置姿态参数使用默认值 0。因此，最后只剩下物体名称和文件名两个作为加载物体参数，物体对象指针作为函数返回值。函数声明如下：

```
static vpObject * CTS_AddObject001(CString ObjectName, CString FileName);
```

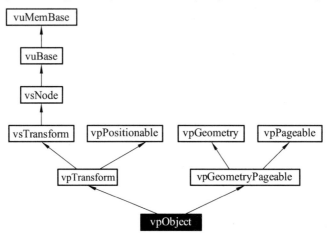

图 6.6.1 物体对象类继承图

加载物体函数的具体代码如图 6.5.2 所示。

```
// 函数声明，应在 PublicMember.h 文件中完成
static vpObject * CTS_AddObject001(CString ObjectName, CString FileName);

//--------------------------------------------------------------
//加载物体函数实现，应在 PublicMember.cpp 文件中完成
vpObject * PublicMember::CTS_AddObject001(CString ObjectName, CString FileName)
{
    //设置物体特性
    vpObject *obj = new vpObject();
    obj->setName(ObjectName );                              //物体名称
    obj->setCullMask( 0x0FFFFFFFF );
    obj->setRenderMask( 0x0FFFFFFFF );
    obj->setIsectMask( 0x0FFFFFFFF );
    obj->setStrategyEnable( true );
    obj->setTranslate(0.000000 ,   0.000000 ,   0.000000 );
    obj->setRotate( 0.000000 ,   0.000000 ,   0.000000 );
    obj->setScale( 1.00000 , 1.000 , 1.000 );
    obj->setStaticEnable( false );

    obj->setFileName( FileName);                            //物体文件名
```

```cpp
obj->setAutoPage( vpObject::AUTO_PAGE_SYNCHRONOUS );
obj->setManualLODChild( -1 );
obj->setLoaderOption( vsNodeLoader::Data::LOADER_OPTION_COMBINE_LIGHT_POINTS ,   true );
obj->setLoaderOption(vsNodeLoader::Data::LOADER_OPTION_SHARE_LIGHT_POINT_CONTROLS
                                ,true );
obj->setLoaderOption( vsNodeLoader::Data::LOADER_OPTION_COMBINE_LODS ,true );
obj->setLoaderOption(     vsNodeLoader::Data::LOADER_OPTION_IGNORE_DOF_CONSTRAINTS    ,
                                true );
obj->setLoaderOption( vsNodeLoader::Data::LOADER_OPTION_PRESERVE_EXTERNAL_REF
                                _FLAGS,true );

obj->setLoaderOption( vsNodeLoader::Data::LOADER_OPTION_PRESERVE_GENERIC_NAMES ,false );

obj->setLoaderOption( vsNodeLoader::Data::LOADER_OPTION_PRESERVE_GENERIC_NODES ,true );
obj->setLoaderOption( vsNodeLoader::Data::LOADER_OPTION_PRESERVE_QUADS ,   false );
obj->setLoaderOption( vsNodeLoader::Data::LOADER_OPTION_ALL_GEOMETRIES_LIT ,   false );
obj->setLoaderOption( vsNodeLoader::Data::LOADER_OPTION_USE_MATERIAL_DIFFUSE_COLOR
                                ,false );
obj->setLoaderOption( vsNodeLoader::Data::LOADER_OPTION_MONOCHROME ,   false );
obj->setLoaderOption( vsNodeLoader::Data::LOADER_OPTION_CREATE_ANIMATIONS ,   true );

obj->setLoaderOption( vsNodeLoader::Data::LOADER_OPTION_SHARE_LIGHT_POINT_ANIMATIONS
                                ,true );
obj->setLoaderOption( vsNodeLoader::Data::LOADER_OPTION_SHARE_LIGHT_POINT
                                _APPEARANCES,true );
obj->setLoaderDetailMultiTextureStage( -1 );
obj->setLoaderBlendTolerance( 0.050000f );
obj->setLoaderUnits( vsNodeLoader::Data::LOADER_UNITS_METERS );
obj->setBuilderOption( vsNodeLoader::Data::BUILDER_OPTION_OPTIMIZE_GEOMETRIES ,true );
obj->setBuilderOption( vsNodeLoader::Data::BUILDER_OPTION_COLLAPSE_BINDINGS ,   true );
obj->setBuilderOption( vsNodeLoader::Data::BUILDER_OPTION_COLLAPSE_TRIANGLE_STRIPS
                                ,true );
obj->setBuilderNormalMode( vsNodeLoader::Data::BUILDER_NORMAL_MODE_PRESERVE );
obj->setBuilderColorTolerance( 0.001000f );
obj->setBuilderNormalTolerance( 0.860000f );
obj->setBuilderVertexTolerance( 0.000100f );
obj->setGeometryOption( vsNodeLoader::Data::GEOMETRY_OPTION_GENERATE_DISPLAY
                                _LISTS ,true );
obj->setGeometryFormat( vrGeometryBase::FORMAT_VERTEX_ARRAY ,   0x0FFF );
obj->setPostLoadOption( vpGeometryPageable::POST_LOAD_OPTION_FLATTEN, true );
obj->setPostLoadOption( vpGeometryPageable::POST_LOAD_OPTION_CLEAN , true );
obj->setPostLoadOption( vpGeometryPageable::POST_LOAD_OPTION_MERGE_GEOMETRIES ,true );
```

```
    obj->setPostLoadOption( vpGeometryPageable::POST_LOAD_OPTION_COLLAPSE_BINDINGS ,true );
    obj->setPostLoadOption( vpGeometryPageable::POST_LOAD_OPTION_COLLAPSE_TRIANGLE
                            _STRIPS,true );

    obj->setPostLoadOption( vpGeometryPageable::POST_LOAD_OPTION_VALIDATE ,true );
    obj->setTextureSubloadEnable( false );
    obj->setTextureSubloadRender( vpGeometry::TEXTURE_SUBLOAD_RENDER_DEFERRED );
    //
    PublicMember::CTS_s_pInstancesToUnref->push_back( obj ); //压进容器，但未加入场景
    return (obj);
}
```

图 6.5.2　加载物体对象函数

定义的加载物体函数如何使用呢？在本质上加载物体与 PublicMember::CTS_Define()函数是一体的，所以加载物体的函数使用应该在 PublicMember::CTS_Define()之后，并且在 vpKernel::instance()->configure()函数之前，在第 5.2.3 节的主线程中按以下方式使用该函数：

```
    //定义场景
    PublicMember::CTS_Define();
    //加载物体
    PublicMember::CTS_pfarmhouse=PublicMember::CTS_AddObject("farmhouse"," farmhouse.flt");
    //加入场景
    PublicMember::CTS_pScene_myScene->addChild(PublicMember::CTS_pfarmhouse);

    //配置场景
    vpKernel::instance()->configure();
```

当然，在使用前必须先定义房屋指针：

```
    static vpObject *CTS_pfarmhouse;
```

并进行初始化：

```
    vpObject * PublicMember::CTS_pfarmhouse=NULL;
```

只要掌握了基本方法，开发者可以灵活应用。例如，可以增加更多的参数，使加载物体函数可以有更多的控制量。

6.7　设置运动模式

引入驾驶模式类的头文件为：#include "vpMotionDrive.h"。
驾驶模式类视图如图 6.7.1 所示。启用运动模式一定要在程序开头进行初始化：

```
    //初始化运动模式模块
    vpModule::initializeModule( "vpMotion" );
```

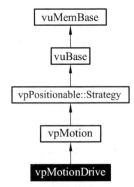

图 6.7.1　驾驶模式类继承图

对于运动模式，每一种模式都有自己特有的属性。使用了驾驶模式，其中的设置主要包含：设置模式名字，设置模式的初始速度，设置模式的速度限制范围，设置模式的加速度，设置模式的方向变化量。具体定义驾驶模式的代码如图 6.7.2 所示。

```
//建立驾驶模式
vpMotionDrive* pMotionDrive_myMotion = new vpMotionDrive();
pMotionDrive_myMotion->setName( "myMotion" );
pMotionDrive_myMotion->setSpeed( 0.000000 );
pMotionDrive_myMotion->setSpeedLimits( -10.000000 ,    10.000000 );
pMotionDrive_myMotion->setSpeedDelta( 1.000000 ,    1.000000 );
pMotionDrive_myMotion->setHeadingDelta( 1.000000 );

PublicMember::CTS_s_pInstancesToUnref->push_back( pMotionDrive_myMotion );
```

图 6.7.2　定义驾驶模式

对于驾驶模式的具体使用，在 void PublicMember::CTS_Define(void)函数中，可以看到这样的配置代码：

　　　　pObserver_myObserver->setStrategy(pMotionDrive_myMotion);
　　　　pObject_Hummer->setStrategy(pMotionDrive_myMotion);

这两句代码就是把观察者和汽车的策略都设置为定义的驾驶模式 pMotionDrive_myMotion。后续章节将使用代码把观察者与汽车联系到一起。

6.8　建立转换

引入转换类的头文件为：#include "vpTransform.h"。

转换类视图如图 6.8.1 所示。每个 Vega Prime 应用都可以建立转换对象实例。在这个应用中，建立了转换实例 pTransform_hummerTransform，然后使用转换实例，把观察者添加为汽车的"孩子"，设置了转换名字，设置了转换剪切掩码，设置了转换的渲染掩码，设置了转换的碰撞掩码，设置了转换的策略选项，设置了转换相对于父亲的位置，设置了转换的姿态，设置了转换的缩放比例，设置了转换的静态选项。最后，把转换放进容器队列中：

　　s_pInstancesToUnref->push_back(pTransform_hummerTransform);

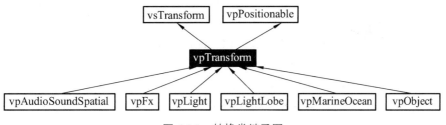

图 6.8.1　转换类继承图

具体代码如图 6.8.2 所示。

```
//建立转换
vpTransform* pTransform_hummerTransform = new vpTransform();
pTransform_hummerTransform->setName( "hummerTransform" );
pTransform_hummerTransform->setCullMask( 0x0FFFFFFFF );
pTransform_hummerTransform->setRenderMask( 0x0FFFFFFFF );
pTransform_hummerTransform->setIsectMask( 0x0FFFFFFFF );
pTransform_hummerTransform->setStrategyEnable( true );
pTransform_hummerTransform->setTranslate( 0.000000, -30.000000, 5.000000 );
pTransform_hummerTransform->setRotate( 0.000000, 0.000000, 0.000000 );
pTransform_hummerTransform->setScale( 1.000000, 1.000000, 1.000000 );
pTransform_hummerTransform->setStaticEnable( false );

PublicMember::CTS_s_pInstancesToUnref->push_back( pTransform_hummerTransform );
```

图 6.8.2　建立转换

6.9　控制观察者

引入观察者类的头文件为：#include "vpObserver.h"。
观察者类视图如图 6.9.1 所示。本节通过编码设置获取对观察者的控制。首先，要定义观察者，其代码如图 6.9.2 所示。

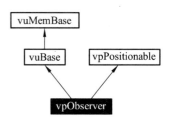

图 6.9.1　观察者类继承图

```
//建立观察者
vpObserver* pObserver_myObserver = new vpObserver();
pObserver_myObserver->setName( "myObserver" );
pObserver_myObserver->setStrategyEnable( false );
pObserver_myObserver->setTranslate( 2300.000000, 2500.000000, 15.000000 );
pObserver_myObserver->setRotate( -90.000000, 0.000000, 0.000000 );
pObserver_myObserver->setLatencyCriticalEnable( false );

PublicMember::CTS_s_pInstancesToUnref->push_back( pObserver_myObserver );
```

图 6.9.2　定义观察者

通过一个转换 vpTransform 把观察者绑定在车上，这样就可以跟随车子一起运动，观察场景。在 void PublicMember::CTS_Define(void) 函数中，可以看到这样的配置代码：

```
    pObserver_myObserver->setStrategy( pMotionDrive_myMotion );
    pObserver_myObserver->addChannel( pChannel_myChannel );
    pObserver_myObserver->addAttachment( pEnv_myEnv );
    pObserver_myObserver->setScene( PublicMember::CTS_pScene_myScene );
    pObserver_myObserver->setLookFrom( pTransform_hummerTransform );

    pObject_Hummer->addChild( pTransform_hummerTransform );
```

pObject_Hummer->setStrategy(pMotionDrive_myMotion);

仔细分析这 7 行代码：第 1 行为观察者设置策略，设置为驾驶模式；第 2 行为观察者添加通道；第 3 行为观察者添加附属物，添加了环境；第 4 行为观察者设置场景；第 5 行为观察者设置观察方向，就是以 pTransform_hummerTransform 转换为观察起点；第六行添加 pTransform_hummerTransform 转换为汽车的"孩子"；第 7 行为汽车设置策略，设置为驾驶模式，与观察者的策略一致。其中，第 5 行与第 6 行通过转换就把观察者绑定到了汽车上，实现了要求。下一节将通过键盘来控制汽车的运动。

6.10 配置键盘函数

在很多时候，使用键盘控制场景中的物体也是一项基本要求。本节将编码实现键盘对观察者的控制。

引入键盘类的头文件为：#include "vpInputKeyboard.h"。

键盘类视图如图 6.10.1 所示。Vega Prime 提供了键盘对应的全部功能键，比如小写字母 f 对应 vpWindow::KEY_f，大写字母 F 对应 vpWindow::KEY_F，上下键分别对应 vpWindow:: KEY_UP 和 vpWindow::KEY_DOWN，左右键分别对应 vpWindow::KEY_LEFT 和 vpWindow:: KEY_RIGHT。

首先，为了使用方便，像定义声明主线程函数一样，定义声明键盘函数：

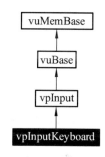

图 6.10.1 键盘类继承图

```
//运动场景键盘操作函数
static void    CTS_Keyboard(vpWindow *window,vpWindow::Key key, int modifier,void *);
```

把该函数定义为 PublicMeber 类的静态成员函数，其中最关心的是该函数的第 2 个参数 key，它用来识别用户的按键。该函数的具体代码如图 6.10.2 所示，其中的静态变量 PublicMember::CTS_pObject_observer 是定义设置的观察者对象指针，定义方法与前面定义静态成员变量一样。

```
//控制观察者位置
void   PublicMember::CTS_Keyboard(vpWindow *window,vpWindow::Key key, int modifier,void *)
{
    switch(key)
    {
//使观察者向前
        case vpWindow::KEY_UP:
            {
                PublicMember::CTS_pObject_observer->setTranslateY(0.2,true);
            }
            break;
//使观察者后退
        case vpWindow::KEY_DOWN:
```

```
                {
                    PublicMember::CTS_pObject_observer->setTranslateY(-0.2,true);
                }
                break;
//使观察者向左转
        case vpWindow::KEY_LEFT:
                {
                    PublicMember::CTS_pObject_observer->setTranslateX(-0.2,true);
                }
                break;
//使观察者向右转
        case vpWindow::KEY_RIGHT:
                {
                    PublicMember::CTS_pObject_observer->setTranslateX(0.2,true);
                }
                break;
        default:
                ;
        }//end of switch
}//end of function
```

图 6.10.2　键盘函数

定义完键盘函数 PublicMember::CTS_Keyboard 后，还需要把它设置成对应窗口的键盘函数才能真正起作用。在主线程中，将添加以下代码：

//设置窗体

vpWindow * vpWin= * vpWindow::begin();

vpWin->setParent(PublicMember::CTS_RunningWindow);

vpWin->setBorderEnable(false);

vpWin->setFullScreenEnable(true);

//设置键盘

vpWin->setInputEnable(true);

vpWin->setKeyboardFunc((vrWindow::KeyboardFunc)PublicMember::CTS_Keyboard,NULL);

vpWin->open();

::SetFocus(vpWin->getWindow());

其中，第 2 行、第 3 行、第 4 行代码分别设置了窗体的位置、边界、全屏，对键盘函数的设定，实际上包含以下两行代码：

vpWin->setInputEnable(true);

vpWin->setKeyboardFunc((vrWindow::KeyboardFunc)PublicMember::CTS_Keyboard,NULL);

在主线程中对观察者对象指针 PublicMember::CTS_pObject_observer 进行了赋值，其中的代码如下：

//设置观察者

PublicMember::CTS_pObject_observer=vpObject::find("hummer");

PublicMember::CTS_pObject_observer->ref();

在前面已经用代码把观察者绑定到了汽车上，实际上，并没有直接获取观察者指针的值，而是根据名称获取了汽车对象指针的值，操作汽车就能跟随汽车观察场景了。图 6.10.3 所示为改造后的完整主线程程序，请仔细查看其中加粗了字体的代码。

```
//VP 运行主线程。
UINT PublicMember::CTS_RunBasicThread(LPVOID)
{
    //初始化
    vp::initialize(__argc,__argv);
    //定义场景
    PublicMember::CTS_Define();
    //配置场景
    vpKernel::instance()->configure();

    //设置观察者
    PublicMember::CTS_pObject_observer=vpObject::find("hummer");
    PublicMember::CTS_pObject_observer->ref();

    //设置窗体
    vpWindow * vpWin= * vpWindow::begin();
    vpWin->setParent(PublicMember::CTS_RunningWindow);
    vpWin->setBorderEnable(false);
    vpWin->setFullScreenEnable(true);

    //设置键盘
    vpWin->setInputEnable(true);
    vpWin->setKeyboardFunc((vrWindow::KeyboardFunc)PublicMember::CTS_Keyboard,NULL);

    vpWin->open();
    ::SetFocus(vpWin->getWindow());     //必须包含此语句，否则键盘函数无效
    //帧循环
    while(vpKernel::instance()->beginFrame()!=0)
    {
        vpKernel::instance()->endFrame();
        if(!PublicMember::CTS_continueRunVP)
        {
            vpKernel::instance()->unconfigure();
            vp::shutdown();
            return 0;
        }
    }
    return 0;
}
```

图 6.10.3　Vega Prime 应用程序主线程函数

6.11 控制物体缩放比例及透明

控制物体的放大缩小是经常需要使用的操作，而操作方法也非常简单。对于 vpObject 对象，使用函数 setScale(x,y,z)就能实现物体在 x 轴、y 轴和 z 轴上缩放比例控制。例如：

  vpObject * pObject;

  pObject->setScale(0.500000 , 1.000000 , 2.000000);

就能实现对物体 pObject 在 x 轴上的尺寸缩小一半、在 y 轴上的尺寸保持原样和在 z 轴上的尺寸放大到两倍。

控制物体透明的最简单的方法是设置其渲染掩码为 0x0000000：

  PublicMember::pObject001->setRenderMask(0x0000000);

实现对物体的透明控制，需要引入 3 个头文件依次为：#include "vsGeometry.h"（几何体头文件），#include "vrMode.h"（方式头文件），#include "vrState.h"（状态头文件）。

几何类视图如图 6.11.1 所示，状态类视图如图 6.11.2 所示。

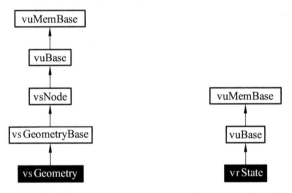

图 6.11.1 几何体类继承图   图 6.11.2 状态类继承图

物体的透明控制可以实现物体的"隐身"，它比缩放比例的控制更复杂，不能采用单一函数实现。在物体的透明控制中，首先要使用结构体 vrAlphaTest::Element，它包含 3 个数据成员：bool m_enable，默认值为 false,如果需要修改，必须把值设置为 true；Mode m_mode,默认值为 vrAlphaTest::MODE_ALWAYS;float m_ref,值为 1.0 时设置为透明，值为 0.0 时设置为不透明。其次，要使用物体对象的几何体，获取几何体的状态，利用状态对象修改元素的值。

为了便于使用，把控制物体透明的功能编写成一个静态函数：

  void PublicMember::CTS_SetObjectTrasnparent(vpObject *pObject,bool transparent)

其中，参数 vpObject *pObject 表示物体对象指针，参数 bool transparent 表示是否设置为透明，true 时为透明，false 时为不透明。

控制物体对象透明的完整代码如图 6.11.3 所示。

```
//控制物体透明
  void PublicMember::CTS_SetObjectTrasnparent(vpObject *pObject,bool transparent)
  {
      vrAlphaTest::Element ate;
```

```
    if(transparent)
        ate.m_ref=1.0f;//设置为透明
    else
        ate.m_ref=0.0f;//设置为不透明

    ate.m_mode=vrAlphaTest::MODE_GREATER;
    ate.m_enable=true;
    vpObject::const_iterator_geometry nit,nite=pObject->end_geometry();
    for(nit=pObject->begin_geometry();nit!=nite;++nit)
    {
        vrState *state=(*nit)->getState();

        state->setElement(vrAlphaTest::Element::Id,&ate);
        (*nit)->setState(state);
    }
}
```

图 6.11.3　Vega Prime 应用程序透明控制函数

透明控制函数的调用位置位于主线程中帧循环判断中，在这里通过控制汽车的透明性来决定是否需要隐身。前面已经定义了汽车对象的静态指针变量 PublicMember::CTS_pObject_observer，这里再定义一个静态逻辑值：

　　bool　PublicMember::CTS_SettingHide;

当 PublicMember::CTS_SettingHide 取值为 true 时，m_ref=1.0f 就把汽车设置为透明；当 PublicMember::CTS_SettingHide 取值为 false 时，m_ref=0.0f 就把汽车设置为不透明。在主线程中，具体的透明控制函数在主线程中的调用代码如下：

```
//帧循环
while(vpKernel::instance()->beginFrame()!=0)
{
    vpKernel::instance()->endFrame();

    //透明控制
    PublicMember::CTS_SetObjectTrasnparent(PublicMember::CTS_pObject_observer,
                            PublicMember::CTS_SettingHide);

    if(!PublicMember::CTS_continueRunVP)
    {
      vpKernel::instance()->unconfigure();
      vp::shutdown();
      return 0;
    }
}
```

6.12 控制碰撞检测

实现对物体的碰撞检测,需要引入两个头文件,依次为:

#include "vpIsectorBump.h"(碰撞检测头文件),#include "vpIsectorServiceInline.h"(碰撞检测服务头文件)。

碰撞类视图如图 6.12.1 所示,碰撞检测服务类视图如图 6.12.2 所示。

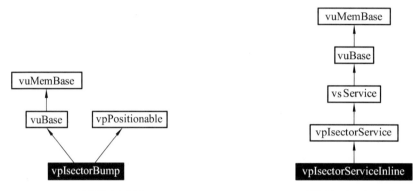

图 6.12.1 碰撞检测类继承图　　　　图 6.12.2 碰撞检测服务类继承图

碰撞检测主要包括建立碰撞模式、设置碰撞检测、碰撞处理 3 个部分。恰当地使用碰撞,能够让虚拟的场景更加真实。

建立碰撞模式的代码如图 6.12.3 所示,其中包含两个对象:

一个是碰撞检测对象:pIsectorBump_bumpIsector,另一个是碰撞检测服务对象:pIsectorServiceInline_myIsectorService。

```
//建立碰撞模式
vpIsectorBump* pIsectorBump_bumpIsector = new vpIsectorBump();
pIsectorBump_bumpIsector->setName( "bumpIsector" );
pIsectorBump_bumpIsector->setEnable( true );
pIsectorBump_bumpIsector->setRenderEnable( false );
pIsectorBump_bumpIsector->setTranslate( 0.000000,  0.000000,  0.000000 );
pIsectorBump_bumpIsector->setRotate( 0.000000,  0.000000,  0.000000 );
pIsectorBump_bumpIsector->setMode( 0x02 );
pIsectorBump_bumpIsector->setIsectMask( 0x0FFFFFFFF );
pIsectorBump_bumpIsector->setStrategyEnable( true );
pIsectorBump_bumpIsector->setDimensions( 3.000000f,  3.000000f,  3.000000f );

PublicMember::CTS_s_pInstancesToUnref->push_back( pIsectorBump_bumpIsector );

//建立碰撞服务对象
vpIsectorServiceInline* pIsectorServiceInline_myIsectorService = new vpIsectorServiceInline();
pIsectorServiceInline_myIsectorService->setName( "myIsectorService" );

PublicMember::CTS_s_pInstancesToUnref->push_back(
                              pIsectorServiceInline_myIsectorService );
```

图 6.12.3 Vega Prime 应用程序建立碰撞模式

设置碰撞检测的代码如下：

    pIsectorBump_bumpIsector->setTarget(pObject_farmhouse);

    pIsectorBump_bumpIsector->setPositionReference(pObject_Hummer);

    pIsectorServiceInline_myIsectorService->addIsector(pIsectorBump_bumpIsector);

第 1 句代码设置碰撞检测的目标是房子，第 2 句代码设置碰撞的位置参考为汽车，第 3 句把碰撞检测添加到碰撞服务对象里面。第 1 句与第 2 句实际上是设置了产生碰撞的两个对象，即房子和车子，第 3 句就是为这个碰撞提供了一个服务对象。

怎么发现碰撞已经发生呢？有一句关键的代码如下：

    if(pIsectorBump_bumpIsector->getHit()==true)
      {
        AfxMessageBox("collision   happened!");
      }

在这里发现碰撞后只是弹出一个消息框，提示碰撞已经发生。当然，也可以做自己的处理，例如，让汽车停下来或者用另外一个破损的房屋来代替原来完好的房屋。在主线程中获取碰撞检测及弹出消息框的具体代码如图 6.12.4 所示。

```
//VP 运行主线程。
UINT PublicMember::CTS_RunBasicThread(LPVOID)
{
    //初始化
    vp::initialize(__argc,__argv);
    //定义场景
    PublicMember::CTS_Define();
    //配置场景
    vpKernel::instance()->configure();

    //设置观察者
    PublicMember::CTS_pObject_observer=vpObject::find("hummer");
    PublicMember::CTS_pObject_observer->ref();

    //设置窗体
    vpWindow * vpWin= * vpWindow::begin();
    vpWin->setParent(PublicMember::CTS_RunningWindow);
    vpWin->setBorderEnable(false);
    vpWin->setFullScreenEnable(true);

    //设置键盘
    vpWin->setInputEnable(true);
    vpWin->setKeyboardFunc((vrWindow::KeyboardFunc)PublicMember::CTS_Keyboard,NULL);

    vpWin->open();
    vpWin->setFocus();

    //帧循环
    while(vpKernel::instance()->beginFrame()!=0)
    {
```

```
        vpKernel::instance()->endFrame();

            //碰撞检测
        if(pIsectorBump_bumpIsector->getHit()==true)
        {
            AfxMessageBox("collision happened!");
            //
        }
        //
        if(!PublicMember::CTS_continueRunVP)
        {
          vpKernel::instance()->unconfigure();
          vp::shutdown();
          return 0;
        }
    }
    return 0;
}
```

图 6.12.4　Vega Prime 应用程序碰撞检测主线程

6.13　控制特效

引入键盘类的头文件为：#include "vpFxDebris.h"。
特效类视图如图 6.13.1 所示。启用特效，一定要在程序开头初始化特效模块：
　　//初始化特效模块
　　vpModule::initializeModule("vpFx");

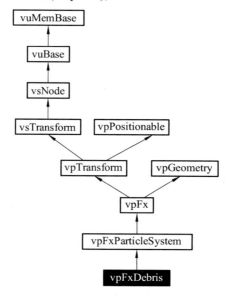

图 6.13.1　粒子特效类继承图

恰当地使用特效才能使开发的虚拟现实更加逼真。特效的使用主要包括建立特效、配置特效和触发特效 3 个步骤。

在应用中，为房屋建立一个碎片特效，一旦汽车碰撞到房屋，就能看到房屋有碎片抛落。建立碎片特效的代码如图 6.13.2 所示，其中这 3 句代码值得留意：

  pFxDebris_Debris->setName("Debris");

  pFxDebris_Debris->setTextureFile("debris.inta");

  pFxDebris_Debris->setOverallDuration(4.000000f);

其中，第 1 句代码设置碎片特效的名称，第 2 句代码设置了碎片特效的材质文件，第 3 句设置了碎片特效的持续时间。建立房屋碎片特效的具体代码如图 6.13.2 所示。

```
vpFxDebris* pFxDebris_Debris = new vpFxDebris();
pFxDebris_Debris->setName( "Debris" );
pFxDebris_Debris->setCullMask( 0x0FFFFFFFF );
pFxDebris_Debris->setRenderMask( 0x0FFFFFFFF );
pFxDebris_Debris->setIsectMask( 0x0FFFFFFFF );
pFxDebris_Debris->setStrategyEnable( true );
pFxDebris_Debris->setTranslate( 0.000000 ,   0.000000 ,   0.000000 );
pFxDebris_Debris->setRotate( 0.000000 ,   0.000000 ,   0.000000 );
pFxDebris_Debris->setScale( 20.000000 ,   20.000000 ,   20.000000 );
pFxDebris_Debris->setStaticEnable( false );
pFxDebris_Debris->setRepeatEnable( false );
pFxDebris_Debris->setOverallColor( 1.000000f ,   1.000000f ,   1.000000f ,   1.000000f );
pFxDebris_Debris->setTextureBlendColor( 1.000000f , 1.000000f , 1.000000f , 1.000000f );
pFxDebris_Debris->setTextureMode( vpFx::TEXTURE_MODE_MODULATE );
pFxDebris_Debris->setTextureFile( "debris.inta" );
pFxDebris_Debris->setOverallDuration( 4.000000f );
pFxDebris_Debris->setFadeDuration( 0.000000f ,   0.000000f );
pFxDebris_Debris->setTransparencyDepthOffset( 0.000000f );
pFxDebris_Debris->setTransparencyDepthRadiusScaler( 0.000000f );
pFxDebris_Debris->setWindSource( vpFxParticleSystem::WIND_SOURCE_TABLE );
PublicMember::CTS_s_pInstancesToUnref->push_back( pFxDebris_Debris );
```

图 6.13.3　建立房屋碎片特效

建立完特效后，需要配置特效。其中：

  pObject_farmhouse->addChild(pFxDebris_Debris);

  pFxDebris_Debris->setTriggerIsector(pIsectorBump_bumpIsector);

第 1 句代码把碎片特效设置为房屋的孩子，第 2 句代码设置了特效的触发者为碰撞对象。

6.14　配置灯光效果

引入灯光类的头文件为：#include "vpLight.h"。

灯光类视图如图 6.14.1 所示。配置灯光主要是为汽车建立聚光灯效果，使汽车能在黑暗

的环境下也能顺利行驶。在建立聚光灯效果的代码中,以下代码为关键代码:

　　pLight_leftHeadlight->setName("leftHeadlight");
　　pLight_leftHeadlight->setTranslate(-0.050000,2.240000,0.900000);
　　pLight_leftHeadlight->setEnable(true);
　　pLight_leftHeadlight->setType(vpLight::TYPE_DIRECTIONAL);
　　pLight_leftHeadlight->setSpotCone(2.000000f ,4.000000f ,0.500000f);

第 1 句代码为灯设置名字,第 2 句设置了灯相对于父物体的位置,第 3 句代码设置灯的开关,第 4 句代码设置了灯的类型,第 5 句设置了灯的光点。具体代码如图 6.14.2 所示。

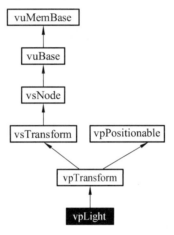

图 6.14.1　灯光类继承图

```
//建立聚光灯效果
  vpLight* pLight_leftHeadlight = new vpLight();
  pLight_leftHeadlight->setName( "leftHeadlight" );
  pLight_leftHeadlight->setCullMask( 0x00 );
  pLight_leftHeadlight->setRenderMask( 0x0FFFFFFFF );
  pLight_leftHeadlight->setStrategyEnable( true );
  pLight_leftHeadlight->setTranslate( -0.050000,  2.240000,  0.900000 );
  pLight_leftHeadlight->setRotate( 0.000000,  0.000000,  0.000000 );
  pLight_leftHeadlight->setScale( 1.000000,  1.000000,  1.000000 );
  pLight_leftHeadlight->setStaticEnable( false );
  pLight_leftHeadlight->setType( vpLight::TYPE_DIRECTIONAL );
  pLight_leftHeadlight->setEnable( true );
  pLight_leftHeadlight->setAttenuation( 0.0000000f,  0.000000f,  0.000000f );
  pLight_leftHeadlight->setSpotCone( 2.000000f,  4.000000f,  0.500000f );
  pLight_leftHeadlight->setColor( vpLight::COLOR_AMBIENT,  1.000000f,
                                  1.000000f,  1.000000f,  1.000000f );
  pLight_leftHeadlight->setColor( vpLight::COLOR_DIFFUSE,  0.1000000f,
                                  0.1000000f,  0.1000000f,  1.000000f );
  pLight_leftHeadlight->setColor( vpLight::COLOR_SPECULAR,  0.1000000f,
                                  0.1000000f,  0.1000000f,  0.100000f );
  pLight_leftHeadlight->setRenderEnable( true );
```

PublicMember::CTS_s_pInstancesToUnref->push_back(pLight_leftHeadlight);

图 6.14.2　建立聚光灯效果

建立完成后，需要把灯"安装"到汽车上，就是用下面这句配置代码，把灯安装到汽车正前面：

pObject_hummer->addChild(pLight_leftHeadlight);

6.15　制造幻影效果

幻影效果本质上就是物体对象每移动一次产生一个复制对象，让原物体留在原地，而产生的复制对象修改位置参数后往前移动。

在开发中，左移时就可以产生幻影效果。首先，重新实例化一个对象 object，然后利用 object 对原物体汽车对象 PublicMember::CTS_pObject 进行复制，接着对复制对象进行位置改变，产生的新的复制对象也必须放进容器中，设置为场景对象的"孩子"。关键代码如下：

```
//复制物体对象
vpObject * object=new vpObject();
object->setCopySource(PublicMember::CTS_pObject);
object->autoPage();
//修改复制的位置
double x,y,z;
PublicMember::CTS_pObject->getTranslate(&x,&y,&z);
    y=y+3;
object->setTranslate(x,y,z);
//修改复制的姿态
PublicMember::CTS_pObject->getRotate(&x,&y,&z);
object->setRotate(x,y,z);
//设置复制为场景的孩子
vpScene *scene=*vpScene::begin();
scene->addChild(object);
//原物体对象指针指向产生的复制对象
PublicMember::CTS_pObject =object;
//把复制放进容器中
PublicMember::CTS_s_pInstancesToUnref->push_back( object );
```

完整代码为键盘函数的一部分，具体代码如图 6.15.1 所示。

```
//下面的代码为键盘函数的一部分，表示按下方向键的左键后，汽车在左移过程中
//产生幻影效果
case vpWindow::KEY_LEFT:
    {    //
```

```
            vpObject * object=new vpObject();
            object->setCopySource(PublicMember::CTS_pObject_observer);
            object->autoPage();
        //
             double x,y,z;
            PublicMember::CTS_pObject_observer->getTranslate(&x,&y,&z);
y=y+3;
            object->setTranslate(x,y,z);

            PublicMember::CTS_pObject_observer->getRotate(&x,&y,&z);
            object->setRotate(x,y,z);

            vpScene *scene=*vpScene::begin();
            scene->addChild(object);
            PublicMember::CTS_pObject_observer =object;
            PublicMember::CTS_s_pInstancesToUnref->push_back( object );
        }
        break;
```

图 6.15.1 制造幻影效果

运行程序，按下键盘的左方向键，其效果如图 6.15.2 所示。

图 6.15.2 汽车的幻影效果

6.16 控制声音

控制声音需要首先引入两个头文件，依次为：
　　#include "vpAudioListener.h"
　　#include "vpAudioSoundSpatial.h"
聆听类继承图如图 6.16.1 所示，它使观察者可以倾听场景中正在播放的声音。声音类继承图如图 6.16.2 所示，它负责声音的播放管理，依据观察者与声源之间的相对位置和速度，它负责调节声音的音量和音调。

第 6 章　Vega Prime 编程对象的实例使用

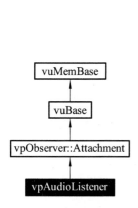

图 6.16.1　聆听类继承图

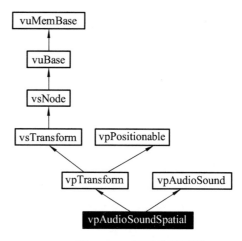

图 6.16.2　声音类继承图

需要特别注意，在 Vega Prime 应用程序中播放声音，需要对声卡进行特殊的驱动，可以到网络上找到 OpenAL，其安装界面如图 6.16.3 所示。

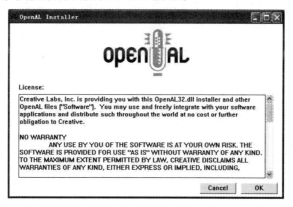

图 6.16.3　OpenAL 安装界面

对于声音的控制，首先需要准备好声音文件。在应用中，把"C:/Program Files/MultiGen-Paradigm/resources/data"目录下的"audio/wav"目录复制到应用程序的 data 目录下，将使用其中的"chopper.wav"文件。这样，就准备好了要播放的声音文件。

在接下来的程序设计中，须先对声音模块进行初始化，其代码如下：

vpModule::initializeModule("vpAudio");

接着，利用路径搜索对象添加声音文件的路径：

CTS_pSearchPath_mySearchPath->append(CTS_RunPath+"//data//audio//wav");

然后，需要定义 vpAudioListener 对象和 vpAudioSoundSpatial 对象。为了便于在程序的其他地方控制声音的播放，定义了一个静态成员变量来控制声音的播放，其代码如下：

static vpAudioSoundSpatial* pAudioSoundSpatial_mySound1;

vpAudioSoundSpatial* PublicMember::pAudioSoundSpatial_mySound1=NULL;

这样，在建立对象时，代码有一些不同，分别为：

vpAudioListener* pAudioListener_myAudioListener = new vpAudioListener();

pAudioSoundSpatial_mySound1 = new vpAudioSoundSpatial();

其中 pAudioListener_myAudioListener 比较简单，不需要特殊考虑。而 pAudioSoundSpatial_mySound1 相对复杂，以下关键代码需要注意：

  pAudioSoundSpatial_mySound1->setFileName("chopper.wav");

  pAudioSoundSpatial_mySound1->setEnable(false);

第 1 句代码设置具体的声音文件；第 2 句设置是否播放声音。建立两个对象的具体代码如图 6.16.4 所示。

```
vpAudioListener* pAudioListener_myAudioListener = new vpAudioListener();
pAudioListener_myAudioListener->setName( "myAudioListener" );
pAudioListener_myAudioListener->setVolume( 1.000000f );
pAudioListener_myAudioListener->setEnable( true );

PublicMember::CTS_s_pInstancesToUnref->push_back( pAudioListener_myAudioListener );

pAudioSoundSpatial_mySound1 = new vpAudioSoundSpatial();
pAudioSoundSpatial_mySound1->setName( "mySound1" );
pAudioSoundSpatial_mySound1->setCullMask( 0x0FFFFFFFF );
pAudioSoundSpatial_mySound1->setRenderMask( 0x0FFFFFFFF );
pAudioSoundSpatial_mySound1->setIsectMask( 0x0FFFFFFFF );
pAudioSoundSpatial_mySound1->setStrategyEnable( true );
pAudioSoundSpatial_mySound1->setTranslate( 0.000000 , 0.000000 , 0.000000 );
pAudioSoundSpatial_mySound1->setRotate( 0.000000 , 0.000000 , 0.000000 );
pAudioSoundSpatial_mySound1->setScale( 1.000000 , 1.000000 , 1.000000 );
pAudioSoundSpatial_mySound1->setStaticEnable( false );
pAudioSoundSpatial_mySound1->setFileName( "chopper.wav" );
pAudioSoundSpatial_mySound1->setVolume( 1.000000f );
pAudioSoundSpatial_mySound1->setEnable(false);
pAudioSoundSpatial_mySound1->setLooping( false );
pAudioSoundSpatial_mySound1->setPauseEnable( false );
pAudioSoundSpatial_mySound1->setMinAttenuationDistance( 20.000000f );
pAudioSoundSpatial_mySound1->setMaxAttenuationDistance( 2000.000000f );
pAudioSoundSpatial_mySound1->setAttenuationFactor( 0.100000f );
pAudioSoundSpatial_mySound1->setPitchShift( 1.000000f );
pAudioSoundSpatial_mySound1->setVelPitchShiftFactor( 0.800000f );

PublicMember::CTS_s_pInstancesToUnref->push_back( pAudioSoundSpatial_mySound1 );
```

图 6.16.4　建立 vpAudioListener 对象和 vpAudioSoundSpatial 对象

建立 vpAudioListener 对象和 vpAudioSoundSpatial 对象完成后，需要特殊的配置才能完成声音的控制，关键代码如下：

  pObserver_myObserver->addAttachment(pAudioListener_myAudioListener);

  pObject_Hummer->addChild(pAudioSoundSpatial_mySound1);

第 1 句添加声音的倾听者到观察者上，第 2 句添加声音播放为汽车的"孩子"，这样就完成了声音播放的特殊配置。具体控制声音的播放与否完全取决于应用需要，可以自由实现。在开发中，利用键盘控制声音的播放。在汽车前进时，声音播放；而在汽车后退时，停止播放声音。作为键盘函数的部分功能，具体代码如下：
```
case vpWindow::KEY_UP://汽车前进
        {
            pAudioSoundSpatial_mySound1->setEnable(true);//播放声音
            PublicMember::CTS_pObject_observer->setTranslateX(1.0,true);
        }
            break;
case vpWindow::KEY_DOWN://汽车后退
        {
            PublicMember::CTS_pObject_observer->setTranslateX(-1.0,true);
            pAudioSoundSpatial_mySound1->setEnable(false);//停止播放
        }
            break;
```

6.17 控制父子关系

对于 Vega Prime 应用程序中的父子关系，首先要认识到这样一个基本的事实：场景中所有的对象都是场景的后代，场景是节点容器，场景是场景图形的根节点。另外，对于子节点，它的位置姿态都是相对于父节点而言的。如果父节点产生了位置姿态的变化，子节点将跟随父节点产生同样的位置姿态变化。但相对于父节点，子节点是静止不动的。同时，也可以单独对子节点进行操作，而父节点保持原样不变。

对于父子关系的设置，代码非常灵活自由，主要有：

（1）直接利用 addChild 方法。

（2）直接利用 push_back_child 方法。

（3）利用 vpTransform 对象，把转换作为父节点的儿子，把子节点作为转换的儿子，这样也可以建立父子关系。

在前面的代码中，可以看到以下代码：

```
pScene_myScene->addChild( pObject_terrain );
pScene_myScene->addChild( pObject_farmhouse );
pScene_myScene->addChild( pObject_Hummer );
pScene_myScene->addChild( pObject_gainstore );
pObject_farmhouse->addChild( pFxDebris_Debris );
```

其中，前 4 句分别是把地形、房屋、汽车和粮仓添加为场景的"孩子"，而第 5 句则是把碎片特效添加为房屋的"孩子"。

在应用开发中，将加入一辆坦克作为汽车的子节点。这样，坦克将跟随汽车一起运动，但坦克可以有自己的操作。

首先，需要准备好素材。与前面动态添加物体一样，把"C:/Program Files/MultiGen-

Paradigm/resources/data/models"目录下的"m1_tank"文件夹复制到应用程序的 data 目录下，将使用其中的文件。

接着，利用路径搜索对象添加坦克文件的路径：
　　　　pSearchPath_mySearchPath->append(CTS_RunPath+"//data//m1_tank");
具体定义坦克对象的代码与动态加载物体的绝大多数代码相同，只是要注意最后一句"pObject_Hummer->addChild(pObject_m1_tank);"，把坦克添加为汽车的"孩子"，而不是添加为场景的"孩子"。详细代码如图 6.17.1 所示。

```
//添加坦克作为汽车的子物体
vpObject* pObject_m1_tank = new vpObject();
pObject_m1_tank->setName( "m1_tank" );
pObject_m1_tank->setCullMask( 0x0FFFFFFFF );
pObject_m1_tank->setRenderMask( 0x0FFFFFFFF );
pObject_m1_tank->setIsectMask( 0x0FFFFFFFF );
pObject_m1_tank->setStrategyEnable( true );
pObject_m1_tank->setTranslate( 4.000000 , 0.000000 , 0.000000 );
pObject_m1_tank->setRotate( 0.000000 , 0.000000 , 0.000000 );
pObject_m1_tank->setScale( 1.000000 , 1.000000 , 1.000000 );
pObject_m1_tank->setStaticEnable( false );
pObject_m1_tank->setFileName( "m1_tank.flt" );
pObject_m1_tank->setAutoPage( vpObject::AUTO_PAGE_SYNCHRONOUS );
pObject_m1_tank->setManualLODChild( -1 );
pObject_m1_tank->setLoaderOption( vsNodeLoader::Data::LOADER_OPTION
                                 _COMBINE_LIGHT_POINTS , true );
pObject_m1_tank->setLoaderOption( vsNodeLoader::Data::LOADER_OPTION
                                 _COMBINE_LODS , true );
pObject_m1_tank->setLoaderOption( vsNodeLoader::Data::LOADER_OPTION
                                 _IGNORE_DOF_CONSTRAINTS , false );
pObject_m1_tank->setLoaderOption( vsNodeLoader::Data::LOADER_OPTION
                                 _PRESERVE_EXTERNAL_REF_FLAGS , true );
pObject_m1_tank->setLoaderOption( vsNodeLoader::Data::LOADER_OPTION
                                 _PRESERVE_GENERIC_NAMES , false );
pObject_m1_tank->setLoaderOption( vsNodeLoader::Data::LOADER_OPTION
                                 _PRESERVE_GENERIC_NODES , false );
pObject_m1_tank->setLoaderOption( vsNodeLoader::Data::LOADER_OPTION
                                 _PRESERVE_QUADS , false );
pObject_m1_tank->setLoaderOption( vsNodeLoader::Data::LOADER_OPTION
                                 _ALL_GEOMETRIES_LIT , false );
pObject_m1_tank->setLoaderOption( vsNodeLoader::Data::LOADER_OPTION
                                 _USE_MATERIAL_DIFFUSE_COLOR , false );
pObject_m1_tank->setLoaderOption( vsNodeLoader::Data::LOADER_OPTION
                                 _MONOCHROME , false );
pObject_m1_tank->setLoaderOption( vsNodeLoader::Data::LOADER_OPTION
                                 _CREATE_ANIMATIONS , true );
```

```
pObject_m1_tank->setLoaderOption( vsNodeLoader::Data::LOADER_OPTION
                    _SHARE_LIGHT_POINT_CONTROLS， true );
pObject_m1_tank->setLoaderOption( vsNodeLoader::Data::LOADER_OPTION
                    _SHARE_LIGHT_POINT_ANIMATIONS， true );
pObject_m1_tank->setLoaderOption( vsNodeLoader::Data::LOADER_OPTION
                    _SHARE_LIGHT_POINT_APPEARANCES， true );
pObject_m1_tank->setLoaderDetailMultiTextureStage( -1 );
pObject_m1_tank->setLoaderBlendTolerance( 0.050000f );
pObject_m1_tank->setLoaderUnits( vsNodeLoader::Data::LOADER_UNITS_METERS );
pObject_m1_tank->setBuilderOption( vsNodeLoader::Data::BUILDER_OPTION
                    _OPTIMIZE_GEOMETRIES， true );
pObject_m1_tank->setBuilderOption( vsNodeLoader::Data::BUILDER_OPTION
                    _COLLAPSE_BINDINGS， true );
pObject_m1_tank->setBuilderOption( vsNodeLoader::Data::BUILDER_OPTION
                    _COLLAPSE_TRIANGLE_STRIPS， true );
pObject_m1_tank->setBuilderNormalMode( vsNodeLoader::Data::BUILDER
                    _NORMAL_MODE_PRESERVE );
pObject_m1_tank->setBuilderColorTolerance( 0.001000f );
pObject_m1_tank->setBuilderNormalTolerance( 0.860000f );
pObject_m1_tank->setBuilderVertexTolerance( 0.000100f );
pObject_m1_tank->setGeometryOption( vsNodeLoader::Data::GEOMETRY_OPTION
                    _GENERATE_DISPLAY_LISTS， false );
pObject_m1_tank->setGeometryFormat( vrGeometryBase::FORMAT_VERTEX
                    _ARRAY， 0x0FFF );
pObject_m1_tank->setPostLoadOption( vpGeometryPageable::POST_LOAD_OPTION
                    _FLATTEN， true );
pObject_m1_tank->setPostLoadOption( vpGeometryPageable::POST_LOAD_OPTION
                    _CLEAN， true );
pObject_m1_tank->setPostLoadOption( vpGeometryPageable::POST_LOAD_OPTION
                    _MERGE_GEOMETRIES， true );
pObject_m1_tank->setPostLoadOption( vpGeometryPageable::POST_LOAD_OPTION
                    _COLLAPSE_BINDINGS， true );
pObject_m1_tank->setPostLoadOption( vpGeometryPageable::POST_LOAD_OPTION
                    _COLLAPSE_TRIANGLE_STRIPS， true );
pObject_m1_tank->setPostLoadOption( vpGeometryPageable::POST_LOAD_OPTION
                    _VALIDATE， true );
pObject_m1_tank->setTextureSubloadEnable( false );
pObject_m1_tank->setTextureSubloadRender( vpGeometry::TEXTURE_SUBLOAD
                    _RENDER_DEFERRED );
PublicMember::CTS_s_pInstancesToUnref->push_back( pObject_m1_tank );
**pObject_Hummer->addChild(pObject_m1_tank);**
```

图 6.17.1　定义坦克为汽车的子物体

运行程序，点击运行按钮，将出现图 6.17.2 所示的画面，键盘控制汽车的同时，坦克将随汽车一起运动。

图 6.17.2 添加坦克为汽车的孩子

6.18 操作 Switch

Switch 本身是一个节点，它能够包含多个子节点，其功能主要在于控制子节点的渲染显示。如图 6.18.1 所示，图中的房子来自系统自带的 farmhouse.flt 文件，该文件在 creator 中的层次结构明确显示，名字为"damage"的 Switch 节点包含两个子节点，一个是 g1，一个是 house_1。左图中的 g1 表示一个完好的房子，其掩码索引为 0；右图中的 house_1 表示一个被损坏了的房子，其掩码索引为 1。

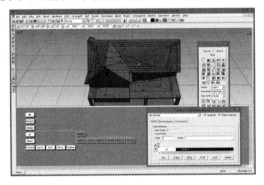 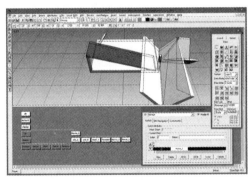

图 6.18.1 Switch 图层结构

Switch 节点可以包含多个子节点，每个子节点可以表示出三维对象不同的状态，如图 6.18.1 所示的完好房子和被破坏的房子两种状态。实际上，一个 Switch 节点可以包含多个子节点，其中 0 是第一个子节点的索引，1 是第二个子节点的索引，以此类推。但是，在任何一个特定的时刻只有一个掩码索引是激活的，表示场景中只渲染显示该 Switch 节点中掩码索引为激活状态的子节点，而其他节点则不渲染显示。这样，利用 Switch 节点可以很方便地显示同一个物体在不同情境下的形状。

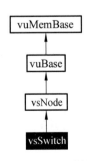

图 6.18.2 Switch 类继承图

vsSwitch 类视图如图 6.18.2 所示，开发者可查看系统的自带文档，详细了解其所有功能。

本节首先把 farmhouse 加载进场景，然后利用代码激活掩码索引，显示房子的不同状态，其效果如图 6.18.3 所示，左图为完好的房子，右图为被损坏了的房子。

图 6.18.3　Switch 效果图

对于 Switch 的编程使用，需要注意的是，对于 Switch 的命名，不能采用默认名称，即 sw1 等名称。否则，可能依据 Switch 的名称不能查找到该 Switch。在 farmhouse.flt 文件中，Switch 命名为 damage。另外，必须引入 Switch 类的头文件为：#include <vsSwitch.h>。

为了应用开发方便，需要定义一个 vsSwitch 静态变量：

static vsSwitch* HouseSwitch;　　//房子 Switch 指针

定义完成后，应该记得要初始化：

vsSwitch * PublicMember:: HouseSwitch =NULL;

操作 Switch 之前，必须首先获得 Switch 指针。而应用中的 Switch 在房子上，则首先应得到房子的指针：

vpObject * house = vpObject::find("farmhouse");

house->ref();

接着获取房子上的 Switch：

//获取房子 Switch

PublicMember::HouseSwitch = static_cast<vsSwitch *>(house->find_named("damage"));

PublicMember::HouseSwitch->ref();

详细代码位于主线程中，具体如图 6.18.4 所示。

```
UINT PublicMember::CTS_RunBasicThread(LPVOID)
{
    //初始化
    vp::initialize(__argc,__argv);
    //定义场景
    PublicMember::CTS_Define();
    //绘制场景
    vpKernel::instance()->configure();
    //设置观察者
    PublicMember::CTS_pObject_observer=vpObject::find("Hummer");
    PublicMember::CTS_pObject_observer->ref();

    vpObject * house = vpObject::find("farmhouse");
```

```
house ->ref();

PublicMember::HouseSwitch = static_cast<vsSwitch *>( house ->find_named("damage"));
PublicMember::HouseSwitch->ref();

//设置窗体
vpWindow * vpWin= * vpWindow::begin();
vpWin->setParent(PublicMember::CTS_RunningWindow);
vpWin->setBorderEnable(false);
vpWin->setFullScreenEnable(true);
//设置键盘
vpWin->setInputEnable(true);
vpWin->setKeyboardFunc((vrWindow::KeyboardFunc)PublicMember::CTS_Keyboard,NULL);
vpWin->open();
::SetFocus(vpWin->getWindow());

//帧循环
while(vpKernel::instance()->beginFrame()!=0)
{
    vpKernel::instance()->endFrame();
    if(!PublicMember::CTS_continueRunVP)
    {
        vpKernel::instance()->endFrame();
        vpKernel::instance()->unconfigure();
        vp::shutdown();
        return 0;
    }
}
return 0;
}
```

图 6.18.4 包含 Switch 操作的主线程

对于 Switch 的编程使用，主要通过设置其活动掩码索引来完成，如图 6.18.3 所示，左图是按钮"完好"执行的效果，其代码为：

PublicMember::HouseSwitch->setActiveMask(0);//设置活动掩码索引为 0

右图是按钮"破坏"执行的效果，其代码为：

PublicMember::HouseSwitch->setActiveMask(1); //设置活动掩码索引为 1

6.19 操作 DOF

需要注意的是，DOF 的命名不能采用默认名称，即 d1 等名称。否则，不能依据 DOF 的名称查找到该 DOF。在此例中，命名为 turretDOF。其次，必须引入 DOF 类的头文件为：#include "vsDOF.h"。

第 6 章　Vega Prime 编程对象的实例使用

　　类视图如图 6.19.1 所示。上一节把坦克作为汽车的"孩子"加进了场景，本节将建立一个火焰特效，利用 DOF 操作坦克的炮管和炮塔，运行实例 VPTestDialogDOF，其效果如图 6.19.2 所示。

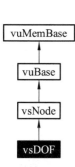

图 6.19.1　DOF 类继承图　　　　　　　　图 6.19.2　效果图

首先，建立火焰特效，其具体代码如图 6.19.3 所示。

```
vpFxParticleSystem* pFxParticleSystem_main_gun_flash = new vpFxParticleSystem();
pFxParticleSystem_main_gun_flash->setName( "maingunflash" );
pFxParticleSystem_main_gun_flash->setCullMask( 0x0FFFFFFFF );
pFxParticleSystem_main_gun_flash->setRenderMask( 0x0FFFFFFFF );
pFxParticleSystem_main_gun_flash->setIsectMask( 0x0FFFFFFFF );
pFxParticleSystem_main_gun_flash->setStrategyEnable( true );
pFxParticleSystem_main_gun_flash->setTranslate( 0.000000 , 0.000000 , 0.000000 );
pFxParticleSystem_main_gun_flash->setRotate( 0.000000 , 0.000000 , 0.000000 );
pFxParticleSystem_main_gun_flash->setScale( 4.000000 , 4.000000 , 4.000000 );
pFxParticleSystem_main_gun_flash->setStaticEnable( false );
pFxParticleSystem_main_gun_flash->setRepeatEnable( false );
pFxParticleSystem_main_gun_flash->setOverallColor( 1.000000f , 1.000000f ,
                                                   1.000000f , 1.000000f );
pFxParticleSystem_main_gun_flash->setTextureBlendColor( 1.000000f , 1.000000f ,
                                                        1.000000f , 1.000000f );
pFxParticleSystem_main_gun_flash->setTextureMode( vpFx::TEXTURE
                                                  _MODE_MODULATE );
pFxParticleSystem_main_gun_flash->setTextureFile( "flash.inta" );
pFxParticleSystem_main_gun_flash->setOverallDuration( 0.200000f );
pFxParticleSystem_main_gun_flash->setFadeDuration( 0.000000f , 0.000000f );
pFxParticleSystem_main_gun_flash->setTransparencyDepthOffset( 0.000000f );
pFxParticleSystem_main_gun_flash->setTransparencyDepthRadiusScaler( 0.000000f );
pFxParticleSystem_main_gun_flash->setParticleReleaseCutoff( 0.000000f );
pFxParticleSystem_main_gun_flash->setSourceDomain( vpFxParticleSystem::
                                                   SOURCE_DOMAIN_CIRCLE );
pFxParticleSystem_main_gun_flash->setSourceDomainSize( 0.000000f );
pFxParticleSystem_main_gun_flash->setBoundingDomainBehavior( vpFxParticleSystem::
                                                   BOUNDING_DOMAIN_BEHAVIOR_NONE );
pFxParticleSystem_main_gun_flash->setBounceCoefficient( 1.000000 );
```

```cpp
pFxParticleSystem_main_gun_flash->setMaxNumParticles( 1 );
pFxParticleSystem_main_gun_flash->setNumParticlesToRelease( 50 );
pFxParticleSystem_main_gun_flash->setReleaseInterval( 10.000000f );
pFxParticleSystem_main_gun_flash->setFlowType( vpFxParticleSystem::
                                               FLOW_TYPE_BURST );
pFxParticleSystem_main_gun_flash->setParticleLifeCycle( 0.200000f );
pFxParticleSystem_main_gun_flash->setParticleTextureAnimationDuration( 0.000000f );
pFxParticleSystem_main_gun_flash->setVelocityDistribution( vpFxParticleSystem::
                                               VELOCITY_DISTRIBUTION_PLANE );
pFxParticleSystem_main_gun_flash->setGravitationalConstant( 0.000000f );
pFxParticleSystem_main_gun_flash->setVelocitySource( vpFxParticleSystem::
                                               VELOCITY_SOURCE_TABLE );
pFxParticleSystem_main_gun_flash->setRelativeEmissionVelocity( 0.000000f );
pFxParticleSystem_main_gun_flash->addVelocity( 0.000000f ,
                                vuVec3d(0.000000,0.000000,0.000000) );
pFxParticleSystem_main_gun_flash->addSphericalVelocity( 0.000000f ,  0.000000f );
pFxParticleSystem_main_gun_flash->addRandomVelocity( 0.000000f ,  0.000000f );
pFxParticleSystem_main_gun_flash->setWindSource( vpFxParticleSystem::
                                               WIND_SOURCE_TABLE );
pFxParticleSystem_main_gun_flash->addWind(0.000000f,
                                vuVec3d(0.000000,0.000000,0.000000) );
pFxParticleSystem_main_gun_flash->addColor( 0.000000f ,
                                vuVec4f(1.000000f, 0.333333f, 0.000000f, 0.498039f) );
pFxParticleSystem_main_gun_flash->addColor( 0.200000f ,
                                vuVec4f(1.000000f, 0.333333f, 0.000000f, 1.000000f) );
pFxParticleSystem_main_gun_flash->addColor( 1.000000f ,
                                vuVec4f(1.000000f, 0.333333f, 0.000000f, 0.000000f) );
pFxParticleSystem_main_gun_flash->addSize( 0.000000f ,   0.000000f );
pFxParticleSystem_main_gun_flash->addSize( 0.200000f ,   5.000000f );
pFxParticleSystem_main_gun_flash->addSize( 1.000000f ,   0.000000f );
pFxParticleSystem_main_gun_flash->addScaleAlongVelocity( 0.000000f ,   1.000000f );
pFxParticleSystem_main_gun_flash->setColorVariation( 0.000000f );
pFxParticleSystem_main_gun_flash->setSizeVariation( 0.000000f );
pFxParticleSystem_main_gun_flash->setEmissivity( 1.000000f );
PublicMember::CTS_s_pInstancesToUnref->push_back( pFxParticleSystem_main_gun_flash );
```

图 6.19.3 建立火焰特效

坦克拥有两个 DOF：一个为 turretDOF，控制炮塔水平旋转；另一个为 barrelDOF，控制炮管上下旋转。为了使场景更加逼真，把火焰特效放到炮管末端，在发射炮弹时，火焰特效发生作用。为了应用开发方便，需要定义 3 个静态变量：2 个 DOF，1 个火焰特效。

```cpp
static vsDOF * m_turret;//炮塔 DOF 指针
static vsDOF * m_barrel;//炮管 DOF 指针
static vpFxParticleSystem *m_mainGunFlash;//火焰特效指针
```

定义完成后，记得要初始化：

```cpp
vsDOF * PublicMember::m_turret=NULL;
```

```
vsDOF * PublicMember::m_barrel=NULL;
vpFxParticleSystem *PublicMember::m_mainGunFlash=NULL;
```
操作 DOF 之前，须先获得 DOF 指针。而应用中的两个 DOF 都在物体坦克上，则先应得到坦克的指针：

```
vpObject * tank = vpObject::find("m1_tank");
```
接着，获取坦克上的 DOF：

```
//获取炮塔 DOF
PublicMember::m_turret = static_cast<vsDOF *>(tank->find_named("turretDOF"));
PublicMember::m_turret->ref();
//获取炮管 DOF
PublicMember::m_barrel = static_cast<vsDOF *>(tank->find_named("barrelDOF"));
PublicMember::m_barrel->ref();
```
炮管的上下旋转应该在一定范围之内，应该加上限制，才符合实际情况。在这里，限制炮管向下 10°，向上 25°。

```
// 对炮管进行控制
vsDOF::Constraint constraint;
constraint.m_enable = true;
constraint.m_min = -10.0;
constraint.m_max = 25.0;
m_barrel->setConstraint(vsDOF::COMPONENT_ROTATE_P, constraint);
```
现在，建立一个转换，把火焰特效放到炮管的末端。其方法是让转换成为炮管的"孩子"，然后让火焰特效成为转换的"孩子"。

```
// 创建一个转换，并且让转换成为炮管的"孩子"
vsTransform * m_mainGunTransform = new vsTransform();
m_mainGunTransform->setTranslate(0.14, 3.22, 0.0);
m_barrel->push_back_child(m_mainGunTransform);
// 找到火焰特效，让火焰特效成为转换的"孩子"，这样就把火焰特效放到了炮管尾部
m_mainGunFlash = vpFxParticleSystem::find("maingunflash");
m_mainGunFlash->ref();
m_mainGunTransform->push_back_child(m_mainGunFlash);
```
详细代码位于主线程中，具体如图 6.19.4 所示。

```
UINT PublicMember::CTS_RunBasicThread(LPVOID)
{
    //初始化
    vp::initialize(__argc,__argv);

    //定义场景
    PublicMember::CTS_Define();

    //绘制场景
```

```cpp
vpKernel::instance()->configure();

//设置观察者
PublicMember::CTS_pObject_observer=vpObject::find("Hummer");
PublicMember::CTS_pObject_observer->ref();

//控制 DOF
vpObject * tank = vpObject::find("m1_tank");

PublicMember::m_turret = static_cast<vsDOF *>(tank->find_named("turretDOF"));
PublicMember::m_turret->ref();

PublicMember::m_barrel = static_cast<vsDOF *>(tank->find_named("barrelDOF"));
    PublicMember::m_barrel->ref();

    // 对炮管进行控制
    vsDOF::Constraint constraint;
    constraint.m_enable = true;
    constraint.m_min = -10.0;
    constraint.m_max = 25.0;
    PublicMember::m_barrel->setConstraint(vsDOF::COMPONENT_ROTATE_P, constraint);

    // 创建一个转换，把火焰特效放到炮管尾部
    vsTransform * m_mainGunTransform = new vsTransform();
    m_mainGunTransform->setTranslate(0.14, 3.22, 0.0);
    m_barrel->push_back_child(m_mainGunTransform);

    // 利用转换，把火焰特效放到炮管尾部
    m_mainGunFlash = vpFxParticleSystem::find("maingunflash");
    m_mainGunFlash->ref();
    m_mainGunTransform->push_back_child(m_mainGunFlash);

    //设置窗体
    vpWindow * vpWin= * vpWindow::begin();
    vpWin->setParent(PublicMember::CTS_RunningWindow);
    vpWin->setBorderEnable(false);
    vpWin->setFullScreenEnable(true);

    //设置键盘
    vpWin->setInputEnable(true);
    vpWin->setKeyboardFunc((vrWindow::KeyboardFunc)PublicMember::CTS_Keyboard,NULL);

    vpWin->open();
    ::SetFocus(vpWin->getWindow());
    //帧循环
```

```
while(vpKernel::instance()->beginFrame()!=0)
{
    vpKernel::instance()->endFrame();
    if(!PublicMember::CTS_continueRunVP)
    {
        vpKernel::instance()->unconfigure();
        vp::shutdown();
        return 0;
    }
}
return 0;
}
```

图 6.19.4 包含 DOF 操作的主线程

主线程中的键盘函数需要进一步改造,定义了 5 个键:
(1) vrWindow::KEY_a: 炮台左转;
(2) vrWindow::KEY_d: 炮台右转;
(3) vrWindow::KEY_w: 炮管向上;
(4) vrWindow::KEY_s: 炮管向下;
(5) vrWindow::KEY_SPACE: 发射炮弹。
具体的键盘函数如图 6.19.5 所示。

```
void   PublicMember::CTS_Keyboard(vpWindow *window,vpWindow::Key key, int modifier,void *)
{
    switch(key)
    {
        //炮台左转
        case vrWindow::KEY_a:
        m_turret->setRotateH(1.0f, true);
break;

        //炮台右转
        case vrWindow::KEY_d:
        m_turret->setRotateH(-1.0f, true);
break;

        //炮管向上
        case vrWindow::KEY_w:
        m_barrel->setRotateP(1.0f, true);
break;

        //炮管向下
        case vrWindow::KEY_s:
        m_barrel->setRotateP(-1.0f, true);
break;
```

```
                //发射炮弹
                case vrWindow::KEY_SPACE:
                m_mainGunFlash->setEnable(true);
                break;


                default: ;
        }
}
```

图 6.19.5　键盘函数的一部分

6.20　获取 DOF 的坐标

DOF 主要用来控制围绕某一点进行旋转，当然也可以进行平移操作。在很多时候，都希望得到某个 DOF 点的坐标。首先要认识到，对于坐标来说，包含本地坐标和世界坐标。本地坐标，就是 DOF 相对于物体本身而言；绝对坐标，是 DOF 相对于场景而言。

对于 DOF 的操作，首先要找到 DOF 所在的物体，在上一节已经使用了这样的代码，找到名称为"m1_tank"的物体"tank"：

　　vpObject * tank = vpObject::find("m1_tank");

找到"tank"上名称为"barrelDOF"的 DOF，同时加以引用：

　　m_barrel = static_cast<vsDOF *>(tank->find_named("barrelDOF"));

　　m_barrel->ref();

创建一个转换 m_mainGunTransform：

　　vsTransform * m_mainGunTransform = new vsTransform();

　　m_mainGunTransform->setTranslate(0.14, 3.22, 0.0);

使转换 m_mainGunTransform 成为 m_barrel 的"孩子"：

　　m_barrel->push_back_child(m_mainGunTransform);

找到名称为"maingunflash"的特效，同时加以引用：

　　m_mainGunFlash = vpFxParticleSystem::find("maingunflash");

　　m_mainGunFlash->ref();

使特效 m_mainGunFlash 成为转换 m_mainGunTransform 的孩子：

　　m_mainGunTransform->push_back_child(m_mainGunFlash);

这样，实际上就完成了利用转换把火焰特效放到炮管尾部的功能。

编写了一个静态函数获取 DOF 的绝对坐标：

　　void PublicMember::CTS_GetAbsolutePositionFromDOF(vsDOF * pDOF,
　　　　　　　　　　　　　　　　double *x, double *y, double *z);

其中，pDOF 为目标 DOF 指针，另外分别设置了 3 个 double 类型的指针变量，用来返回获取的坐标值。其详细代码如图 6.20.1 所示。

```
//获取DOF的绝对坐标
void PublicMember::CTS_GetAbsolutePositionFromDOF(vsDOF * pDOF,double *x, double *y, double *z)
{
    //获DOF的绝对坐标
    vsTraversalLocate    trav;
    trav.setMode(vsTraversalLocate::MODE_MATRIX_STACK
                vsTraversalLocate::MODE_NODE_STACK);
    trav.visit(PublicMember::CTS_pDOF8);
    vuMatrix <double> m=trav.getTopMatrixStack();
    m.getTranslate(x,y,z);
}
```

图 6.20.1　获取 DOF 的绝对坐标

编写了一个静态函数，获取 DOF 的相对坐标：

void PublicMember::CTS_GetRelativePositionFromDOF(vsDOF * pDOF,
double *x,double *y, double *z, double *h,double *p, double *r);

其中，pDOF 为目标 DOF 指针，另外分别设置了 6 个 double 类型的指针变量，用来返回获取的坐标值。其详细代码如图 6.20.2 所示。

```
//获取外挂物DOF的相对坐标
void PublicMember::CTS_GetRelativePositionFromDOF(vsDOF * pDOF,
        double *x,double *y, double *z, double *h,double *p, double *r)

{
    vuMatrixAffine<double> v;
     v=pDOF->getLocalOrigin();
    *x=v.getTranslateX(),
    *y=v.getTranslateY(),
    *z=v.getTranslateZ();
    *h=v.getRotateH();
    *p=v.getRotateP();
    *r=v.getRotateR();
}
```

图 6.20.2　获取 DOF 的相对坐标

6.21　配置多通道

引入通道类的头文件为：#include "vpChannel.h"。

通道类视图如图 6.21.1 所示。一般情况下，只需要一个通道，从一个地方观察场景。有些时候，需要多个通道，从不同角度和方位观察场景的不同地方。

在这里，定义了 4 个通道：

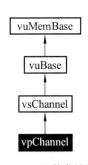

图 6.21.1　通道类继承图

默认通道：pChannel_myChannel ;
房屋通道：pChannel_houseChannel ;
转角通道：pChannel_portchChannel;
俯视通道：pChannel_orthoChannel ;
建立了4个观察者：
默认观察者：pObserver_myObserver;
房屋观察者：pObserver_houseObserver;
转角观察者：pObserver_porchObserver;
俯视观察者：pObserver_orthoObserver;
把4个通道添加到窗体：
pWindow_myWindow->addChannel(pChannel_myChannel);
pWindow_myWindow->addChannel(pChannel_houseChannel);
pWindow_myWindow->addChannel(pChannel_portchChannel);
pWindow_myWindow->addChannel(pChannel_orthoChannel);
分别设置4个观察者：
pObserver_myObserver->setStrategy(pMotionDrive_myMotion);
pObserver_myObserver->addChannel(pChannel_myChannel);
pObserver_myObserver->addAttachment(pEnv_myEnv);
pObserver_myObserver->setScene(pScene_myScene);
Observer_myObserver->setLookFrom(pTransform_hummerTransform);

Observer_houseObserver->addChannel(pChannel_houseChannel);
pObserver_houseObserver->addAttachment(pEnv_myEnv);
pObserver_houseObserver->setScene(pScene_myScene);
pObserver_houseObserver->setLookAt(pObject_farmhouse);

pObserver_porchObserver->addChannel(pChannel_portchChannel);
pObserver_porchObserver->addAttachment(pEnv_myEnv);
pObserver_porchObserver->setScene(pScene_myScene);
pObserver_porchObserver->setLookFrom(pTransform_hummerTransform);

pObserver_orthoObserver->addChannel(pChannel_orthoChannel);
pObserver_orthoObserver->setScene(pScene_myScene);
完整代码如图6.21.2所示。

```
//建立默认通道
vpChannel* pChannel_myChannel = new vpChannel();
pChannel_myChannel->setName( "myChannel" );
pChannel_myChannel->setOffsetTranslate( 0.000000, 0.000000, 0.000000 );
```

```cpp
pChannel_myChannel->setOffsetRotate( 0.000000 ,  0.000000 ,  0.000000 );
pChannel_myChannel->setCullMask( 0x0FFFFFFFF );
pChannel_myChannel->setRenderMask( 0x0FFFFFFFF );
pChannel_myChannel->setClearColor( 0.000000f ,  0.500000f ,  1.000000f ,  0.000000f );
pChannel_myChannel->setClearBuffers( 0x03 );
pChannel_myChannel->setDrawArea( 0.000000 ,  1.000000 ,  0.000000 ,  1.000000 );
pChannel_myChannel->setFOVSymmetric( 45.000000f ,  -1.000000f );
pChannel_myChannel->setNearFar( 1.000000f ,  35000.000000f );
pChannel_myChannel->setLODVisibilityRangeScale( 1.000000 );
pChannel_myChannel->setLODTransitionRangeScale( 1.000000 );
pChannel_myChannel->setCullThreadPriority( vuThread::PRIORITY_NORMAL );
pChannel_myChannel->setCullThreadProcessor( -1 );
pChannel_myChannel->setGraphicsModeEnable( vpChannel::GRAPHICS_MODE
                            _WIREFRAME ,  false );
pChannel_myChannel->setGraphicsModeEnable( vpChannel::GRAPHICS_MODE
                            _TRANSPARENCY ,  true );
pChannel_myChannel->setGraphicsModeEnable( vpChannel::GRAPHICS_MODE
                            _TEXTURE ,  true );
pChannel_myChannel->setGraphicsModeEnable( vpChannel::GRAPHICS_MODE_LIGHT ,
                            true );
pChannel_myChannel->setGraphicsModeEnable( vpChannel::GRAPHICS_MODE_FOG ,
                            true );
pChannel_myChannel->setLightPointThreadPriority( vuThread::PRIORITY_NORMAL );
pChannel_myChannel->setLightPointThreadProcessor( -1 );
pChannel_myChannel->setMultiSample( vpChannel::MULTISAMPLE_OFF );
pChannel_myChannel->setStatisticsPage( vpChannel::PAGE_OFF );
pChannel_myChannel->setCullBoundingBoxTestEnable( false );
pChannel_myChannel->setOpaqueSort( vpChannel::OPAQUE_SORT_TEXTURE ,
                            vpChannel::OPAQUE_SORT_MATERIAL );
pChannel_myChannel->setTransparentSort( vpChannel::TRANSPARENT_SORT_DEPTH );
pChannel_myChannel->setDrawBuffer( vpChannel::DRAW_BUFFER_DEFAULT );
pChannel_myChannel->setStressEnable( false );
pChannel_myChannel->setStressParameters( 1.0000f ,  20.00f ,  0.750f ,  0.50f ,  2.00f );

PublicMember::CTS_s_pInstancesToUnref ->push_back( pChannel_myChannel );

//建立房屋通道
vpChannel* pChannel_houseChannel = new vpChannel();
pChannel_houseChannel->setName( "houseChannel" );
//相同代码省略..............................
pChannel_houseChannel->setStressParameters( 1.00f ,  20.00f ,  0.750f ,  0.50f ,  2.00f );

PublicMember::CTS_s_pInstancesToUnref ->push_back( pChannel_houseChannel );

//建立转角通道
vpChannel* pChannel_portchChannel = new vpChannel();
pChannel_portchChannel->setName( "portchChannel" );
//相同代码省略..............................
```

```cpp
pChannel_portchChannel->setStressParameters( 1.00f,  20.00f,  0.750f,  0.500f,  2.00f );

PublicMember::CTS_s_pInstancesToUnref ->push_back( pChannel_portchChannel );

//建立俯视通道
vpChannel* pChannel_orthoChannel = new vpChannel();
pChannel_orthoChannel->setName( "orthoChannel" );
pChannel_orthoChannel->setOffsetTranslate( 0.000000,  0.000000,  600.000000 );
pChannel_orthoChannel->setOffsetRotate( 0.000000,  -90.000000,  0.000000 );
//相同代码省略…………………………………
pChannel_orthoChannel->setStressParameters( 1.0f,  20.00f,  0.75f,  0.500f,  2.0f );

PublicMember::CTS_s_pInstancesToUnref ->push_back( pChannel_orthoChannel );

//建立默认观察者
vpObserver* pObserver_myObserver = new vpObserver();
pObserver_myObserver->setName( "myObserver" );
pObserver_myObserver->setStrategyEnable( false );
pObserver_myObserver->setTranslate( 2300.000000,  2500.000000,  15.000000 );
pObserver_myObserver->setRotate( -90.000000,  0.000000,  0.000000 );
pObserver_myObserver->setLatencyCriticalEnable( false );

PublicMember::CTS_s_pInstancesToUnref ->push_back( pObserver_myObserver );

//建立房屋观察者
vpObserver* pObserver_houseObserver = new vpObserver();
pObserver_houseObserver->setName( "houseObserver" );
pObserver_houseObserver->setStrategyEnable( true );
pObserver_houseObserver->setTranslate( 2360.000000,  2490.000000,  2.000000 );
pObserver_houseObserver->setRotate( 0.000000,  0.000000,  0.000000 );
pObserver_houseObserver->setLatencyCriticalEnable( false );

PublicMember::CTS_s_pInstancesToUnref ->push_back( pObserver_houseObserver );

//建立转角观察者
vpObserver* pObserver_porchObserver = new vpObserver();
pObserver_porchObserver->setName( "porchObserver" );
pObserver_porchObserver->setStrategyEnable( true );
pObserver_porchObserver->setTranslate( 0.000000,  0.000000,  0.000000 );
pObserver_porchObserver->setRotate( 0.000000,  0.000000,  0.000000 );
pObserver_porchObserver->setLatencyCriticalEnable( false );

PublicMember::CTS_s_pInstancesToUnref ->push_back( pObserver_porchObserver );

//建立俯视观察者
vpObserver* pObserver_orthoObserver = new vpObserver();
pObserver_orthoObserver->setName( "orthoObserver" );
pObserver_orthoObserver->setStrategyEnable( true );
pObserver_orthoObserver->setTranslate( 1500.000000,  1500.000000,  0.000000 );
```

```
pObserver_orthoObserver->setRotate( 0.000000 ,   0.000000 ,   0.000000 );
pObserver_orthoObserver->setLatencyCriticalEnable( false );

PublicMember::CTS_s_pInstancesToUnref ->push_back( pObserver_orthoObserver );

//把四个通道添加到窗体
pWindow_myWindow->addChannel( pChannel_myChannel );
pWindow_myWindow->addChannel( pChannel_houseChannel );
pWindow_myWindow->addChannel( pChannel_portchChannel );
pWindow_myWindow->addChannel( pChannel_orthoChannel );

//设置默认观察者
pObserver_myObserver->setStrategy( pMotionDrive_myMotion );
pObserver_myObserver->addChannel( pChannel_myChannel );
pObserver_myObserver->addAttachment( pEnv_myEnv );
pObserver_myObserver->setScene( pScene_myScene );
pObserver_myObserver->setLookFrom( pTransform_hummerTransform );

//设置房屋观察者
pObserver_houseObserver->addChannel( pChannel_houseChannel );
pObserver_houseObserver->addAttachment( pEnv_myEnv );
pObserver_houseObserver->setScene( pScene_myScene );
pObserver_houseObserver->setLookAt( pObject_farmhouse );

//设置转角观察者
pObserver_porchObserver->addChannel( pChannel_portchChannel );
pObserver_porchObserver->addAttachment( pEnv_myEnv );
pObserver_porchObserver->setScene( pScene_myScene );
pObserver_porchObserver->setLookFrom( pTransform_hummerTransform );

//设置俯视观察者
pObserver_orthoObserver->addChannel( pChannel_orthoChannel );
pObserver_orthoObserver->setScene( pScene_myScene );
```

图 6.21.2　建立多通道

运行程序，其效果如图 6.21.3 所示。

图 6.21.3　多通道效果

6.22 物体平面影子效果

首先引入平面光照影子的头文件：#include "vpShadowPlanar.h"。

在 Define 函数中，首先启动平面光照影子功能模块：

vpModule::initializeModule("vpShadow");

在 VP 的安装目录下找到 MultiGen-Paradigm\resources\data\models\f16cn，复制该文件夹到自己程序的 data 目录下，该目录包含了一个飞机模型。把该目录添加到 VP 程序管理目录下：

pSearchPath_mySearchPath->append(CTS_RunPath+"//data//f16cn");

正常添加飞行物体对象,这里需要注意的是：飞行对象需要离开地面一定的距离，才能实现平面光照影子效果。

vpObject* pObject_flight = new vpObject();

pObject_flight->setName("flight");

pObject_flight->setTranslate(2450.000000 , 2530.000000 , 10.000000);

pObject_flight->setRotate(90.000000 , 0.000000 , 0.000000);

添加平面光照影子的效果对象：

vpShadowPlanar* pShadowPlanar_myShadowPlanar = new vpShadowPlanar();

pShadowPlanar_myShadowPlanar->setName("myShadowPlanar");

pShadowPlanar_myShadowPlanar->setEnable(true);

pShadowPlanar_myShadowPlanar->setEffectiveRange(10000.000000f);

PublicMember::CTS_s_pInstancesToUnref->push_back(pShadowPlanar_myShadowPlanar);

最后，实现平面光照影子效果的配置代码，与观察者关联，设置光源，添加平面光照影子对象：

pObserver_myObserver->addAttachment(pShadowPlanar_myShadowPlanar);

pShadowPlanar_myShadowPlanar->setLightSourceCelestial(pEnvSun_myEnvSun);

pShadowPlanar_myShadowPlanar->addCaster(pObject_flight);

其详细代码如图 6.22.1 所示。

```
#include "vpShadowPlanar.h"
void PublicMember::CTS_Define( void )
{
    //初始化平面光照影子模块
    vpModule::initializeModule( "vpShadow" );

    //添加模型目录
    pSearchPath_mySearchPath->append( CTS_RunPath+"//data//f16cn" );

    //正常添加目标对象
    vpObject* pObject_flight = new vpObject();
    pObject_flight->setName( "flight" );
    pObject_flight->setCullMask( 0x0FFFFFFFF );
```

```
pObject_flight->setRenderMask( 0x0FFFFFFFF );
pObject_flight->setIsectMask( 0x0FFFFFFFF );
pObject_flight->setStrategyEnable( true );
pObject_flight->setTranslate( 2450.000000 , 2530.000000 , 10.000000 );
pObject_flight->setRotate( 90.000000 , 0.000000 , 0.000000 );
//其他代码省略
PublicMember::CTS_s_pInstancesToUnref->push_back( pObject_flight );

//平面光照影子效果对象
vpShadowPlanar* pShadowPlanar_myShadowPlanar = new vpShadowPlanar();
pShadowPlanar_myShadowPlanar->setName( "myShadowPlanar" );
pShadowPlanar_myShadowPlanar->setEnable( true );
pShadowPlanar_myShadowPlanar->setEffectiveRange( 10000.000000f );
PublicMember::CTS_s_pInstancesToUnref->push_back( pShadowPlanar_myShadowPlanar );

//平面光照影子效果配置
pObserver_myObserver->addAttachment( pShadowPlanar_myShadowPlanar );
pShadowPlanar_myShadowPlanar->setLightSourceCelestial( pEnvSun_myEnvSun );
pShadowPlanar_myShadowPlanar->addCaster( pObject_flight );
}
```

图 6.22.1　平面倒影效果实现代码

运行程序，其效果如图 6.22.2 所示。另外，可以添加多个平面光照影子对象，实现平面光照影子效果。

图 6.22.2　飞行物体平面影子效果

6.23　物体颜色控制

在某些场景中，物体颜色需要动态控制。例如，在危险情况下，物体改变为红色；在正常情况下，物体改变为蓝色。

首先引入材质、状态的头文件：

```
#include " vrMaterial.h"
```

```cpp
#include " vrState.h"
#include " vsGeometry.h"
```

整个设置过程：分别设置状态的材质元素、光照元素和颜色元素，然后依次遍历物体的几何体，并把几何体的状态设置成赋值了材质元素、光照元素和颜色元素的状态。

设置材质元素，包含设置漫反射颜色、散射颜色、镜面反射颜色和发射颜色：

```cpp
//为材质元素作准备
vrMaterial *material = new vrMaterial();
material->setColor(vrMaterial::COLOR_AMBIENT,    r, g, b, 1.0);
material->setColor(vrMaterial::COLOR_DIFFUSE,    r, g, b, 0.0);
material->setColor(vrMaterial::COLOR_SPECULAR,   r, g, b, 1.0);
material->setColor(vrMaterial::COLOR_EMISSIVE,   r, g, b, 1.0);
material->setSpecularExponent(50.0);

//建立材质元素并赋值
vrMaterial::Element materialElement;
materialElement.m_side = vrMaterial::SIDE_FRONT_AND_BACK;
materialElement.m_material = material;
```

设置光照模型，包含设置光照的漫反射颜色：

```cpp
//为光照模式元素作准备
vrLightModel *lightModel = new vrLightModel();
lightModel->setLocalViewerEnable(false);
lightModel->setTwoSidedLightingEnable(false);
lightModel->setAmbientColor(0.2f, 0.2f, 0.2f, 1.0);

//建立光照模式元素并赋值
vrLightModel::Element lightModelElement;
lightModelElement.m_lightModel = lightModel;
```

设置跟随颜色元素：

```cpp
//建立颜色轨迹元素并赋值
vrColorTrack::Element colorTrackElement;
colorTrackElement.m_enable = false;
```

把材质、光照模型和跟随元素颜色赋值于状态：

```cpp
//建立状态对象，并对其相关元素并赋值
vrState* pNonTexState = new vrState();
pNonTexState->setElement(   vrMaterial::Element::Id, &materialElement );
pNonTexState->setElement(vrLightModel::Element::Id, &lightModelElement);
pNonTexState->setElement(vrColorTrack::Element::Id, &colorTrackElement);
```

最后，遍历物体对象的几何体，设置几何体的状态值：

```cpp
vsGeometry *geometry;
```

```
vpObject::const_iterator_geometry it, ite =pObject->end_geometry();
for (it=pObject->begin_geometry();it!=ite;++it)
{
            if ((*it)->isOfClassType(vsGeometry::getStaticClassType()))
    {
        // 获取几何体，并设置材质
            geometry = static_cast<vsGeometry *>(*it);
       geometry->setState(pNonTexState);
        }
}
```

整个详细代码如图 6.23.1 所示，其中，r、g、b 分别代表红、绿、蓝，取值方位均为[0,1]。

```
//控制物体颜色
void    PublicMember::SetObjectColor(vpObject *pObject,float r,float g,float b)
{
    //为材质元素作准备
    vrMaterial *material = new vrMaterial();
    material->setColor(vrMaterial::COLOR_AMBIENT,    r, g, b, 1.0);
    material->setColor(vrMaterial::COLOR_DIFFUSE,    r, g, b, 0.0);
    material->setColor(vrMaterial::COLOR_SPECULAR,   r, g, b, 1.0);
    material->setColor(vrMaterial::COLOR_EMISSIVE,   r, g, b, 1.0);
    material->setSpecularExponent(50.0);

    //建立材质元素并赋值
    vrMaterial::Element materialElement;
    materialElement.m_side = vrMaterial::SIDE_FRONT_AND_BACK;
    materialElement.m_material = material;

    //为光照模式元素作准备
    vrLightModel *lightModel = new vrLightModel();
    lightModel->setLocalViewerEnable(false);
    lightModel->setTwoSidedLightingEnable(false);
    lightModel->setAmbientColor(0.2f, 0.2f, 0.2f, 1.0);

    //建立光照模式元素并赋值
    vrLightModel::Element lightModelElement;
    lightModelElement.m_lightModel = lightModel;

    //建立颜色轨迹元素并赋值
    vrColorTrack::Element colorTrackElement;
    colorTrackElement.m_enable = false;

    //建立状态对象，并对其相关元素并赋值
    vrState* pNonTexState = new vrState();
    pNonTexState->setElement(   vrMaterial::Element::Id, &materialElement );
    pNonTexState->setElement(vrLightModel::Element::Id, &lightModelElement);
```

```
pNonTexState->setElement(vrColorTrack::Element::Id, &colorTrackElement);

vsGeometry *geometry;
vpObject::const_iterator_geometry it, ite =pObject->end_geometry();
for (it=pObject->begin_geometry();it!=ite;++it)
{
    if ((*it)->isOfClassType(vsGeometry::getStaticClassType()))
    {
        // 获取几何体,并设置材质
        geometry = static_cast<vsGeometry *>(*it);
        geometry->setState(pNonTexState);
    }
}
```

图 6.23.1 物体颜色控制实现代码

如图 6.23.2 所示,分别调用函数,实现功能:
PublicMember::SetObjectColor(pObjectHummer,1.0,0.0,0.0);//设置成红色
PublicMember::SetObjectColor(pObjectHummer,0.0,1.0,0.0);//设置成绿色
PublicMember::SetObjectColor(pObjectHummer,0.0,0.0,1.0);//设置成蓝色
另外,可以设置 RGB 的混合调色值,呈现出不同的颜色,满足不同的需求。

(a) 原色

(b) 红色

(c) 绿色

(d) 蓝色

图 6.22.1 物体颜色控制

6.24 雨雪天气控制

环境的控制，主要是对环境及其特效的控制，主要包含环境（vpEnv）、太阳(EnvSun)、月亮(EnvMoon)、天幕(EnvSkyDome)、云层(EnvCloudLayer)、星星(EnvStars)、云柱体(EnvCloudVolume)、雪(EnvSnow)、雨(EnvRain)、风(EnvWind)、风层(EnvWindLayer)和风柱体(EnvWindVolume)，如图 6.24.1 所示，这些特效都可以在 LynX Prime 界面中完成，然后再导出相关代码即可。

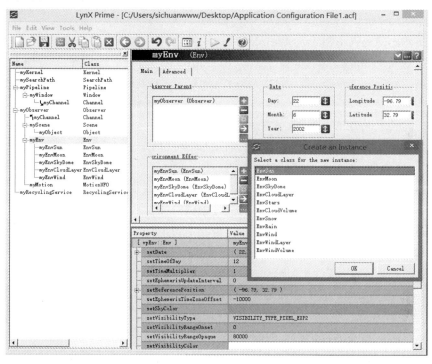

图 6.24.1　环境特效控制

本节只讨论环境、太阳、天幕、云层、风、雪和雨，其他特效读者可自行查询相关资料。在导出的 cpp 代码中，主要涉及环境及其相关特效的声明：

vpEnv* pEnv_myEnv = new vpEnv();

vpEnvSun* pEnvSun_myEnvSun = new vpEnvSun();

vpEnvSkyDome* pEnvSkyDome_myEnvSkyDome = new vpEnvSkyDome();

vpEnvCloudLayer* pEnvCloudLayer_myEnvCloudLayer = new vpEnvCloudLayer();

vpEnvWind* pEnvWind_myEnvWind = new vpEnvWind();

vpEnvSnow* pEnvSnow_myEnvSnow = new vpEnvSnow();

vpEnvRain* pEnvRain_myEnvRain = new vpEnvRain();

另外，包含把以下特效添加进环境的代码：

pEnv_myEnv->addEnvFx(pEnvSun_myEnvSun);

pEnv_myEnv->addEnvFx(pEnvMoon_myEnvMoon);

pEnv_myEnv->addEnvFx(pEnvSkyDome_myEnvSkyDome);

```cpp
pEnv_myEnv->addEnvFx( pEnvCloudLayer_myEnvCloudLayer );
pEnv_myEnv->addEnvFx( pEnvWind_myEnvWind );
pEnv_myEnv->addEnvFx( pEnvRain_myEnvRain );
pEnv_myEnv->addEnvFx( pEnvSnow_myEnvSnow );
```

把这些特效的相关代码放进 define( ) 函数即可。

以下是下雪特效的相关代码：

```cpp
vpEnvSnow* pEnvSnow_myEnvSnow = new vpEnvSnow();
pEnvSnow_myEnvSnow->setName( "myEnvSnow" );
pEnvSnow_myEnvSnow->setEnable( false );
pEnvSnow_myEnvSnow->setColor( 1.00f, 1.00f, 1.00f, 1.00f );
pEnvSnow_myEnvSnow->setTextureBlendColor( 1.00f, 1.00f, 1.00f, 1.00f );
pEnvSnow_myEnvSnow->setTextureBlendMode( vpEnvFx::TEXTURE_BLEND
                                        _MODE_MODULATE );
pEnvSnow_myEnvSnow->setTextureFile( "" );
pEnvSnow_myEnvSnow->setNumParticles( 50000);
pEnvSnow_myEnvSnow->setEmitterBoxSize( 10 );
pEnvSnow_myEnvSnow->setInternalClipBoxEnable( false );
pEnvSnow_myEnvSnow->setInternalClipBoxSize( 2 );
PublicMember::CTS_s_pInstancesToUnref->push_back( pEnvSnow_myEnvSnow );
```

其中，使能开关的代码为：

```cpp
pEnvSnow_myEnvSnow->setEnable( false );
```

为了能够控制其使能，开始设置为关闭，也就是设置其使能值为 false。

设置特效颗粒数的代码：

```cpp
pEnvSnow_myEnvSnow->setNumParticles( 50000);
```

默认值为 50，其值太小，效果不明显。读者也可以根据需要调整数字。

以下是下雨特效的相关代码：

```cpp
vpEnvRain* pEnvRain_myEnvRain = new vpEnvRain();
pEnvRain_myEnvRain->setName( "myEnvRain" );
pEnvRain_myEnvRain->setEnable( false );
pEnvRain_myEnvRain->setColor( 1.00f, 1.00f, 1.00f, 1.00f );
pEnvRain_myEnvRain->setTextureBlendColor( 1.00f, 1.00f, 1.00f, 1.00f );
pEnvRain_myEnvRain->setTextureBlendMode( vpEnvFx::TEXTURE_BLEND
                                        _MODE_MODULATE );
pEnvRain_myEnvRain->setTextureFile( "" );
pEnvRain_myEnvRain->setNumParticles( 50000 );
pEnvRain_myEnvRain->setEmitterBoxSize( 10 );
pEnvRain_myEnvRain->setInternalClipBoxEnable( false );
pEnvRain_myEnvRain->setInternalClipBoxSize( 2 );
PublicMember::CTS_s_pInstancesToUnref->push_back( pEnvRain_myEnvRain );
```

其中，使能开关的代码为：

pEnvRain_myEnvRain->setEnable(false);;

为了能够控制其使能，开始设置为关闭，也就是设置其使能值为 false。

设置特效颗粒数的代码：

pEnvRain_myEnvRain->setNumParticles(50000);

默认值为 50，其值太小，效果不明显。读者也可以根据需要调整数字。

把相关环境特效加入到 define 函数后，就可以通过代码来控制其状态和效果。设计了一个 PublicMember 的静态成员函数：

void   setWeather(int   WeatherType);

便于设置天气状况，其中的参数 WeatherType 为整数类型：其值为 1 时，设置天气为晴天；其值为 2 时，设置天气为阴天；其值为 3 时，设置天气为雾天；其值为 4 时，设置天气为雨天；其值为 5 时，设置天气为雪天。

整个详细代码如图 6.24.2 所示。其中，声明函数：

static void   setWeather(int   WeatherType=1);

首先设置为静态成员，然后让 WeatherType 的默认值为 1，即为晴天。

```cpp
//函数声明，并设置WeatherType的默认值为1
// 天气类型：  1, 晴天;2, 阴天 ;3, 雾天 ;4, 雨天 ;5, 雪天
static void    setWeather(int    WeatherType=1);

//函数实现
// 天气类型:1, 晴天;2, 阴天 ;3, 雾天 ;4, 雨天 ;5, 雪天
void    PublicMember::setWeather(int    WeatherType)
{
    char str[256], *cp = getenv("MPI_LOCATE_VEGA_PRIME");

    vpEnv*              m_env= * vpEnv::begin();
    vpEnvSun*           m_sun=* vpEnvSun::begin();
    vpEnvSkyDome*       m_skyDome=* vpEnvSkyDome::begin();
    vpEnvCloudLayer*    m_cloudLayer=* vpEnvCloudLayer::begin();
    vpEnvWind*          m_wind=* vpEnvWind::begin();
    vpEnvSnow*          m_snow=* vpEnvSnow::begin();
    vpEnvRain*          m_rain=* vpEnvRain::begin();

    switch (WeatherType)
    {
        case 1: // 默认值，晴天
            m_env->setVisibilityColor(1.0f, 1.0f, 1.0f, 1.0f);
            m_env->setVisibilityRangeOpaque(60000.0f);
            m_sun->setHorizonColor(1.0f, 0.545f, 0.239f, 1.0f);
            m_skyDome->setGroundColor(0.2117f, 0.286f, 0.15f, 1.0f);

            sprintf(str, "%s/config/vegaprime/vpenv/cloud_scattered.inta", cp);
            m_cloudLayer->setTextureFile(str);
            m_cloudLayer->setTextureTiling(3.0f, 3.0f);
```

```cpp
        m_cloudLayer->setColor(0.96f, 0.98f, 0.98f, 1.0f);

        m_cloudLayer->setElevation(3000.0f, 5000.0f);
        m_cloudLayer->setTransitionRange(500.0f, 500.0f);
        m_wind->setSpeed(100.0f);
        m_snow->setEnable(false);
        m_rain->setEnable(false);
        break;

    case 2: //阴天
        m_env->setVisibilityColor(1.0f, 1.0f, 1.0f, 1.0f);
        m_env->setVisibilityRangeOpaque(100000.0f);
        m_sun->setHorizonColor(1.0f, 0.6f, 0.29f, 1.0f);
        m_skyDome->setGroundColor(0.33f, 0.29f, 0.15f, 1.0f);

        sprintf(str, "%s/config/vegaprime/vpenv/cloud_few.inta", cp);
        m_cloudLayer->setTextureFile(str);
        m_cloudLayer->setTextureTiling(5.0f, 5.0f);
        m_cloudLayer->setColor(0.96f, 0.98f, 0.98f, 1.0f);

        m_cloudLayer->setElevation(3000.0f, 5000.0f);
        m_cloudLayer->setTransitionRange(500.0f, 500.0f);
        m_wind->setSpeed(15.0f);
        m_snow->setEnable(false);
        m_rain->setEnable(false);
        break;

    case 3: //雾天
        m_env->setVisibilityColor(0.75f, 0.75f, 0.75f, 1.0f);
        m_env->setVisibilityRangeOpaque(5000.0f);
        m_sun->setHorizonColor(1.0f, 0.6f, 0.29f, 1.0f);
        m_skyDome->setGroundColor(0.8f, 0.82f, 0.83f, 1.0f);

        sprintf(str, "%s/config/vegaprime/vpenv/cloud_overcast.inta", cp);
        m_cloudLayer->setTextureFile(str);
        m_cloudLayer->setTextureTiling(3.0f, 3.0f);
        m_cloudLayer->setColor(0.69f, 0.75f, 0.77f, 1.0f);

        m_cloudLayer->setElevation(500.0f, 3000.0f);
        m_cloudLayer->setTransitionRange(500.0f, 500.0f);
        m_wind->setSpeed(0.0f);
        m_snow->setEnable(false);
        m_rain->setEnable(false);
        break;

    case 4: //雨天
        m_env->setVisibilityColor(0.40f, 0.43f, 0.45f, 1.0f);
```

```
            m_env->setVisibilityRangeOpaque(50000.0f);
            m_sun->setHorizonColor(0.65f, 0.45f, 0.25f, 1.0f);
            m_skyDome->setGroundColor(0.33f, 0.29f, 0.15f, 1.0f);

            sprintf(str, "%s/config/vegaprime/vpenv/cloud_storm.inta", cp);
            m_cloudLayer->setTextureFile(str);
            m_cloudLayer->setTextureTiling(3.0f, 3.0f);
            m_cloudLayer->setColor(0.52f, 0.56f, 0.61f, 1.0f);

            m_cloudLayer->setElevation(1000.0f, 4000.0f);
            m_cloudLayer->setTransitionRange(500.0f, 500.0f);
            m_wind->setSpeed(5.0f);
            m_snow->setEnable(false);
            m_rain->setEnable(true);
            break;

    case 5: // 雪天
            m_env->setVisibilityColor(0.40f, 0.43f, 0.45f, 1.0f);
            m_env->setVisibilityRangeOpaque(50000.0f);
            m_sun->setHorizonColor(0.65f, 0.45f, 0.25f, 1.0f);
            m_skyDome->setGroundColor(0.33f, 0.29f, 0.15f, 1.0f);

            sprintf(str, "%s/config/vegaprime/vpenv/cloud_storm.inta", cp);
            m_cloudLayer->setTextureFile(str);
            m_cloudLayer->setTextureTiling(3.0f, 3.0f);
            m_cloudLayer->setColor(0.52f, 0.56f, 0.61f, 1.0f);

            m_cloudLayer->setElevation(1000.0f, 4000.0f);
            m_cloudLayer->setTransitionRange(500.0f, 500.0f);
            m_wind->setSpeed(5.0f);
            m_snow->setEnable(true);
            m_rain->setEnable(false);
            break;

    default:
            break;
    }
}
```

图 6.24.2　环境控制实现代码

整个设计大体分为两类：晴天、阴天和雾天基本是依靠纹理进行设计；雨天和下雪天依靠粒子特效完成。

如图 6.24.3 所示，分别调用函数，实现功能：

　　PublicMember::setWeather(1) ;//晴天
　　PublicMember::setWeather(2) ;//阴天
　　PublicMember::setWeather(3) ;//雾天
　　PublicMember::setWeather(4) ;//雨天

PublicMember::setWeather(5) ;//雪天

读者可以对其中的参数进行修改设置，以达到用户的需求。

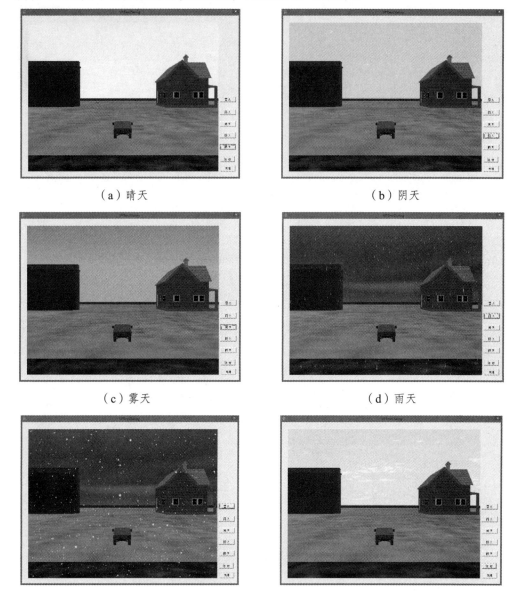

图 6.24.3 环境特效控制

6.25 场景能见度控制

仿真环境中的雾天控制效果不是很明显。但是，可以通过控制仿真环境的能见度实现更为明显的雾天效果。

在其他因素完全相同的条件下,定义环境对象:

vpEnv* pEnv_myEnv = new vpEnv();

然后设置环境的能见度范围。在图 6.25.1 中,分别设置了不同的能见度范围,可得到不同的能见度控制效果。

在图 6.25.1(a)中,设置能见度为 50 m 的效果:

pEnv_myEnv->setVisibilityRangeOpaque(50.00f);

在图 6.25.1(b)中,设置能见度为 500 m 的效果:

pEnv_myEnv->setVisibilityRangeOpaque(500.00f);

在图 6.25.1(c)中,不设置能见度的效果:

pEnv_myEnv->setVisibilityRangeOpaque(-1.00f);

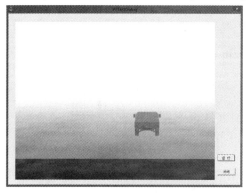
(a)50 m 能见度

(b)500 m 能见度

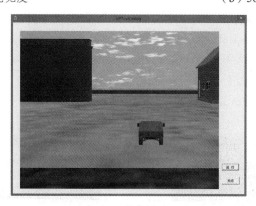
(c)正常能见度

图 6.25.1 环境能见度控制

本书只讨论设置了能见度的可设置性,用户可以根据需要,动态设置能见度。

6.26 纯色场景控制

在集中展示某个对象的时候,需要排除其他影响因素,这个时候就需要比较纯色的背景,如环境中使用纯蓝色背景。

在其他因素完全相同的条件下，定义环境对象：

  vpEnv* pEnv_myEnv = new vpEnv();

在环境特效中，只添加太阳特效，其他特效一概不要使用：

  pEnv_myEnv->addEnvFx(pEnvSun_myEnvSun);
  // pEnv_myEnv->addEnvFx(pEnvMoon_myEnvMoon);
  // pEnv_myEnv->addEnvFx(pEnvSkyDome_myEnvSkyDome);
  // pEnv_myEnv->addEnvFx(pEnvCloudLayer_myEnvCloudLayer);
  // pEnv_myEnv->addEnvFx(pEnvWind_myEnvWind);

同时，在场景中也最好不使用地形：

  //pScene_myScene->addChild(pObject_terrain);

最后，设置场景的颜色：

  pEnv_myEnv->setSkyColor(0.513725f,0.701961f ,0.941176f,1.00f);
  pEnv_myEnv->setVisibilityRangeOpaque(5000.000000f);
  pEnv_myEnv->setVisibilityColor(0.00f ,0.00f ,1.00f ,1.00f);

在能见度范围设置为 5 000 m 时，可见颜色起作用，如图 6.26.1(a)所示。

  pEnv_myEnv->setSkyColor(0.513725f,0.701961f ,0.941176f,1.00f);
  pEnv_myEnv->setVisibilityRangeOpaque(500000.000000f);
  pEnv_myEnv->setVisibilityColor(0.00f ,0.00f ,1.00f ,1.00f);

在能见度范围设置为 500 000 m 时，可见颜色不起作用，而是天空颜色起作用，如图 6.26.1(b)所示。

当然，对于色彩本身的配置，用户可以根据 RGB 值进行自由调配，调配出自己需要的环境色彩。

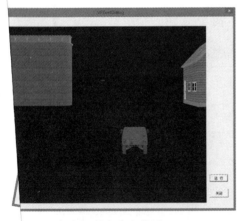 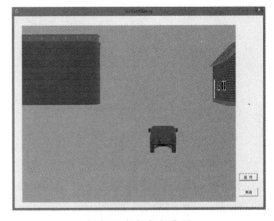

（a）可见颜色起作用   （b）天空颜色起作用

图 6.26.1 纯色环境控制

6.27 仿真场景全屏设计

对于仿真环境的控制，某些时候需要单独全屏显示仿真场景。仿真场景的全屏显示，主要包含两个步骤：

一是 vpWindow 的属性设置，这里主要是关闭其全屏属性，以便于后面根据屏幕尺寸改变 vpWindow 的尺寸。

  vpWindow * vpWin= * vpWindow::begin();
  vpWin->setFullScreenEnable(false);

二是根据需要获得屏幕的尺寸，设置仿真场景的尺寸：

  int width=::GetSystemMetrics(SM_CXSCREEN);
  int height=::GetSystemMetrics(SM_CYSCREEN);

  ::MoveWindow(PublicMember::CTS_MainWindow,0,0,width,height,true);
  ::MoveWindow(PublicMember::CTS_RunningWindow,0,0,width,height,true);

  vpWindow * vpWin= * vpWindow::begin();
  vpWin->setSize(width,height);

有了这两个步骤，就可以自由控制仿真场景的尺寸大小了。

实际设计中，还包含相关的辅助设计。首先在 PublicMember 中定义了两个静态窗口指针：

  //场景窗口句柄
  static HWND CTS_RunningWindow ;
  //程序主窗口句柄
  static HWND CTS_MainWindow;

然后进行初始化：

  HWND PublicMember::CTS_RunningWindow =NULL;
  HWND PublicMember::CTS_MainWindow=NULL;

并在适当的地方获取真实的值，本节是在"运行"按钮下进行赋值：

  PublicMember::CTS_MainWindow=theApp.GetMainWnd()->GetSafeHwnd();

  CWnd *pWnd=GetDlgItem(IDC_grScene);
  PublicMember::CTS_RunningWindow=pWnd->GetSafeHwnd();

接着，在窗口函数 PublicMember::CTS_Keyboard(vpWindow *window,vpWindow::Key k int modifier,void *)中添加代码，按下 M 键时，窗口全屏显示，按下 N 键时，恢复到 800×(的窗口：

//全屏显示
case vpWindow::KEY_M:
case vpWindow::KEY_m:
  {

```cpp
            int width=::GetSystemMetrics(SM_CXSCREEN);
            int height=::GetSystemMetrics(SM_CYSCREEN);

            ::MoveWindow(PublicMember::CTS_MainWindow,0,0,width,height,true);
            ::MoveWindow(PublicMember::CTS_RunningWindow,0,0,width,height,true);

        vpWindow * vpWin= * vpWindow::begin();
        vpWin->setSize(width,height);
          }
        break;

                //恢复原状
case vpWindow::KEY_N:
case vpWindow::KEY_n:
          {
          int width=800;
          int height=600;

            ::MoveWindow(PublicMember::CTS_MainWindow,0,0,width+100,height+100,true);
            ::MoveWindow(PublicMember::CTS_RunningWindow,0,0,width,height,true);

        vpWindow * vpWin= * vpWindow::begin();
        vpWin->setSize(width,height);
          }
        break;
```

其中的关键代码，首先是设置宽度和高度值，其次是设置主窗口的尺寸，接着是设置仿[真窗]口控件的尺寸，最后才是设置仿真窗口的尺寸。

最后，需要在 VP 主线程中进行相关的配置：

```cpp
//绘制场景
vpKernel::instance()->configure();
//设置观察者
PublicMember::CTS_pObject_observer=vpObject::find("Hummer");
PublicMember::CTS_pObject_observer->ref();

//设置窗体
vpWindow * vpWin= * vpWindow::begin();
vpWin->setParent(PublicMember::CTS_RunningWindow);
vpWin->setFullScreenEnable(false);

//设置键盘函数
vpWin->setInputEnable(true);
vpWin->setKeyboardFunc((vrWindow::KeyboardFunc)PublicMember::CTS_Keyboard,NULL);
```

vpWin->open();
::SetFocus(vpWin->getWindow());

在图 6.27.1（a）中，按下 M 键，仿真场景是单独全屏；在图 6.26.1（b）中，按下 M 键，仿真场景恢复原状。

（a）仿真场景全屏

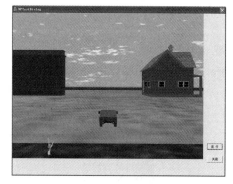
（b）恢复原状

图 6.27.1　仿真场景全屏控制

当然，用户还可以根据需要做更多的控制。

6.28　加快仿真场景物体加载速度

对于虚拟现实仿真来说，最开始的环境加载和渲染非常重要。按照经验，一个场景的加载超过 10 s，用户体验就比较糟糕了；如果超过 30 s，基本不能接受了。虽然可以通过提供加载进度条改善用户体验，但这是"治标不治本"的方法。Vega Prime 的加载，对于不超过 10 M 的物体对象来说，速度还是很快的，一般超过 20 M 的单个物体，加载速度就很慢了。

其实，Vega Prime 提供了一种文件格式，Vega Scenegraph Binary(VSB)不仅可以大大减小 FLT 文件的尺寸，还可以大大提高加载速度。VSB 文件是序列化的场景对象格式，不再需要对场景对象进行转换和优化，因此，在加载和内存分页管理时可以节省很多时间，这样就可以大大提高物体对象的加载速度。

Vega Prime 提供了一种转换工具可以把 FLT 模型文件转换为 VSB 文件格式，VSB 文件格式为系统尽可能快地加载模型到场景中提供了一种有效的途径。VSB 文件是预先采用工具从 FLT 格式转换而来的，其数据方式与 Vega Prime 场景里的需要的方式非常相似，不用在加载时再进行优化处理，所以可以非常快地载入。FLT 文件格式在加入 Vega Prime 场景时都需要做大量的处理工作，这在处理少量数据或静态调入时对系统运行时间没有什么影响，但是在需要动态实时调入模型数据（尤其是大大规模数据时）就有很大的延迟，影响浏览效果。

通常情况下，VSB 文件里不储存纹理文件，因为有可能几个 VSB 文件用同样的纹理文件，这样就使 VSB 文件变得更小，加载速度比更快。场景的实时调入也包含纹理的实时调入，如果模型的纹理总量超过硬件纹理内存的容量，甚至超过系统内存的容量，在浏览过程中，频繁交换内存或硬盘中的纹理，就很难保证其实时性。因此，可以把纹理文件存储在 VSB 文件里，这样也可以大大地提高载入的实时性。

VSB 文件格式具有很多好处：第一，创建好的 FLT 模型文件，转化成 VSB 格式的二进制文件，并将模型纹理包含其中，可以大大加快系统的加载速度，提高运行效率（特别是对于大型的模型数据库而言）；第二，Creator 中不能打开 VSB 格式的模型，VSB 模型由于具有不可逆性，所以从另一个方面看，又同时具备了保护自己模型的知识产权的目的；第三，包含纹理后，VSB 文件的模型成为一个单独的文件，不再受到路径的困扰，发布仿真应用程序时也不用再附带纹理文件夹了。

把 FLT 文件转换为 VSB 文件，既可以通过 Vega Prime 的 API 函数来完成，也可以通过 Vega Prime 提供的工具来完成。

如图 6.28.1 所示，启动 VSB Generation Utility，其界面如图 6.28.2 所示。

图 6.28.1 启动 VSB 转换工具

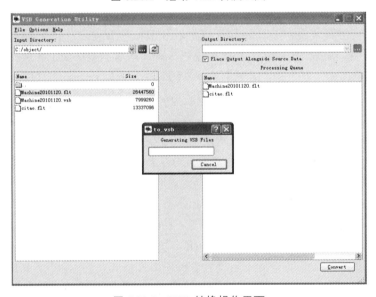

图 6.28.2 VSB 转换操作界面

在图 6.28.2 中左上部的"Input Directory"下选择模型所在的目录，这里是"C:/object/"，其中有两个 FLT 文件，双击其中的 FLT 文件，其自动添加到图中右侧列表中，也就是"Processing Queue"处理队列中，勾选右上部分的"Place Output Alongside Source Data"，让转换完成的 VSB 文件和源文件放到一起。最后，点击图中右下角的"Convert"按钮，开始转换。

如果需要转换的 FLT 文件很大或者 FLT 文件较多，可能需要较长的时间。在这里，要重点强调一下，FLT 文件或 VSB 文件最好不要使用中文名称，文件路径也最好不要包含空格、中文等。

转换完成后，结果如图 6.28.3 所示。原来大小约为 13 M 的 FLT 文件变成了大小约为 5 M 的 VSB 文件，原来大小约为 26 M 的 FLT 文件变成了大小约为 8 M 的 VSB 文件，其尺寸变化非常明显。其实，最大的变化是物体对象的加载时间，对于 25 M 的 FLT 文件，大概需要 30 s 时间，而 8 M 的 VSB 文件，加载时间大概只需要 5 s，优越性非常明显。

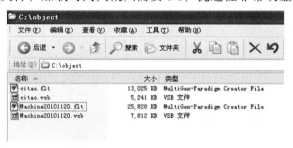

图 6.28.3　VSB 转换结果

刚才对 VSB 转换工具的使用，采用的是默认选项，其实，用户可以根据自己的需要对转换选项进行设置。在图 6.28.2 中，选择菜单"Options"，点击其中的"VSB Generation Options"，将出现如图 6.28.4 所示的选项。

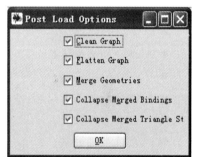
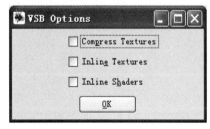
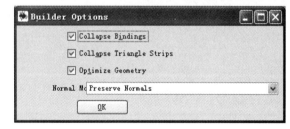

图 6.28.4　VSB 选项

图 6.28.4 中的 VSB 转换选项主要包含 4 个选项板："VSB Options"主要包含纹理压缩、内联纹理、内联阴影等；"Builder Options"主要包含折叠包围盒、折叠三角条、优化几何体等；"Loading Options"主要包含合并光照点、合并 LODS、创建动画、忽略 DOF 限制等；"Post Load Options"主要包含清理图形、平滑图形、合并结合体等。用户可以仔细研究，按照自己需求使用该工具，进行 VSB 转换。

第 7 章 Vega Prime 自绘图形设计

VSG(Vega Scene Graph)是跨平台的场景渲染 API，是 Vega Prime 的基础。Vega Prime 包括 VSG 提供的所有功能，并在易用性和生产效率上做了相应改进。前面章节都是利用 Creator 进行建模，然后利用 Vega Prime 进行驱动渲染，在图形构造上稍欠灵活性。现在利用 VSG 的强大绘图功能，灵活构造所需的图形。本章主要内容包括工具 VSG 的特点、VSG 图形绘制过程、VSG 简单图形绘制、VSG 纹理设置、VSG 材质设置等。整个过程充分展示了利用 VSG 的强大功能在 Vega Prime 应用程序中绘制图形的方法和步骤。

【本章重点】

- VSG 的特点；
- VSG 图形绘制过程；
- VSG 绘制箱体；
- VSG 绘制球体；
- VSG 绘制平面；
- VSG 简单图形绘制；
- VSG 字符输出；
- VSG 纹理设置；
- VSG 材质设置；
- VSG 在场景显示中文。

7.1 认识 VSG

7.1.1 VSG 的特点

在为视景仿真和可视化应用提供的各种低成本商业开发软件中，VSG 具有强大的功能，它为仿真、训练和可视化等高级三维应用开发人员提供了极佳的可扩展的基础。VSG 具有以下特点：

- 帧频率控制；
- 内存分配；

- 内存泄漏跟踪；
- 基于帧的纹理调用；
- 异步光线点处理；
- (优化的)分布式渲染；
- 跨平台可扩展的开发环境，支持 Windows、Irix、Linux 和 Solaris；
- 与 C++ STL 相兼容的体系结构；
- 强大的可扩展性，允许最大限度的定制，使得用户可调整 VSG 来满足应用需求，而不是根据产品的限制来调整应用需求；
- 支持多处理器多线程的定制与配置；
- 应用程序也具有跨平台性，用户在任意一种平台上开发的应用程序无须进行修改即可在另一个平台上运行；
- 支持 OpenGL 和 Direct3D 的优化渲染功能，应用程序能基于 OpenGL 或 Direct3D 运行，其间无须改动程序代码；
- 支持双精度浮点数，使几何物体和地形在场景中能够精确地放置并表示；
- 支持虚拟纹理、软件实现图像的动态查阅，使高级功能与平台无关。

7.1.2 VSG 的功能模块

VSG 包含的功能模块：
- Vega Scene Graph Font Loaders，图形字体加载控制模块。
- Vega Scene Graph Shader Loaders，图形着色加载控制模块。
- Vega Scene Graph Texture Loaders，图形纹理加载控制模块。
- Vega Scene Graph Geo-Builder，图形建造控制模块。
- Vega Scene Graph Rendering Library，图形渲染控制库。
- Vega Scene Graph Scene Library，图形场景控制库。
- Vega Scene Graph Statistics，图形统计数据控制模块。
- Vega Scene Graph Utility Library，图形实用工具库。
- Vega Scene Graph Virtual Texture，图形虚拟纹理模块。
- Vega Scene Graph Node Loaders，图形节点加载控制模块。
- Vega Scene Graph Image Loaders，图形图像加载控制模块。

7.1.3 VSG 的图形绘制过程

可把 VSG 物体绘制分为 5 个步骤：
（1）状态设置，由 vrState 类完成。主要代码为：

    vrState *pState = new vrState;
    vrAlphaBlend::Element alphaBlendElement;
    alphaBlendElement.m_enable = true;

pState->setElement(vrAlphaBlend::Element::Id,&alphaBlendElement);
(2) 物体的绘制，由 vrGeometry 类完成。主要代码为：
```
vrGeometry *pRGrom = new vrGeometry;
vuVec4<float> *color = vuAllocArray<vuVec4<float> >::malloc(1);
color[0].set(1.0f, 1.0f, 0.0f, 1.0f);
vuVec3<float> *vertex = vuAllocArray<vuVec3<float> >::malloc(4);
vertex[0].set(-0.5f, -0.5f, 0.0f);
vertex[1].set(0.5f, -0.5f, 0.0f);
vertex[2].set(-0.5f, -0.3f, 0.0f);
vertex[3].set(0.5f, -0.3f, 0.0f);
pRGrom->setPrimitive(vrGeometry::PRIMITIVE_LINE);
pRGrom->setNumPrimitives(2);
pRGrom->setColors(color, vrGeometry::BINDING_OVERALL);
pRGrom->setVertices(vertex);
```
(3) 用 vsGeometry 包装 vrGeometry。主要代码为：
```
vsGeometry *pSGeom = new vsGeometry;
pSGeom->setGeometry(pRGrom);
pSGeom->setState(pState);
```
(4) 将 vsGeometry 作为子物体加到 vpTransform 中。主要代码为：
```
vpTransform *pTrans =new vpTransform();
pTrans->setTranslate( 2360.000000 ,2490.000000 ,3.000000 );
pTrans->setRotate( 0.000000 ,0.000000 ,0.000000 );
pTrans->setScale( 10.000000 ,10.000000 ,10.000000 );
pTrans->insert_child(pSGeom,pTrans->end_child());
pTrans->ref();
```
(5) vpTransform 作为孩子添加到场景中。主要代码为：
```
vpScene* Scene = (vpScene*)vpScene::find("myScene");
Scene->addChild(pTrans);
```
下一节将会详细展示 VSG 绘图的完整过程。

7.2 VSG 图形绘制

VSG 的图形类包括盒子类 vrBox、椎体类 vrFrustum、几何体类 vrGeometry、光点类 vrLightPoint、平面类 vrPlane、球体类 vrSphere 和字符串类 vrString 等。VSG 图形类的继承关系如图 7.2.1 所示。

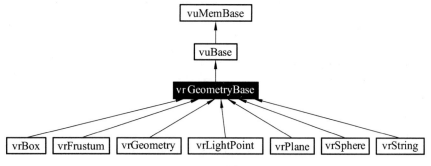

图 7.2.1　VSG 图形类的继承关系

7.2.1　简单几何体绘制

根据实际需求，利用 VSG 绘制不同图形。采用 VSG 图形绘制的 5 个步骤。在 VSG 图形绘制的 5 个步骤中，第 2 个步骤主要涉及几何体的设计问题，其余 4 个步骤的内容基本没有变化。在第 2 个步骤中，首先定义了一系列的顶点向量，然后设置这些顶点的处理方式："pRGrom->setPrimitive(vrGeometry::PRIMITIVE_LINE);"即设置为线段方式处理顶点。实际上，VSG 包含以下处理方式：

PRIMITIVE_POINT：独立点方式，只包含离散的点集合。

PRIMITIVE_LINE：线段方式，相邻两个顶点产生一条线段，第 0 个顶点与第 1 个顶点产生一条线段；第 2 个顶点与第 3 个顶点产生一条线段；以此类推。

PRIMITIVE_LINE_STRIP：闭环线方式，产生一系列线段闭环图形。

PRIMITIVE_TRIANGLE：三角形方式，每 3 个顶点产生一个三角形，以此类推。

PRIMITIVE_TRIANGLE_FAN：三角扇形方式，每 3 个顶点产生一个包含一条弧形边的三角形，以此类推。

PRIMITIVE_QUAD：四边形方式，每 4 个顶点产生一个四边形，以此类推。

接下来，按照 VSG 绘制物体的五个步骤，进行简单图形绘制。

在主线程中，在 vpKernel::instance()->configure();之后，加入了 VSG 绘制图形的代码。

为了便于操作绘制的图形，在 PublicMember 类中定义了一个静态变量：

  static vpObject * PublicMember::CTS_pObject_other;

在第 4 步中，使用了代码：

  PublicMember::CTS_pObject_other=static_cast<vpObject *> (pTrans);

  PublicMember::CTS_pObject_other->ref();

用 PublicMember::CTS_pObject_other 指向了绘制的物体，这样就可以在键盘函数中操作绘制的物体。

主线程中详细代码如图 7.2.2 所示，有关 VSG 绘图的代码用黑体字体进行了标识。

```
//VP 运行主线程
UINT PublicMember::CTS_RunBasicThread(LPVOID)
{
    //初始化
    vp::initialize(__argc,__argv);
```

```cpp
//定义场景
PublicMember::CTS_Define();
//绘制场景
vpKernel::instance()->configure();
//VSG 绘图开始,绘制线段//1 状态设置,由 vrState 类完成
vrAlphaBlend::Element alphaBlendElement;
alphaBlendElement.m_enable = true;
vrState *pState = new vrState;
pState->setElement(vrAlphaBlend::Element::Id,&alphaBlendElement);

//2 物体的绘制,由 vrGeometry 类完成

vrGeometry *pRGrom = new vrGeometry;
vuVec4<float> *color = vuAllocArray<vuVec4<float> >::malloc(1);
color[0].set(1.0f, 1.0f, 0.0f, 1.0f);
vuVec3<float> *vertex = vuAllocArray<vuVec3<float> >::malloc(24);

vertex[0].set(-5.0f, -5.0f, 0.0f);
vertex[1].set(5.0f, -5.0f, 0.0f);

vertex[2].set(-5.0f, -3.0f, 0.0f);
vertex[3].set(5.0f, -3.0f, 0.0f);

vertex[4].set(-5.0f, -1.0f, 0.0f);
vertex[5].set(5.0f, -1.0f, 0.0f);

vertex[6].set(-5.0f, 1.0f, 0.0f);
vertex[7].set(5.0f, 1.0f, 0.0f);

vertex[8].set(-5.0f, 3.0f, 0.0f);
vertex[9].set(5.0f, 3.0f, 0.0f);

vertex[10].set(-5.0f, 5.0f, 0.0f);
vertex[11].set(5.0f, 5.0f, 0.0f);

pRGrom->setPrimitive(vrGeometry::PRIMITIVE_LINE);
pRGrom->setNumPrimitives(6);
pRGrom->setColors(color, vrGeometry::BINDING_OVERALL);
pRGrom->setVertices(vertex);

//3 用 vsGeometry 包装 vrGeometry
vsGeometry *pSGeom = new vsGeometry;
pSGeom->setGeometry(pRGrom);
pSGeom->setState(pState);

//4 将 vsGeometry 作为子物体加到 vpTransform 中
vpTransform *pTrans =new   vpTransform();
```

```
pTrans->setTranslate( 2346.000000 , 2490.000000 , 3.000000 );
pTrans->setRotate( 0.000000 , 0.000000 , 0.000000 );
pTrans->setScale( 1.000000 , 1.000000 , 1.000000 );

pTrans->insert_child(pSGeom,pTrans->end_child());
pTrans->ref();

PublicMember::CTS_pObject_other=static_cast<vpObject *> (pTrans);
PublicMember::CTS_pObject_other->ref();

//5   vpTransform 作为孩子添加到场景中
vpScene* Scene = (vpScene*)vpScene::find("myScene");
Scene->addChild(pTrans);

//设置观察者
PublicMember::CTS_pObject_observer=vpObject::find("Hummer");
PublicMember::CTS_pObject_observer->ref();

//设置窗体
vpWindow * vpWin= * vpWindow::begin();
vpWin->setParent(PublicMember::CTS_RunningWindow);
vpWin->setBorderEnable(false);
vpWin->setFullScreenEnable(true);

//设置键盘
vpWin->setInputEnable(true);
vpWin->setKeyboardFunc((vrWindow::KeyboardFunc)PublicMember::CTS_Keyboard,NULL);
vpWin->open();
::SetFocus(vpWin->getWindow());

//帧循环
while(vpKernel::instance()->beginFrame()!=0)
{
    vpKernel::instance()->endFrame();

    if(!PublicMember::CTS_continueRunVP)
    {
      vpKernel::instance()->endFrame();
      vpKernel::instance()->unconfigure();
      vp::shutdown();
      return 0;
    }
}
return 0;
}
```

图 7.2.2   VSG 绘制简单图形的主线程

运行主线程，将在汽车旁绘制 6 条线段，效果如图 7.2.3 所示。

图 7.2.3　VSG 绘制线段

7.2.2　箱体绘制

如果所有的图形都需要从点开始绘制，将会非常麻烦。VSG 提供了常见图形的定义类，这样，就可以直接使用 VSG 的图形类定义图形。

接下来，按照 VSG 绘制物体的 5 个步骤，进行简单箱体绘制。

在主线程中，在 vpKernel::instance()->configure();之后，加入了 VSG 绘制图形的代码。

在第 2 步中，不再需要按照点的方式来定义一个几何体，而是直接使用 VSG 的箱体类 vrBox 来定义一个箱体，代码如下：

vrBox* pBox = new vrBox(vuVec3f(1.0,1.0,0.0),vuVec3f(2.0,2.0,1.0));

其中的两个顶点，代表箱体的两个斜对角的顶点。

主线程中详细代码如图 7.2.4 所示，有关 VSG 绘图的代码用黑体字体进行了标识。

```
//VP 运行主线程
UINT PublicMember::CTS_RunBasicThread(LPVOID)
{
    //初始化
    vp::initialize(__argc,__argv);
    //定义场景
    PublicMember::CTS_Define();
    //绘制场景
    vpKernel::instance()->configure();
    //VSG 绘图开始
    //1 状态设置，由 vrState 类完成
    vrAlphaBlend::Element alphaBlendElement;
    alphaBlendElement.m_enable = true;
    vrState *pState = new vrState;
    pState->setElement(vrAlphaBlend::Element::Id,&alphaBlendElement);

    //2 物体的绘制，由 vrGeometry 类完成
    //绘制箱体
    vrBox* pBox = new vrBox( vuVec3f(1.0,1.0,0.0),vuVec3f(2.0,2.0,1.0) );
    pBox->setColor( 0.0,0.0,0.6f,0.5f );
```

```cpp
//3 用 vsGeometry 包装 vrGeometry
vsGeometry *pSGeom = new vsGeometry;
pSGeom->setGeometry(pBox);//注意几何体变量的名称          *
pSGeom->setState(pState);

//4 将 vsGeometry 作为子物体加到 vpTransform 中
vpTransform *pTrans =new   vpTransform();
pTrans->setTranslate( 2366.000000 ,  2490.000000 ,  3.000000 );
pTrans->setRotate( 0.000000 ,  0.000000 ,  0.000000 );
pTrans->setScale( 1.000000 ,  1.000000 ,  1.000000 );

pTrans->insert_child(pSGeom,pTrans->end_child());
pTrans->ref();

PublicMember::CTS_pObject_other=static_cast<vpObject *> (pTrans);
PublicMember::CTS_pObject_other->ref();

//5 vpTransform 作为孩子添加到场景中
vpScene* Scene = (vpScene*)vpScene::find("myScene");
Scene->addChild(pTrans);

//设置观察者
PublicMember::CTS_pObject_observer=vpObject::find("Hummer");
PublicMember::CTS_pObject_observer->ref();

//设置窗体
vpWindow * vpWin= * vpWindow::begin();
vpWin->setParent(PublicMember::CTS_RunningWindow);
vpWin->setBorderEnable(false);
vpWin->setFullScreenEnable(true);

//设置键盘
vpWin->setInputEnable(true);
vpWin->setKeyboardFunc((vrWindow::KeyboardFunc)PublicMember::CTS_Keyboard,NULL);
vpWin->open();
::SetFocus(vpWin->getWindow());

//帧循环
while(vpKernel::instance()->beginFrame()!=0)
{
    vpKernel::instance()->endFrame();

    if(!PublicMember::CTS_continueRunVP)
    {
        vpKernel::instance()->endFrame();
        vpKernel::instance()->unconfigure();
        vp::shutdown();
```

```
            return 0;
        }
    }
    return 0;
}
```

图 7.2.4 VSG 绘制箱体的主线程

运行主线程，将在汽车旁绘制一个箱体，效果如图 7.2.5 所示。

图 7.2.5 VSG 绘制箱体

7.2.3 梯形平台绘制

接下来，按照 VSG 绘制物体的 5 个步骤，进行简单梯形平台绘制。

在主线程中，在 vpKernel::instance()->configure();之后，加入了 VSG 绘制图形的代码。

在第 2 步中，不再需要按照点的方式来定义一个几何体，而是直接使用 VSG 的梯形平台类 vuFrustum 来定义一个梯形平台，代码如下：

```
vuFrustum <float> vuf;
vuf.makeSymmetric(45.0,45.0,2.0,4.0);
vrFrustum *pFrustum=new vrFrustum();
pFrustum->setFrustum(vuf);
```

其中的两个顶点，代表箱体的两个斜对角的顶点。

主线程中详细代码如图 7.2.6 所示，有关 VSG 绘图的代码用黑体字体进行了标识。

```
//VP 运行主线程
UINT PublicMember::CTS_RunBasicThread(LPVOID)
{
    //初始化
    vp::initialize(__argc,__argv);
    //定义场景
    PublicMember::CTS_Define();
    //绘制场景
    vpKernel::instance()->configure();
    //VSG 绘图开始，绘制线段//1 状态设置，由 vrState 类完成
    vrAlphaBlend::Element alphaBlendElement;
    alphaBlendElement.m_enable = true;
```

```cpp
vrState *pState = new vrState;
pState->setElement(vrAlphaBlend::Element::Id,&alphaBlendElement);


//2 物体的绘制，由 vrGeometry 类完成
/绘制圆锥体
vuFrustum <float> vuf;
vuf.makeSymmetric(45.0,45.0,2.0,4.0);


vrFrustum *pFrustum=new vrFrustum();
pFrustum->setFrustum(vuf);
pFrustum->setColor( 1.0,0.0,0.0f,0.5f );


//3 用 vsGeometry 包装 vrGeometry
pSGeom->setGeometryBase(pFrustum);
pSGeom->setState( pState );


//4 将 vsGeometry 作为子物体加到 vpTransform 中
vpTransform *pTrans =new   vpTransform();
pTrans->setTranslate( 2365.000000， 2485.000000， 3.000000 );
pTrans->setRotate( 0.000000， -90.000000， 0.000000 );
pTrans->setScale( 1.000000， 1.000000， 1.000000 );


pTrans->insert_child(pSGeom,pTrans->end_child());
pTrans->ref();


PublicMember::CTS_pObject_other=static_cast<vpObject *> (pTrans);
PublicMember::CTS_pObject_other->ref();


//5  vpTransform 作为孩子添加到场景中
vpScene* Scene = (vpScene*)vpScene::find("myScene");
Scene->addChild(pTrans);


//设置观察者
PublicMember::CTS_pObject_observer=vpObject::find("Hummer");
PublicMember::CTS_pObject_observer->ref();


//设置窗体
vpWindow * vpWin= * vpWindow::begin();
vpWin->setParent(PublicMember::CTS_RunningWindow);
vpWin->setBorderEnable(false);
vpWin->setFullScreenEnable(true);


//设置键盘
vpWin->setInputEnable(true);
vpWin->setKeyboardFunc((vrWindow::KeyboardFunc)PublicMember::CTS_Keyboard,NULL);
vpWin->open();
::SetFocus(vpWin->getWindow());
```

```
    //帧循环
    while(vpKernel::instance()->beginFrame()!=0)
    {
        vpKernel::instance()->endFrame();

        if(!PublicMember::CTS_continueRunVP)
        {
            vpKernel::instance()->endFrame();
            vpKernel::instance()->unconfigure();
            vp::shutdown();
            return 0;
        }
    }
    return 0;
}
```

图 7.2.6 VSG 绘制圆柱体的主线程

运行主线程,将在汽车旁绘制一个梯形平台,效果如图 7.2.7 所示。

图 7.2.7 平台绘制

7.2.4 平面绘制

接下来,按照 VSG 绘制物体的 5 个步骤,进行平面绘制。

在主线程中,在 vpKernel::instance()->configure();之后,加入了 VSG 绘制图形的代码。

在第 2 步中,不再需要按照点的方式来定义一个几何体,而是直接使用 VSG 的箱体类 vrPlane 来定义一个箱体,代码如下:

    vrPlane *pPlan=new vrPlane(vuVec3f(0.0,0.0,3.0), 0.50, 10.0f);

其中的第 1 个参数顶点表示垂直于平面的法线,第 2 个参数表示法线方向上的偏移量,第 3 个参数表示平面的尺寸大小。

主线程中详细代码如图 7.2.8 所示,有关 VSG 绘图的代码用黑体字体进行了标识。

```
//VP 运行主线程
UINT PublicMember::CTS_RunBasicThread(LPVOID)
{
```

```cpp
//初始化
vp::initialize(__argc,__argv);
//定义场景
PublicMember::CTS_Define();
//绘制场景
vpKernel::instance()->configure();
//VSG 绘图开始，绘制线段//1 状态设置，由 vrState 类完成
vrAlphaBlend::Element alphaBlendElement;
alphaBlendElement.m_enable = true;
vrState *pState = new vrState;
pState->setElement(vrAlphaBlend::Element::Id,&alphaBlendElement);

//2 物体的绘制，由 vrGeometry 类完成
//绘制平面
vrPlane *pPlan=new vrPlane( vuVec3f(0.0,0.0,3.0), 0.50, 10.0f);
pPlan ->setColor( 0.0,0.0,0.6f,0.5f );   //

//3 用 vsGeometry 包装 vrGeometry
vsGeometry *pSGeom = new vsGeometry;
pSGeom->setGeometry(pPlan); //注意几何体变量的名称
pSGeom->setState(pState);

//4 将 vsGeometry 作为子物体加到 vpTransform 中
vpTransform *pTrans =new   vpTransform();
pTrans->setTranslate( 2356.000000 ,  2490.000000 ,  3.000000 );
pTrans->setRotate( 0.000000 ,   0.000000 ,   0.000000 );
pTrans->setScale( 1.000000 ,   1.000000 ,   1.000000 );

pTrans->insert_child(pSGeom,pTrans->end_child());
pTrans->ref();

PublicMember::CTS_pObject_other=static_cast<vpObject *> (pTrans);
PublicMember::CTS_pObject_other->ref();

//5  vpTransform 作为孩子添加到场景中
vpScene* Scene = (vpScene*)vpScene::find("myScene");
Scene->addChild(pTrans);

//设置观察者
PublicMember::CTS_pObject_observer=vpObject::find("Hummer");
PublicMember::CTS_pObject_observer->ref();

//设置窗体
vpWindow * vpWin= * vpWindow::begin();
vpWin->setParent(PublicMember::CTS_RunningWindow);
vpWin->setBorderEnable(false);
vpWin->setFullScreenEnable(true);
```

```
//设置键盘
vpWin->setInputEnable(true);
vpWin->setKeyboardFunc((vrWindow::KeyboardFunc)PublicMember::CTS_Keyboard,NULL);
vpWin->open();
::SetFocus(vpWin->getWindow());

//帧循环
while(vpKernel::instance()->beginFrame()!=0)
{
    vpKernel::instance()->endFrame();

    if(!PublicMember::CTS_continueRunVP)
    {
      vpKernel::instance()->endFrame();
      vpKernel::instance()->unconfigure();
      vp::shutdown();
      return 0;
    }
}
return 0;
}
```

图 7.2.8　VSG 绘制平面的主线程

运行主线程，将在汽车旁绘制一个平面，效果如图 7.2.9 所示。

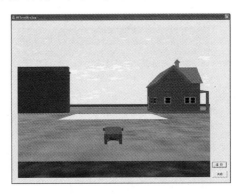

图 7.2.9　VSG 绘制平面

7.2.5　球体绘制

接下来，按照 VSG 绘制物体的 5 个步骤，进行简单球体绘制。

在主线程中，在 vpKernel::instance()->configure();之后，加入了 VSG 绘制图形的代码。

在第 2 步中，不再需要按照点的方式来定义一个几何体，而是直接使用 VSG 的箱体类 vrSphere 来定义一个球体，代码如下：

　　　　vrSphere* pSphere = new vrSphere(vuVec3f(0.0,0.0,0.0), 1.0,100);

第 1 个参数顶点表示圆心，第 2 个参数表示圆半径，第 3 个参数表示球体分片度数。主线程中详细代码如图 7.2.10 所示，有关 VSG 绘图的代码用黑体字体进行了标识。

```
//VP 运行主线程
UINT PublicMember::CTS_RunBasicThread(LPVOID)
{
    //初始化
    vp::initialize(__argc,__argv);
    //定义场景
    PublicMember::CTS_Define();
    //绘制场景
    vpKernel::instance()->configure();
    //VSG 绘图开始，绘制线段
    //1 状态设置，由 vrState 类完成
    vrAlphaBlend::Element alphaBlendElement;
    alphaBlendElement.m_enable = true;
    vrState *pState = new vrState;
    pState->setElement(vrAlphaBlend::Element::Id,&alphaBlendElement);

    //2 物体的绘制，由 vrGeometry 类完成
    //绘制球体
    vrSphere* pSphere = new vrSphere( vuVec3f(0.0,0.0,0.0), 1.0,100);
    pSphere ->setColor( 0.0,1.0,0.0,1.0 );

    //3 用 vsGeometry 包装 vrGeometry
    vsGeometry *pSGeom = new vsGeometry;
    pSGeom->setGeometry(pSphere); //注意几何体变量的名称         *
    pSGeom->setState(pState);

    //4 将 vsGeometry 作为子物体加到 vpTransform 中
    vpTransform *pTrans =new   vpTransform();
    pTrans->setTranslate( 2366.000000 ，  2490.000000 ，  3.000000 );
    pTrans->setRotate( 0.000000 ，   0.000000 ，   0.000000 );
    pTrans->setScale( 1.000000 ，   1.000000 ，   1.000000 );

    pTrans->insert_child(pSGeom,pTrans->end_child());
    pTrans->ref();

    PublicMember::CTS_pObject_other=static_cast<vpObject *> (pTrans);
    PublicMember::CTS_pObject_other->ref();

    //5 vpTransform 作为孩子添加到场景中
    vpScene* Scene = (vpScene*)vpScene::find("myScene");
    Scene->addChild(pTrans);

    //设置观察者
    PublicMember::CTS_pObject_observer=vpObject::find("Hummer");
    PublicMember::CTS_pObject_observer->ref();
```

```
//设置窗体
vpWindow * vpWin= * vpWindow::begin();
vpWin->setParent(PublicMember::CTS_RunningWindow);
vpWin->setBorderEnable(false);
vpWin->setFullScreenEnable(true);

//设置键盘
vpWin->setInputEnable(true);
vpWin->setKeyboardFunc((vrWindow::KeyboardFunc)PublicMember::CTS_Keyboard,NULL);
vpWin->open();
::SetFocus(vpWin->getWindow());

//帧循环
while(vpKernel::instance()->beginFrame()!=0)
{
    vpKernel::instance()->endFrame();

    if(!PublicMember::CTS_continueRunVP)
    {
        vpKernel::instance()->endFrame();
        vpKernel::instance()->unconfigure();
        vp::shutdown();
        return 0;
    }
}
return 0;
}
```

图 7.2.10 VSG 绘制球体的主线程

运行主线程,将在汽车旁绘制一个球体,效果如图 7.2.11 所示。

图 7.2.11 VSG 绘制球体

7.2.6 字符输出

字符输出相对需要更多步骤,除了前面提到的绘制图形的步骤,另外主要包含以下步骤:
(1)定义字体。主要代码为:

vrFont* pFont2D = new vrFont2D("ariel", 18, 16);

（2）设置字符内容和字体，由 vrString 负责。主要代码为：

vrString* pvrBoxString = new vrString();
pvrBoxString->setString("This is a   string over box!");
pvrBoxString->setPosition(0.0, 0.0, 1.0);
pvrBoxString->setFont(pFont2D);

（3）用 vsString 包装 vrString。主要代码为：

vsString * pBoxString = new vsString();
pBoxString->setString(pvrBoxString);
pBoxString->setState(pState);

（4）定义 vsTransform，并添加 vsString 对象为孩子。主要代码为：

vsTransform* pBoxTextOffset = new vsTransform();
pBoxTextOffset->setTranslate(0.0, 0.0f, 1.0f);
pBoxTextOffset->setRotate(0.0, -90.0, 0.0);
pBoxTextOffset->setScale(1.0f, 1.0f, 1.0f);
pBoxTextOffset->push_back_child(pBoxString);

（5）把 vsTransform 对象添加为绘制对象的孩子。主要代码为：

pSphGeom->push_back_child(pBoxTextOffset);

主线程中详细代码如图 7.2.12 所示。

```
//VP 运行主线程。
UINT PublicMember::CTS_RunBasicThread(LPVOID)
{
    //初始化
    vp::initialize(__argc,__argv);
    //定义场景
    PublicMember::CTS_Define();
    //绘制场景
    vpKernel::instance()->configure();

    //定义箱体
    vrBox* pBox= new vrBox( vuVec3f(1.0,1.0,0.0),vuVec3f(2.0,2.0,1.0) );
    pBox->setColor( 0.0,0.0,0.6f,0.5f );    //

    //定义状态
    vrState *pState = new vrState;
    vrAlphaBlend::Element alphaBlendElement;
    alphaBlendElement.m_enable = true;
    pState->setElement(vrAlphaBlend::Element::Id,&alphaBlendElement);

    //定义几何体
    vsGeometry* pSphGeom = new vsGeometry();
    pSphGeom->setGeometryBase(pBox);
    pSphGeom->setState( pState );
```

```cpp
//定义字体
vrFont* pFont2D = new vrFont2D( "ariel", 18, 16 );

//定义字体内容和使用字体
vrString* pvrBoxString = new vrString();
pvrBoxString->setString("This is a   string over box!");
pvrBoxString->setPosition( 0.0, 0.0, 1.0 );
pvrBoxString->setFont( pFont2D );

//使用 vsString 包装 vrString 对象
vsString * pBoxString = new vsString();
pBoxString->setString( pvrBoxString );
pBoxString->setState( pState );

//定义 vsTransform
vsTransform* pBoxTextOffset = new vsTransform();
pBoxTextOffset->setTranslate( 0.0, 0.0f, 1.0f );
pBoxTextOffset->setRotate( 0.0, -90.0, 0.0 );
pBoxTextOffset->setScale( 1.0f, 1.0f, 1.0f );

// vsTransform 对象添加 vsString 对象为孩子
pBoxTextOffset->push_back_child( pBoxString );

//几何体添加 vsTransform 为孩子
pSphGeom->push_back_child( pBoxTextOffset );

vpTransform *pTrans =new   vpTransform();
pTrans->setTranslate( 2366.000000 ,   2490.000000 ,   2.000000 );
pTrans->setRotate( 0.000000 ,   0.000000 ,   0.000000 );
pTrans->setScale( 1.000000 ,   1.000000 ,   1.000000 );

pTrans->insert_child(pSphGeom,pTrans->end_child());
pTrans->ref();

PublicMember::CTS_pObject_other=static_cast<vpObject *> (pTrans);
PublicMember::CTS_pObject_other->ref();

vpScene* Scene = (vpScene*)vpScene::find("myScene");
Scene->addChild(pTrans);

//设置观察者
PublicMember::CTS_pObject_observer=vpObject::find("Hummer");
PublicMember::CTS_pObject_observer->ref();

//设置窗体
vpWindow * vpWin= * vpWindow::begin();
```

```
vpWin->setParent(PublicMember::CTS_RunningWindow);
vpWin->setBorderEnable(false);
vpWin->setFullScreenEnable(true);

//设置键盘
vpWin->setInputEnable(true);
vpWin->setKeyboardFunc((vrWindow::KeyboardFunc)PublicMember::CTS_Keyboard,NULL);
vpWin->open();
::SetFocus(vpWin->getWindow());
//
    //帧循环
while(vpKernel::instance()->beginFrame()!=0)
{
    vpKernel::instance()->endFrame();
    //
    if(!PublicMember::CTS_continueRunVP)
    {
        vpKernel::instance()->endFrame();
        vpKernel::instance()->unconfigure();
        vp::shutdown();
        return 0;
    }
}
return 0;
}
```

图 7.2.12　字符输出设计

其效果如图 7.2.13 所示。

图 7.2.13　VSG 二维字符输出

7.3　图形纹理控制

在上一节中按照 VSG 绘制物体的 5 个步骤绘制的物体显得非常单调，仅仅简单地设置了颜色。本节将重点讨论 VSG 中的纹理设置。

需要设置图形纹理时，都是通过在状态类里设置元素值来完成的。首先需要找到纹理文件，纹理文件可以为 rgb、rgba、jpg 和 int 等格式。在目录 C:\Program Files\MultiGen- Paradigm\resources\data\databases\town 下可以找到 house1.rgb 文件和 house1.rgb.attr 文件，首先把这两个文件复制到工程所在的目录下。这样，就准备好了纹理文件。

接下来，重点处理怎么把纹理设置到绘制的图形上。纹理的设置，主要分为 3 个步骤：
首先，利用纹理工厂类 vrTextureFactory 对象加载纹理文件，其代码如下：

```
vrTextureFactory* pTexFactory = new vrTextureFactory();
vrTexture* houseTexture = pTexFactory->read("house1.rgb");
pTexFactory->unref();
```

接着，设置纹理元素属性，其代码如下：

```
vrTexture::Element textureElement;
textureElement.m_enable[0]  = true;
textureElement.m_texture[0] = houseTexture;
```

最后，利用状态对象，把纹理元素设置为状态对象的元素，其代码如下：

```
vrState* treeState = new vrState();
treeState->setElement( vrTexture::Element::Id, &textureElement);
```

其他代码与上一节绘制箱体基本相同。主线程中详细代码如图 7.3.1 所示，有关图形纹理的代码用黑体字体进行了标识。

```
UINT PublicMember::CTS_RunBasicThread(LPVOID)
{
    //初始化
    vp::initialize(__argc,__argv);

    //定义场景
    PublicMember::CTS_Define();

    //绘制场景
    vpKernel::instance()->configure();

    //加载纹理
    vrTextureFactory* pTexFactory = new vrTextureFactory();
    vrTexture* houseTexture = pTexFactory->read("house1.rgb");
    pTexFactory->unref();

    // 创建纹理元素和状态
    vrTexture::Element textureElement;
    textureElement.m_enable[0]  = true;
    textureElement.m_texture[0] = houseTexture;

    //1 状态设置
    vrState* treeState = new vrState();
```

```cpp
treeState->setElement( vrTexture::Element::Id, &textureElement);

//2 物体绘制
vrBox* pBox = new vrBox( vuVec3f(0.0,0.0,0.0),vuVec3f(18.0,18.0,18.0) );
pBox->setColor( 0.0,0.0,0.6f,0.5f );   //

//3 vsGeometry 包装 vrGeometry 对象
vsGeometry *pSGeom = new vsGeometry;
pSGeom->setGeometryBase(pBox);
pSGeom->setState(treeState);

//4 将 vsGeometry 作为子物体加到 vpTransform 中
vpTransform *pTrans =new   vpTransform();
pTrans->setTranslate( 2430.000000 , 2480.000000 , 18.000000 );
pTrans->setRotate( 0.000000 , 270.000000 , 0.000000 );
pTrans->setScale( 1.000000 , 1.000000 , 1.000000 );

pTrans->insert_child(pSGeom,pTrans->end_child());
pTrans->ref();

PublicMember::CTS_pObject_other=static_cast<vpObject *> (pTrans);
PublicMember::CTS_pObject_other->ref();
//5  vpTransform 作为孩子添加到场景中
vpScene* Scene = (vpScene*)vpScene::find("myScene");
Scene->addChild(pTrans);

//设置观察者
PublicMember::CTS_pObject_observer=vpObject::find("Hummer");
PublicMember::CTS_pObject_observer->ref();

//设置窗体
vpWindow * vpWin= * vpWindow::begin();
vpWin->setParent(PublicMember::CTS_RunningWindow);
vpWin->setBorderEnable(false);
vpWin->setFullScreenEnable(true);

//设置键盘
vpWin->setInputEnable(true);
vpWin->setKeyboardFunc((vrWindow::KeyboardFunc)PublicMember::CTS_Keyboard,NULL);
vpWin->open();
::SetFocus(vpWin->getWindow());

//帧循环
while(vpKernel::instance()->beginFrame()!=0)
{
    vpKernel::instance()->endFrame();
    if(!PublicMember::CTS_continueRunVP)
```

```
            {
                vpKernel::instance()->endFrame();
                vpKernel::instance()->unconfigure();
                vp::shutdown();
                return 0;
            }
        }
        return 0;
}
```

图 7.3.1 纹理控制代码

运行程序，其效果如图 7.3.2 所示。图中仅仅绘制了一个长宽高都为 8 m 的箱体，但使用纹理设置后完全是一个房屋的视觉效果。

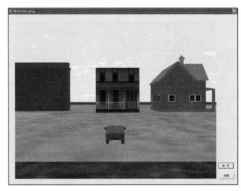

图 7.3.2 纹理效果图

7.4 图形材质控制

除了为图形设置纹理外，还可以为图形设置材质，使图形看起来更加形象逼真。本节将重点讨论 VSG 中的材质控制。

控制图形材质是通过在状态类里设置元素值来完成的。需要设置材质元素、光照模式元素和颜色轨迹元素，然后把这 3 个元素值赋予状态类。

首先，为材质元素做准备，并设置相关的值，其代码为：

  vrMaterial *material = new vrMaterial();
  material->setColor(vrMaterial::COLOR_AMBIENT, 0.2f, 0.0f, 0.0f, 1.0);
  material->setColor(vrMaterial::COLOR_DIFFUSE, 1.0, 0.0, 0.0, 0.0);
  material->setColor(vrMaterial::COLOR_SPECULAR, 1.0, 0.0, 0.0, 1.0);
  material->setColor(vrMaterial::COLOR_EMISSIVE, 0.0, 0.0, 0.0, 1.0);
  material->setSpecularExponent(50.0);

第 2 行代码设置其环绕颜色，第 3 行代码设置其漫反射颜色，第 4 行代码设置其镜面反射颜色，第 5 行代码设置其散发光颜色，第 6 行代码设置其镜面反射的尺寸和亮度。

其次，为光照模式元素做准备，并设置其相关的值，其代码为：

vrLightModel *lightModel = new vrLightModel();

lightModel->setLocalViewerEnable(false);

lightModel->setTwoSidedLightingEnable(false);

lightModel->setAmbientColor(0.2f, 0.2f, 0.2f, 1.0);

第 2 行代码设置其本地光无效，第 3 行代码设置其双面光照无效，第 4 行代码设置其漫反射颜色。

接着，分别建立材质元素、光照模式元素和颜色轨迹元素，并赋值，其代码如下：

//建立材质元素并赋值

vrMaterial::Element materialElement;

materialElement.m_side = vrMaterial::SIDE_FRONT_AND_BACK;

materialElement.m_material = material;

//建立光照模式元素并赋值

vrLightModel::Element lightModelElement;

lightModelElement.m_lightModel = lightModel;

//建立颜色轨迹元素并赋值

vrColorTrack::Element colorTrackElement;

colorTrackElement.m_enable = false;

最后，利用状态对象，分别把材质元素、光照模式元素和颜色轨迹元素赋值给状态对象的元素，其代码如下：

vrState* pNonTexState = new vrState();

pNonTexState->setElement(vrMaterial::Element::Id, &materialElement);

pNonTexState->setElement(vrLightModel::Element::Id, &lightModelElement);

pNonTexState->setElement(vrColorTrack::Element::Id, &colorTrackElement);

其他代码与绘制球体基本相同。主线程中详细代码如图 7.4.1 所示，有关图形材质的代码用黑体字体进行了标识。

```
UINT PublicMember::CTS_RunBasicThread(LPVOID)
{
    //初始化
    vp::initialize(__argc,__argv);

    //定义场景
    PublicMember::CTS_Define();

    //绘制场景
    vpKernel::instance()->configure();

    //为材质元素作准备
    vrMaterial *material = new vrMaterial();
      material->setColor(vrMaterial::COLOR_AMBIENT, 0.2f, 0.0f, 0.0f, 1.0);
      material->setColor(vrMaterial::COLOR_DIFFUSE, 1.0, 0.0, 0.0, 0.0);
```

```cpp
material->setColor(vrMaterial::COLOR_SPECULAR, 1.0, 0.0, 0.0, 1.0);
material->setColor(vrMaterial::COLOR_EMISSIVE, 0.0, 0.0, 0.0, 1.0);
material->setSpecularExponent(50.0);

//为光照模式元素作准备
vrLightModel *lightModel = new vrLightModel();
lightModel->setLocalViewerEnable(false);
lightModel->setTwoSidedLightingEnable(false);
lightModel->setAmbientColor(0.2f, 0.2f, 0.2f, 1.0);

//建立材质元素并赋值
vrMaterial::Element materialElement;
materialElement.m_side = vrMaterial::SIDE_FRONT_AND_BACK;
materialElement.m_material = material;

//建立光照模式元素并赋值
vrLightModel::Element lightModelElement;
lightModelElement.m_lightModel = lightModel;

//建立颜色轨迹元素并赋值
vrColorTrack::Element colorTrackElement;
colorTrackElement.m_enable = false;

//建立状态对象,并对其相关元素并赋值
vrState* pNonTexState = new vrState();
pNonTexState->setElement( vrMaterial::Element::Id, &materialElement );
pNonTexState->setElement(vrLightModel::Element::Id, &lightModelElement);
pNonTexState->setElement(vrColorTrack::Element::Id, &colorTrackElement);

//绘制球体
vrSphere* pSphere = new vrSphere( vuVec3f(0.0,0.0,0.0), 4.0,100);
pSphere->setColor( 0.0,1.0,0.0,1.0 );

vsGeometry *pSGeom = new vsGeometry;
pSGeom->setGeometryBase(pSphere);
pSGeom->setState(pNonTexState);

vpTransform *pTrans =new   vpTransform();
pTrans->setTranslate( 2430.000000 ,  2480.000000 ,  4.000000 );
pTrans->setRotate( 0.000000 ,  0.000000 ,  0.000000 );
pTrans->setScale( 1.000000 ,  1.000000 ,  1.000000 );

pTrans->insert_child(pSGeom,pTrans->end_child());
pTrans->ref();

PublicMember::CTS_pObject_other=static_cast<vpObject *> (pTrans);
PublicMember::CTS_pObject_other->ref();
```

```
vpScene* Scene = (vpScene*)vpScene::find("myScene");
Scene->addChild(pTrans);

//设置观察者
PublicMember::CTS_pObject_observer=vpObject::find("Hummer");
PublicMember::CTS_pObject_observer->ref();

//设置窗体
vpWindow * vpWin= * vpWindow::begin();
vpWin->setParent(PublicMember::CTS_RunningWindow);
vpWin->setBorderEnable(false);
vpWin->setFullScreenEnable(true);

//设置键盘
vpWin->setInputEnable(true);
vpWin->setKeyboardFunc((vrWindow::KeyboardFunc)PublicMember::CTS_Keyboard,NULL);
vpWin->open();
::SetFocus(vpWin->getWindow());

//帧循环
while(vpKernel::instance()->beginFrame()!=0)
{
    vpKernel::instance()->endFrame();
    if(!PublicMember::CTS_continueRunVP)
    {
        vpKernel::instance()->endFrame();
        vpKernel::instance()->unconfigure();
        vp::shutdown();
        return 0;
    }
}
return 0;
}
```

图 7.4.1　图形材质控制主线程

运行程序，其效果如图 7.4.2 所示。图中仅仅绘制了一个半径为 4 m 的球体，为图形设置材质后，这个红色的球体不再随着运动而变换颜色，而是具有了它自己的材料特性。

图 7.4.2　材质效果图

7.5 VSG 在场景中显示中文

本节要解决的问题是在三维可视化仿真中如何实时显示相关采集数据或者 VP 的三维场景中直接显示中文。解决的办法是：先把相关数据绘制到图片上，然后在场景中动态绘制图形并通过切换纹理，以达到实时显示相关信息的目的。如图 7.5.1 所示，在三维场景中绘制了一个白色正方形，并间隔一定的时间在其表面进行纹理贴图，图片采用实时动态绘制，不仅可以显示中文，也可以实时刷新随机数。

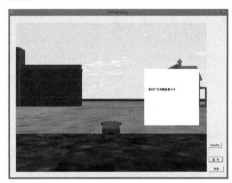

图 7.5.1　动态切换纹理显示实时信息效果图

整个设计分为两个步骤，第一个步骤是产生图片，第二个步骤是在场景中进行纹理切换控制。

7.5.1　图片产生类设计

动态产生图片的方法有很多种，读者可以采用自己最喜欢的一种方式进行。本书设计了一个图片绘制类 DrawHDCpicture，该类使用 Windows 中使用的 DC(绘制上下文)，绘制图片，然后把数据复制到 BITMAP 中，保存为位图。

图片绘制类 DrawHDCpicture 中，除了构造函数和析构函数外，还包含 5 个成员函数。如果对实现细节不感兴趣，直接调用即可。前 3 个函数是基本功能函数：CopyDCToBitmap 函数处理 DC 结构的相关数据，最后复制 DC 数据到 BITMAP 结构中；SaveBmp 函数则是处理 BITMAP 结构的相关数据，并最终保存成硬盘文件，本书把图片文件默认保存为 mm.bmp；OnbuildBitmap 函数则是建立一个 DC 结构，并把用户需要的信息绘制到 DC 结构中，然后调用前面 2 个函数，完成图片文件的保存。GetBitmap 函数是一个线程函数，按一定的时间间隔更新图片文件的内容。布尔变量 running 是控制线程函数的运行。

```
static BOOL SaveBmp(HBITMAP hBitmap, CString FileName);  //保存位图
static HBITMAP CopyDCToBitmap(HDC hScrDC, LPRECT lpRect);
                                                          //从 DC 复制到位图
static void OnbuildBitmap( );                             //绘制 DC
static UINT GetBitmap(LPVOID );                           //固定间隔产生图片线程函数
static bool running;                                      //线程控制变量
```

图片绘制类 DrawHDCpicture 的详细代码如图 7.5.2 所示。

```cpp
#pragma once

class DrawHDCpicture
{
public:
    DrawHDCpicture(void);
    ~DrawHDCpicture(void);

    static BOOL SaveBmp(HBITMAP hBitmap, CString FileName);          //保存位图
    static HBITMAP CopyDCToBitmap(HDC hScrDC, LPRECT lpRect);        //从 DC 复制到位图
    static void OnbuildBitmap( );                                    //绘制 DC
    static UINT GetBitmap(LPVOID );                                  //固定间隔产生图片线程函数
    static bool running;                                             //线程控制变量
};

#include "StdAfx.h"
#include ".\drawhdcpicture.h"

bool DrawHDCpicture::running=false;

DrawHDCpicture::DrawHDCpicture(void){}
DrawHDCpicture::~DrawHDCpicture(void){}
BOOL DrawHDCpicture::SaveBmp(HBITMAP hBitmap, CString FileName)
{
    HDC hDC;
    int iBits;                  //当前分辨率下每象素所占字节数
    WORD wBitCount;             //位图中每象素所占字节数
    //定义调色板大小，位图中像素字节大小，位图文件大小，写入文件字节数
    DWORD dwPaletteSize=0, dwBmBitsSize=0, dwDIBSize=0, dwWritten=0;

    BITMAP Bitmap;                      //位图属性结构
    BITMAPFILEHEADER bmfHdr;            //位图文件头结构
    BITMAPINFOHEADER bi;                //位图信息头结构
    LPBITMAPINFOHEADER lpbi;            //指向位图信息头结构
    HANDLE fh, hDib, hPal,hOldPal=NULL; //定义文件，分配内存句柄，调色板句柄

    //计算位图文件每个像素所占字节数
    hDC = CreateDC("DISPLAY", NULL, NULL, NULL);
    iBits = GetDeviceCaps(hDC, BITSPIXEL) * GetDeviceCaps(hDC, PLANES);
    DeleteDC(hDC);
    if (iBits <= 1) wBitCount = 1;
    else if (iBits <= 4) wBitCount = 4;
    else if (iBits <= 8) wBitCount = 8;
    else wBitCount = 24;

    GetObject( hBitmap, sizeof( Bitmap ), ( LPSTR )&Bitmap );
```

```
bi.biSize = sizeof( BITMAPINFOHEADER );
bi.biWidth = Bitmap.bmWidth;
bi.biHeight = Bitmap.bmHeight;
bi.biPlanes = 1;
bi.biBitCount = wBitCount;
bi.biCompression = BI_RGB;
bi.biSizeImage = 0;
bi.biXPelsPerMeter = 0;
bi.biYPelsPerMeter = 0;
bi.biClrImportant = 0;
bi.biClrUsed = 0;

dwBmBitsSize = ((Bitmap.bmWidth * wBitCount + 31) / 32) * 4 * Bitmap.bmHeight;

//为位图内容分配内存
hDib = GlobalAlloc(GHND,dwBmBitsSize + dwPaletteSize + sizeof(BITMAPINFOHEADER));
lpbi = (LPBITMAPINFOHEADER)GlobalLock(hDib);
*lpbi = bi;

// 处理调色板
hPal = GetStockObject(DEFAULT_PALETTE);
if (hPal)
{
    hDC = ::GetDC(NULL);
    //hDC = m_pDc->GetSafeHdc();
    hOldPal = ::SelectPalette(hDC, (HPALETTE)hPal, FALSE);
    RealizePalette(hDC);
}
// 获取该调色板下新的像素值
GetDIBits(hDC, hBitmap, 0, (UINT) Bitmap.bmHeight, (LPSTR)lpbi + sizeof(BITMAPINFOHEADER)
    +dwPaletteSize, (BITMAPINFO *)lpbi, DIB_RGB_COLORS);

//恢复调色板
if (hOldPal)
{
    ::SelectPalette(hDC, (HPALETTE)hOldPal, TRUE);
    RealizePalette(hDC);
    ::ReleaseDC(NULL, hDC);
}

//创建位图文件
fh = CreateFile(FileName, GENERIC_WRITE,0, NULL, CREATE_ALWAYS,
    FILE_ATTRIBUTE_NORMAL | FILE_FLAG_SEQUENTIAL_SCAN, NULL);

if (fh == INVALID_HANDLE_VALUE) return FALSE;

// 设置位图文件头
```

```
bmfHdr.bfType = 0x4D42; // "BM"
dwDIBSize = sizeof(BITMAPFILEHEADER) + sizeof(BITMAPINFOHEADER) + dwPaletteSize + dwBmBitsSize;
bmfHdr.bfSize = dwDIBSize;
bmfHdr.bfReserved1 = 0;
bmfHdr.bfReserved2 = 0;
bmfHdr.bfOffBits = (DWORD)sizeof(BITMAPFILEHEADER) + (DWORD)sizeof(BITMAPINFOHEADER) + dwPaletteSize;
// 写入位图文件头
WriteFile(fh, (LPSTR)&bmfHdr, sizeof(BITMAPFILEHEADER), &dwWritten, NULL);
// 写入位图文件其余内容
WriteFile(fh, (LPSTR)lpbi, dwDIBSize, &dwWritten, NULL);
//清除
GlobalUnlock(hDib);
GlobalFree(hDib);
CloseHandle(fh);
return TRUE;
}

HBITMAP DrawHDCpicture::CopyDCToBitmap(HDC hScrDC, LPRECT lpRect)
{
    HDC hMemDC;                       // 屏幕和内存设备描述表
    HBITMAP hBitmap,hOldBitmap;       // 位图句柄
    int nX, nY, nX2, nY2;             // 选定区域坐标
    int nWidth, nHeight;              // 位图宽度和高度

    if ( IsRectEmpty( lpRect ) ) return NULL; // 确保选定区域不为空矩形

    // 获得选定区域坐标
    nX=    lpRect->left;
    nY=    lpRect->top;
    nX2 = lpRect->right;
    nY2 = lpRect->bottom;
    nWidth  = nX2 - nX;
    nHeight = nY2 - nY;
    //为屏幕设备描述表创建兼容的内存设备描述表
    hMemDC= CreateCompatibleDC( hScrDC );
    // 创建一个与屏幕设备描述表兼容的位图
    hBitmap = CreateCompatibleBitmap( hScrDC, nWidth, nHeight );
    // 把新位图选到内存设备描述表中
    hOldBitmap = ( HBITMAP )SelectObject( hMemDC, hBitmap );
    // 把屏幕设备描述表复制到内存设备描述表中    BLACKNESS    WHITENESS    SRCCOPY
    StretchBlt( hMemDC, 0, 0, nWidth, nHeight, hScrDC, nX, nY, nWidth, nHeight, SRCCOPY );
    //得到屏幕位图的句柄
    hBitmap = ( HBITMAP )SelectObject( hMemDC, hOldBitmap );
    //清除
    DeleteDC( hMemDC );
    DeleteObject( hOldBitmap );
```

```
   //返回位图句柄
   return hBitmap;
}

void DrawHDCpicture::OnbuildBitmap( )
{
    int iwidth=400;
    int iheight=400;

    CDC *pdc=new CDC();
    pdc->m_hDC=::GetDC(::GetDesktopWindow());
//--------------------------------------------------
// 下面是创建兼容 DC 和兼容 DC 使用的 CBitmap，并规定兼容
// DC 的绘图绘制在创建的 CBitmap 上。
// 经过这一句：MenDC.SelectObject(&bm);以后，不管使用
// MenDC 绘制什么，实际上都是绘制在了 CBitmap bm;这个内
// 存位图上了。
//--------------------------------------------------
    CDC MenDC;
    CBitmap bm;
    MenDC.CreateCompatibleDC( pdc );
    bm.CreateCompatibleBitmap( pdc, iwidth, iheight ); //设定背景位图大小，最好是整个客户区大小
    MenDC.SelectObject( &bm );

//--------------------------------------------------
// 下面使用 MenDC 绘制你想要的任何东西，这里只添加了一个文本
//--------------------------------------------------
    CString a;//,
    a.Format("     我们产生的随机数:%d",::GetTickCount()%1000);//

    MenDC.SetBkColor(RGB(255, 255, 255));
    for(int i=0;i<iheight;i++)
        for(int k=0;k<iwidth;k++)
            MenDC.TextOut( i, k, "    ");

    MenDC.SetTextColor(RGB(0, 0,255));
    MenDC.TextOut( 10, 100, a);
//--------------------------------------------------
// 下面设定你要从 CBitmap bm;上截取哪一部分。
//--------------------------------------------------
    RECT rt;
    rt.left = 0; rt.right  = iwidth;
    rt.top  = 0; rt.bottom = iheight;
//--------------------------------------------------
    HBITMAP hBmp = CopyDCToBitmap( MenDC.GetSafeHdc(), &rt );
    SaveBmp( hBmp, "MM.bmp" );
//--------------------------------------------------
```

```
    bm.DeleteObject();
    MenDC.DeleteDC();
}

UINT DrawHDCpicture::GetBitmap(LPVOID )
{
    while(running)
    {
        OnbuildBitmap();
        PublicMember::DrawPicSpan=0;
        ::Sleep(1000);
    }
    return 0;
}
```

图 7.5.2　图形绘制类 DrawHDCpicture

OnbuildBitmap 函数中的绘制代码已经被设置为黑色倾斜格式，里面的内容可按照 DC 的绘制方法进行绘制。例如，可以使用 TextOut 绘制文本，可以用 LineTo 绘制线段等。当然，也可以设置相关的颜色信息。完全可以根据用户的需求进行设置。

GetBitmap 函数则是按一定的时间间隔调用 OnbuildBitmap 函数，产生新的图片，这里设置的时间间隔为 1 s。PublicMember::DrawPicSpan=0 则是相当于供 VP 的主线程进行通信的一个开关变量，该值为零，则表示 VP 主线程需要重新加载新的纹理图片文件，其他情况则不需要重新加载。

7.5.2　纹理切换控制设计

纹理切换控制的设计，主要包含两个步骤：第一个步骤是关于动态绘制图形并在图形上设置最新纹理图片，本书设计了函数 PublicMember::DrawBoard 来实现该功能；第二个步骤是在 VP 主线程中如何使用 PublicMember::DrawBoard 函数。

实现 PublicMember::DrawBoard 函数的步骤，与本书前面关于材质纹理操作的步骤大体类似。

需要设置图形纹理时，都是通过在状态类 vrState 里设置元素值来完成的。首先需要找到纹理文件，在图片产生类的成员函数 SaveBmp 中把图片文件默认保存为 mm.bmp，在这里就使用该图片文件。

首先，利用纹理工厂类 vrTextureFactory 对象加载纹理文件，其代码如下：

```
vrTextureFactory* pTexFactory = new vrTextureFactory();
vrTexture* treeTex = pTexFactory->read("mm.bmp");
pTexFactory->unref();
```

其次，设置纹理元素属性，其代码如下：

```
vrTexture::Element textureElement;
textureElement.m_enable[0]   = true;
textureElement.m_texture[0] =treeTex;
```

接着，设置与材质有关的元素，否则，新绘制的图形会在观察者移动中产生闪烁。

  vrState* treeState = new vrState();

  treeState->setElement(vrTexture::Element::Id, &textureElement);

  treeState->setElement(vrMaterial::Element::Id, &materialElement);

  treeState->setElement(vrLightModel::Element::Id, &lightModelElement);

  treeState->setElement(vrColorTrack::Element::Id, &colorTrackElement);

其他代码与绘制箱体基本相同，详细代码如图 7.5.3 所示。其中有一个静态成员变量 PublicMember::CTS_pObject_other，它指向绘制物体，后面有关绘制物体的操作都可以使用该指针变量进行操作。

```
void PublicMember::DrawBoard(void)
{
  vrTextureFactory* pTexFactory = new vrTextureFactory();
  // ----------------- Load the textures ----------------------------------
  vrTexture* treeTex = pTexFactory->read("mm.bmp");
  pTexFactory->unref();
  // -------Create a single texture element & corresponding vrState ---------
  vrTexture::Element textureElement;
  textureElement.m_enable[0]  = true;
  textureElement.m_texture[0] =treeTex;

    float r=1.0;      float g=1.0;       float b=1.0; //设置材质颜色

  //为材质元素作准备
  vrMaterial *material = new vrMaterial();
  material->setColor(vrMaterial::COLOR_AMBIENT,    r, g, b, 1.0);
  material->setColor(vrMaterial::COLOR_DIFFUSE,    r, g, b, 0.0);
  material->setColor(vrMaterial::COLOR_SPECULAR,   r, g, b, 1.0);
  material->setColor(vrMaterial::COLOR_EMISSIVE,   r, g, b, 1.0);
  material->setSpecularExponent(50.0);
  //为光照模式元素作准备
  vrLightModel *lightModel = new vrLightModel();
  lightModel->setLocalViewerEnable(true);
  lightModel->setTwoSidedLightingEnable(true);
  lightModel->setAmbientColor(0.2f, 0.2f, 0.2f, 1.0);
  //建立材质元素并赋值
  vrMaterial::Element materialElement;
  materialElement.m_side = vrMaterial::SIDE_FRONT_AND_BACK;
  materialElement.m_material = material;
  //建立光照模式元素并赋值
  vrLightModel::Element lightModelElement;
  lightModelElement.m_lightModel = lightModel;
  //建立颜色轨迹元素并赋值
  vrColorTrack::Element colorTrackElement;
  colorTrackElement.m_enable = false;

    //建立状态对象，并对其相关元素并赋值
```

```
vrState* treeState = new vrState();
treeState->setElement( vrTexture::Element::Id, &textureElement);
treeState->setElement( vrMaterial::Element::Id, &materialElement );
treeState->setElement(vrLightModel::Element::Id, &lightModelElement);
treeState->setElement(vrColorTrack::Element::Id, &colorTrackElement);

vrBox* pBox = new vrBox( vuVec3f(0.0,0.0,0.0),vuVec3f(0.01,10.0,10.0) );
pBox->setColor( 0.0,0.0,0.6f,0.5f );    //

vsGeometry *pSGeom = new vsGeometry;
pSGeom->setGeometryBase(pBox);
pSGeom->setState(treeState);

vpTransform *pTrans =new   vpTransform();
pTrans->setTranslate( 2360.000000 ,   2480.000000 ,   10.000000 );
pTrans->setRotate( 0.000000 ,   -90.000000 ,   0.000000 );
pTrans->setScale( 1.000000 ,   1.000000 ,   1.000000 );
pTrans->insert_child(pSGeom,pTrans->end_child());
pTrans->ref();

PublicMember::CTS_pObject_other=static_cast<vpObject *> (pTrans);
PublicMember::CTS_pObject_other->ref();

vpScene* Scene = (vpScene*)vpScene::find("myScene");
Scene->addChild(PublicMember::CTS_pObject_other);
}
```

图 7.5.3　动态绘制图形并控制纹理代码

使用 PublicMember::DrawBoard 函数绘制了一个长方体，并设置了相关纹理和材质，纹理图片文件使用的是默认图片文件 mm.bmp，同时用指针 PublicMember::CTS_pObject_other 指向了绘制的长方体。

绘制函数设计完成后，还需要在 VP 主线程中进行调用。其核心代码如下：

```
//帧循环
        while(vpKernel::instance()->beginFrame()!=0)
        {
            vpKernel::instance()->endFrame();
            if(PublicMember::DrawPicSpan<1)
            {
                vpScene *scene=*vpScene::begin();
                scene->erase_child(PublicMember::CTS_pObject_other);
                PublicMember::DrawBoard();
                PublicMember::DrawPicSpan++;
            }
        }
```

静态变量 PublicMember::DrawPicSpan 默认值为零，进入帧循环则先从场景中移除原有长方体，重新运行 PublicMember::DrawBoard 函数进行一次崭新的绘制；然后修改 PublicMember::DrawPicSpan 的值，以免每一帧都进行绘制。主线程中详细代码如图 7.5.4 所示，有关动态纹理切换的代码用黑体字体进行了标识。

```cpp
UINT PublicMember::CTS_RunBasicThread(LPVOID)
{
    //初始化
    vp::initialize(__argc,__argv);
    //定义场景
    PublicMember::CTS_Define();
    //绘制场景
    vpKernel::instance()->configure();
    //设置观察者
    PublicMember::CTS_pObject_observer=vpObject::find("Hummer");
    PublicMember::CTS_pObject_observer->ref();
    //设置窗体
    vpWindow * vpWin= * vpWindow::begin();
    vpWin->setParent(PublicMember::CTS_RunningWindow);
    vpWin->setBorderEnable(false);
    vpWin->setFullScreenEnable(true);

    //设置键盘
    vpWin->setInputEnable(true);
    vpWin->setKeyboardFunc((vrWindow::KeyboardFunc)PublicMember::CTS_Keyboard,NULL);
    vpWin->open();
    ::SetFocus(vpWin->getWindow());

    //帧循环
    while(vpKernel::instance()->beginFrame()!=0)
    {
        vpKernel::instance()->endFrame();
        if(PublicMember::DrawPicSpan<1)
        {
            vpScene *scene=*vpScene::begin();
            scene->erase_child(PublicMember::CTS_pObject_other);
            PublicMember::DrawBoard();
            PublicMember::DrawPicSpan++;
        }
        if(!PublicMember::CTS_continueRunVP)
        {
            vpKernel::instance()->endFrame();
            vpKernel::instance()->unconfigure();
            vp::shutdown();
            return 0;
        }
    }
```

```
    return 0;
}
```

图 7.5.4　动态纹理控制代码

VP 主线程中主要在于通过静态变量 PublicMember::DrawPicSpan 判断是否有新图片的产生，从而决定是否移除原有的图形绘制新图形。

新图片的产生，依赖于图片绘制线程的运行。将图片产生类成员函数 DrawHDCpicture::GetBitmap 作为线程运行函数，启动代码如图 7.5.5 所示。

```
void CVPTestDialogDlg::OnBnClickedGetPic()
{
    DrawHDCpicture::running=false;
    ::Sleep(2000);
    DrawHDCpicture::running=true;
    AfxBeginThread(DrawHDCpicture::GetBitmap,this);
}

UINT DrawHDCpicture::GetBitmap(LPVOID )
{
    while(running)
    {
        OnbuildBitmap();
        PublicMember::DrawPicSpan=0;
        ::Sleep(1000);
    }
    return 0;
}
```

图 7.5.5　动态图片绘制线程及其启动

在该线程函数中，每隔 1 s 产生一次新图片，同时修改 PublicMember::DrawPicSpan 的值为零，这样 VP 主线程将会移除原有绘制图形，重新进行绘制，从而实时显示最新的内容。其效果如图 7.5.6 所示。图中白色长方体上的数字将会每隔 1 s 进行更新变化。

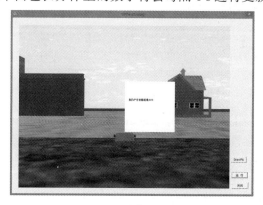

图 7.5.6　动态纹理显示

第 8 章　Vega Prime 和 OpenGL 混合编程

OpenGL 作为当前主流的图形 API 之一，它在一些场合具有比 DirectX 更优越的特性。OpenGL 作为 Vega Prime 的底层图形库，若能灵活应用，将会大大地提高开发能力。本章先充分认识 OpenGL 的强大功能，然后重点研究如何在 Vega Prime 中使用 OpenGL。本章内容主要包括 OpenGL 的特点、OpenGL 程序构成、OpenGL 绘制几何图形、OpenGL 颜色模式、OpenGL 视图变换、理解 Vega Prime 与 OpenGL、定义订阅类、Vega Prime 中使用 OpenGL 绘图等。整个过程充分展示了利用 OpenGL 在 Vega Prime 应用程序中绘制图形的方法和步骤。

【本章重点】

- OpenGL 的特点；
- OpenGL 程序构成；
- OpenGL 绘制几何图形；
- OpenGL 颜色模式；
- OpenGL 视图变换；
- 理解 Vega Prime 与 OpenGL；
- 定义订阅类；
- Vega Prime 中使用 OpenGL 绘图。

8.1　OpenGL 基础

8.1.1　OpenGL 的特点

1. 与 C 语言紧密结合

OpenGL 命令最初就是用 C 语言函数来进行描述的，对于学习过 C 语言的人来说，OpenGL 是容易理解和学习的。如果曾经接触过 TC 的 graphics.h，将会发现，使用 OpenGL 作图甚至比 TC 更加简单。

2. 强大的可移植性

微软的 Direct3D 虽然也是十分优秀的图形 API，但它只能用于 Windows 系统（现在还要加上一个 Xbox 游戏机）。而 OpenGL 不仅用于 Windows，还可以用于 Unix/Linux 等其他系统，它甚至在大型计算机、各种专业计算机（如医疗用显示设备）上都有应用。并且，OpenGL 的基本命令做到了与硬件无关，甚至是与平台无关。

3. 高性能的图形渲染

OpenGL 是一个工业标准，其技术紧跟时代，现今各个显卡厂家均对 OpenGL 提供了强力支持，使得 OpenGL 性能一直领先。

8.1.2 OpenGL 开发环境配置

OpenGL 官方网站为开发提供了最权威的资料。首先在 Visual C++ 下配置 OpenGL 开发环境，以便于能实际体验 OpenGL 的强大功能。

第一步：选择一个编译环境。

现在 Windows 系统的主流编译环境有 Visual Studio，Broland C++ Builder，Dev-C++等，它们均支持 OpenGL。本节选择 VC++ 作为学习 OpenGL 的环境。

第二步：安装 GLUT 工具包。

GLUT 不是 OpenGL 所必需的，但它会给学习带来一定方便，推荐开发者安装。Windows 环境下 GLUT（大小约为 150k）的下载地址为：http://www.opengl.org/resources/libraries/ glut/glutdlls37beta.zip。

Windows 环境下安装 GLUT 的步骤为：

（1）解压下载的压缩包将得到 5 个文件。

（2）以默认的安装目录为例：

① 将解压得到的 glut.h 放到 GL 文件夹里，没有 GL 文件夹可以自己创建一个，"c:\Program Files\Microsoft Visual Studio\VC98\include\GL" 文件夹。

② 将解压得到的 glut.lib 和 glut32.lib 文件放到静态函数库所在文件夹，即 "c:\Program Files\Microsoft Visual Studio\VC98\lib" 文件夹。

③ 将解压得到的 glut.dll 和 glut32.dll 文件放到操作系统目录下面的 system32 文件夹内，典型的位置为 "C:\Windows\System32"，这是非常重要的动态链接库设置。

第三步，创建工程。其步骤如下：

（1）创建一个 Win32 Console Application。

（2）链接 OpenGL libraries:在 Visual C++中先单击 Project，再单击 Settings，再找到 Link 单击，最后在 Object/library modules 的最前面加上 opengl32.lib Glut32.lib Glaux.lib glu32.lib 。

（3）单击 Project Settings 中的 C/C++标签，将 Preprocessor definitions 中的_CONSOLE 改为__WINDOWS，最后单击 OK。

现在，准备工作已经基本上完成，可不要轻视这一步，如果没有设置好，在编译及运行过程中总是会出错。

8.1.3 OpenGL 程序构成

紧接着上一节的设置，现在就可以编写一个简单的 OpenGL 程序，其代码如图 8.1.1 所示。

```
#include <GL/glu.h>
#include <GL/gl.h>
#include <GL/glut.h>
#include <GL/glaux.h>
void myDisplay(void)
{
    glClear(GL_COLOR_BUFFER_BIT);
    glRectf(-0.5f, -0.5f, 0.5f, 0.5f);
    glFlush();
}
int main(int argc, char *argv[])
{
    glutInit(&argc, argv);
    glutInitDisplayMode(GLUT_RGB | GLUT_SINGLE);
    glutInitWindowPosition(100, 100);
    glutInitWindowSize(400, 400);
    glutCreateWindow("第一个 OpenGL 程序");
    glutDisplayFunc(&myDisplay);
    glutMainLoop();
    return 0;
}
```

图 8.1.1　简单的 OpenGL 程序

该程序的作用是在一个黑色的窗口中央画一个白色的矩形。下面对各行语句进行说明。首先，OpenGL 程序一般还要包含<GL/gl.h>和<GL/glu.h>，还有 <GL/glut.h>，这是 GLUT 的头文件。然后是 main 函数，int main(int argc, char *argv[])，这个是带命令行参数的 main 函数。注意 main 函数中的各语句，除了最后的 return 之外，其余全部以 glut 开头。这种以 glut 开头的函数都是 GLUT 工具包所提供的函数，下面对用到的几个函数进行介绍。

（1）glutInit：对 GLUT 进行初始化，这个函数必须在其他的 GLUT 使用之前调用一次。其格式较死板，一般照抄这句 glutInit(&argc, argv)即可。

（2）glutInitDisplayMode：设置显示方式，其中 GLUT_RGB 表示使用 RGB 颜色，与之对应的还有 GLUT_INDEX（表示使用索引颜色）。GLUT_SINGLE 表示使用单缓冲，与之对应的还有 GLUT_DOUBLE（使用双缓冲）。

（3）glutInitWindowPosition：设置窗口在屏幕中的位置。

（4）glutInitWindowSize：设置窗口的大小。

（5）glutCreateWindow：根据前面设置的信息创建窗口。参数将被作为窗口的标题。注意：窗口被创建后，并不立即显示到屏幕上。需要调用 glutMainLoop 才能看到窗口。

（6）glutDisplayFunc：设置一个函数，当需要进行画图时，这个函数就会被调用。

（7）glutMainLoop：进行一个消息循环。可以简单理解为，这个函数可以显示窗口，并且等待窗口关闭后才会返回。在 glutDisplayFunc 函数中，设置了"当需要画图时，请调用 myDisplay 函数"，于是 myDisplay 函数就用来画图。

观察 myDisplay 中 3 个函数的调用，会发现它们都以 gl 开头。这种以 gl 开头的函数都是 OpenGL 的标准函数，下面对用到的函数进行介绍。

（1）glClear：清除。GL_COLOR_BUFFER_BIT 表示清除颜色，glClear 函数还可以清除其他的东西。

（2）glRectf：画一个矩形。4 个参数分别表示位于对角线上的两个点的横纵坐标。

（3）glFlush：保证前面的 OpenGL 命令立即执行，而不是让它们在缓冲区中等待。其作用跟 fflush(stdout)类似。

这样，就创建好了基本的 OpenGL 开发环境和框架。接下来，需要充分认识 OpenG，才能使用它开发出功能强大的程序。

8.1.4 OpenGL 绘制几何图形

在实际绘制之前，先熟悉一些概念。

1. 点、直线和多边形

数学中的几何学有点、直线和多边形的概念，但这些概念在计算机中会有所不同。数学上的点，只有位置，没有大小。但在计算机中，无论计算精度如何提高，始终不能表示一个无穷小的点。另一方面，无论图形输出设备（如显示器）的精度多高，始终不能输出一个无穷小的点。一般情况下，OpenGL 中的点将被画成单个的像素，虽然它可能足够小，但并不会是无穷小。同一像素上，OpenGL 可以绘制许多坐标只有稍微不同的点，但该像素的具体颜色将取决于 OpenGL 的实现。当然，大可不必花费过多的精力去研究"多个点如何画到同一像素上"。同样的，数学上的直线没有宽度，但 OpenGL 的直线则是有宽度的。同时，OpenGL 的直线必须是有限长度，而不是像数学概念那样是无限的。可以认为，OpenGL 的"直线"概念与数学上的"线段"非常接近，它可以由两个端点来确定。多边形是由多条线段首尾相连而形成的闭合区域。OpenGL 规定，一个多边形必须是一个"凸多边形"（其定义为：多边形内任意两点所确定的线段都在多边形内，由此也可以推导出，凸多边形不能是空心的）。多边形可以由其边的端点（这里可称为顶点）来确定。（注意：如果使用的多边形不是凸多边形，则最后输出的效果是未定义的（OpenGL 为了效率，放宽了检查，这可能导致显示错误）。要避免这个错误，尽量使用三角形，因为三角形都是凸多边形）可以想象，通过点、直线和多边形，就可以组合成各种几何图形。甚至，可以把一段弧看成是很多短的直线段相连，这些直线段足够短，以至于其长度小于一个像素的宽度。这样，弧和圆也可以表示出来了。通过位于不同平面的相连的小多边形，还可以组成一个"曲面"。

2. 在 OpenGL 中指定顶点

由以上的讨论可以知道，"点"是一切的基础。如何指定一个点呢？OpenGL 提供了一系列函数。它们都以 glVertex 开头，后面跟一个数字和 1~2 个字母。例如，glVertex2d、glVertex2f、glVertex3f、glVertex3fv 等。数字表示参数的个数，2 表示有 2 个参数，3 表示 3 个，4 表示 4 个。字母表示参数的类型，s 表示 16 位整数（OpenGL 中将这个类型定义为 GLshort），i 表示 32 位整数（OpenGL 中将这个类型定义为 GLint 和 GLsizei），f 表示 32 位浮点数（OpenGL 中将这个类型定义为 GLfloat 和 GLclampf），d 表示 64 位浮点数（OpenGL 中将这个类型定义为 GLdouble 和 GLclampd），v 表示传递的几个参数将使用指针的方式。这些函数除了参数的类型和个数不同外，功能都是相同的。例如，以下 5 个代码段的功能是等效的：

（1）glVertex2i(1, 3);

（2）glVertex2f(1.0f, 3.0f);

（3）glVertex3f(1.0f, 3.0f, 0.0f);

（4）glVertex4f(1.0f, 3.0f, 0.0f, 1.0f);

（5）GLfloat VertexArr3[] = {1.0f, 3.0f, 0.0f};
　　 glVertex3fv(VertexArr3);

以后将用 glVertex*来表示这一系列函数。注意：OpenGL 的很多函数都是采用这样的形式，一个相同的前缀再加上参数说明标记，这一点会随着学习的深入而有更多的体会。

3. 开始绘制

假设现在已经指定了若干顶点，那么 OpenGL 是如何知道想拿这些顶点来做什么呢？是一个一个的画出来，还是连成线？或者是构成一个多边形？或者是做其他什么事情？为了解决这一问题，OpenGL 要求：指定顶点的命令必须包含在 glBegin 函数之后，glEnd 函数之前（否则指定的顶点将被忽略），并由 glBegin 来指明如何使用这些点。

例如：

glBegin(GL_POINTS);

glVertex2f(0.0f, 0.0f);

glVertex2f(0.5f, 0.0f);glEnd();

则这两个点将分别被画出来。如果将 GL_POINTS 替换成 GL_LINES，则两个点将被认为是直线的两个端点，OpenGL 将会画出一条直线。还可以指定更多的顶点，然后画出更复杂的图形。另一方面，glBegin 支持的方式除了 GL_POINTS 和 GL_LINES，还有 GL_LINE_STRIP、GL_LINE_LOOP、GL_TRIANGLES、GL_TRIANGLE_STRIP、GL_TRIANGLE_FAN 等。

可以自己尝试改变 glBegin 的方式和顶点的位置，生成一些有趣的图案。程序代码：

void myDisplay(void)

{

　　glClear(GL_COLOR_BUFFER_BIT);

　　glBegin(/* 在这里填上所希望的模式 */);

　　 /* 在这里使用 glVertex*系列函数 */

　　 /* 指定所希望的顶点位置 */

　　glEnd();

　　glFlush();

}

把这段代码改成希望的样子，然后用它替换 myDisplay 函数，编译后即可运行。

【例】画一个圆。

正四边形、正五边形、正六边形、......，直到正 n 边形，当 n 越大时，这个图形就越接近圆，当 n 大到一定程度后，人眼将无法把它跟真正的圆相区别。这时已经成功的画出了一个"圆"（注：画圆的方法很多，这里使用的是比较简单，但效率较低的一种）试修改下面的 const int n 的值，观察当 n=3,4,5,8,10,15,20,30,50 等不同数值时输出的变化情况。将 GL_POLYGON

改为 GL_LINE_LOOP，GL_POINTS 等其他方式，观察输出的变化情况。

```
#include <math.h>
const int n = 20;
const GLfloat R = 0.5f;
const GLfloat Pi = 3.1415926536f;
void myDisplay(void)
{   int i;
    glClear(GL_COLOR_BUFFER_BIT);
    glBegin(GL_POLYGON);
    for(i=0; i<n; ++i)
    glVertex2f(R*cos(2*Pi/n*i), R*sin(2*Pi/n*i));
    glEnd();
    glFlush();
}
```

前面介绍了关于点、直线和多边形的概念及如何使用 OpenGL 来描述点，并使用点来描述几何图形。可以发挥自己的想象，画出各种几何图形，当然，也可以用 GL_LINE_STRIP 把很多位置相近的点连接起来，构成函数图像。如果有兴趣，也可以去找一些图像比较美观的函数，自己用 OpenGL 把它画出来。

虽然学习了如何绘制几何图形，但是多写几个程序就会发现其实还是存在不方便之处。例如：点太小，难以看清楚；直线也太细，不舒服；或者想画虚线，但不知道方法只能用许多短直线，甚至用点组合而成。这些问题将在后续章节解决。下面就点、直线、多边形分别讨论。

1. 关于点

点的大小默认为 1 个像素，可以改变，改变的命令为 glPointSize，其函数原型如下：

void glPointSize(GLfloat size);size 必须大于 0.0f，默认值为 1.0f，单位为"像素"。注意：对于具体的 OpenGL 实现，点的大小都是限度的，如果设置的 Size 超过最大值，则设置可能会有问题。

示例：

```
void myDisplay(void)
{
  glClear(GL_COLOR_BUFFER_BIT);
  glPointSize(5.0f);
  glBegin(GL_POINTS);
    glVertex2f(0.0f, 0.0f);
    glVertex2f(0.5f, 0.5f);
  glEnd();
  glFlush();
}
```

2. 关于直线

（1）直线可以指定宽度"void glLineWidth(GLfloat width);"，其用法与 glPointSize 类似。

（2）画虚线。首先，使用"glEnable(GL_LINE_STIPPLE);"来启动虚线模式，使用 glDisable(GL_LINE_STIPPLE)可以关闭。然后，使用 glLineStipple 来设置虚线的样式。"void glLineStipple(GLint factor, GLushort pattern);pattern"是由 1 和 0 组成的长度为 16 的序列，从最低位开始看，如果为 1，则直线上接下来应该画的 factor 个点将被画为实的；如果为 0，则直线上接下来应该画的 factor 个点将被画为虚的。

示例：

```
void myDisplay(void)
{
    glClear(GL_COLOR_BUFFER_BIT);
    glEnable(GL_LINE_STIPPLE);
    glLineStipple(2, 0x0F0F);
    glLineWidth(2.0f);
    glBegin(GL_LINES);
        glVertex2f(0.0f, 0.0f);
        glVertex2f(0.5f, 0.5f);
    glEnd();
    glFlush();
}
```

3. 关于多边形

（1）多边形的两面以及绘制方式。

虽然目前还没有真正的使用三维坐标来画图，但是建立一些三维的概念还是有必要的。从三维的角度来看，一个多边形具有两个面。每一个面都可以设置不同的绘制方式：填充、只绘制边缘轮廓线、只绘制顶点。其中"填充"是默认的方式。可以为两个面分别设置不同的方式。

设置正面为填充方式：

　　glPolygonMode(GL_FRONT, GL_FILL);

设置反面为边缘绘制方式：

　　glPolygonMode(GL_BACK, GL_LINE);

设置两面均为顶点绘制方式：

　　glPolygonMode(GL_FRONT_AND_BACK, GL_POINT);

（2）反转。

一般约定为"顶点以逆时针顺序出现在屏幕上的面"为"正面"，另一个面即为"反面"。生活中常见的物体表面，通常都可以用这样的"正面"和"反面"合理地被表现出来（找一个较透明的矿泉水瓶，在正对的一面沿逆时针画一个圆，并标明画的方向，然后将背面转为正面，画一个类似的圆，体会一下"正面"和"反面"。这时会发现正对你的方向，瓶的外侧是正面，而背对你的方向，瓶的内侧才是正面。正对你的内侧和背对你的外侧则是反面。这样一来，同样属于"瓶的外侧"这个表面，但某些地方算是正面，某些地方却算是反面了）。但也有一些表面比较特殊，如"麦比乌斯带"可以全部使用"正面"或全部使用"背面"来表示。可以通过 glFrontFace 函数来交换"正面"和"反面"的概念。

glFrontFace(GL_CCW)函数设置 CCW（CounterClockWise，逆时针）方向为"正面"。

glFrontFace(GL_CW)函数设置 CW（ClockWise，顺时针）方向为"正面"。

下面是一个示例程序，请用它替换 myDisplay 函数，并将 glFrontFace(GL_CCW)修改为 glFrontFace(GL_CW)，并观察结果的变化。

```
void myDisplay(void)
{
glClear(GL_COLOR_BUFFER_BIT);
glPolygonMode(GL_FRONT, GL_FILL);     // 设置正面为填充模式
glPolygonMode(GL_BACK, GL_LINE);  // 设置反面为线形模式
glFrontFace(GL_CCW);     // 设置逆时针方向为正面
glBegin(GL_POLYGON); // 按逆时针绘制一个正方形，在左下方
    glVertex2f(-0.5f, -0.5f);
    glVertex2f(0.0f, -0.5f);
    glVertex2f(0.0f, 0.0f);
    glVertex2f(-0.5f, 0.0f);
glEnd();
glBegin(GL_POLYGON); // 按顺时针绘制一个正方形，在右上方
    glVertex2f(0.0f, 0.0f);
    glVertex2f(0.0f, 0.5f);
    glVertex2f(0.5f, 0.5f);
    glVertex2f(0.5f, 0.0f);
glEnd();
glFlush();
}
```

（3）剔除多边形表面。

在三维空间中，一个多边形虽然有两个面，但无法看见背面的那些多边形，而一些多边形虽然是正面的，但被其他多边形所遮挡。如果将无法看见的多边形和可见的多边形同等对待，无疑会降低处理图形的效率。在这种时候，可以将不必要的面剔除。首先，使用 glEnable(GL_CULL_FACE)来启动剔除功能；使用 glDisable(GL_CULL_FACE)可以关闭该功能。然后，使用 glCullFace 来进行剔除。glCullFace 的参数可以是 GL_FRONT，GL_BACK 或者 GL_FRONT_AND_BACK，分别表示剔除正面、剔除反面、剔除正反两面的多边形。注意：剔除功能只影响多边形,而对点和直线无影响。例如,使用 glCullFace(GL_FRONT_AND_BACK)后，所有的多边形都将被剔除，所以看见的就只有点和直线。

（4）镂空。

多边形直线可以被画成虚线，而多边形则可以进行镂空。首先，使用 glEnable

(GL_POLYGON_STIPPLE)来启动镂空模式；使用 glDisable(GL_POLYGON_STIPPLE)可以关闭该模式。

然后，使用 glPolygonStipple 来设置镂空的样式。"void glPolygonStipple(const GLubyte *mask);"中的参数 mask 指向一个长度为 128 字节的空间，它表示了一个 32×32 的矩形应该如何镂空。其中：第一个字节表示了最左下方的从左到右（也可以是从右到左，这个可以修改）8 个像素是否镂空（1 表示不镂空，显示该像素；0 表示镂空，显示其后面的颜色），最后一个字节表示了最右上方的 8 个像素是否镂空。

这样一堆数据非常缺乏直观性，需要很费劲的去分析，才会发现它表示的竟然是一只苍蝇。如果将这样的数据保存成图片，并用专门的工具进行编辑，显然会方便很多。下面介绍如何做到这一点。首先，用 Windows 自带的画笔程序新建一张图片，取名为 mask.bmp，注意保存时应该选择"单色位图"。在"图象"→"属性"对话框中，设置图片的高度和宽度均为 32。用放大镜观察图片，并编辑。黑色对应二进制 0（镂空），白色对应二进制 1（不镂空），编辑完后保存。然后，就可以使用以下代码来获得这个 Mask 数组了。

```
static GLubyte Mask[128];
FILE *fp;
fp = fopen("mask.bmp", "rb");
if( !fp )      exit(0);
/* 移动文件指针到这个位置，使得再读 sizeof(Mask)个字节就会遇到文件结束。 注意 -(int)sizeof(Mask)虽然不是什么好的写法，但这里它确实是正确有效的。如果直接写 -sizeof(Mask)，由于 sizeof 取得的是一个无符号数，取负号会有问题*/
if( fseek(fp, -(int)sizeof(Mask), SEEK_END) )      exit(0);
// 读取 sizeof(Mask)个字节到 Mask
if( !fread(Mask, sizeof(Mask), 1, fp) )      exit(0);
fclose(fp);
```

现在请自己编辑一个图片作为 mask，并用上述方法取得 Mask 数组，运行后观察效果。说明：绘制虚线时可以设置 factor 因子，但多边形的镂空无法设置 factor 因子。请用鼠标改变窗口的大小，观察镂空效果的变化情况。

```
#include <stdio.h>#include <stdlib.h>void myDisplay(void)
{
  static GLubyte Mask[128];
  FILE *fp;        fp = fopen("mask.bmp", "rb");
  if( !fp )
     exit(0);
   if( fseek(fp, -(int)sizeof(Mask), SEEK_END) )
       exit(0);
  if( !fread(Mask, sizeof(Mask), 1, fp) )
     exit(0);
  fclose(fp);
  glClear(GL_COLOR_BUFFER_BIT);
```

```
glEnable(GL_POLYGON_STIPPLE);
glPolygonStipple(Mask);
glRectf(-0.5f, -0.5f, 0.0f, 0.0f);    // 在左下方绘制一个有镂空效果的正方形
glDisable(GL_POLYGON_STIPPLE);
glRectf(0.0f, 0.0f, 0.5f, 0.5f);      // 在右上方绘制一个无镂空效果的正方形
glFlush();}
```

学习了绘制几何图形的一些细节：点可以设置大小；直线可以设置宽度；可以将直线画成虚线；多边形的两个面的绘制方法可以分别设置；在三维空间中，不可见的多边形可以被剔除；可以将填充多边形绘制成镂空的样式。了解这些细节后，在一些图像绘制时会更加得心应手。另外，把一些数据写到程序之外的文件中，并用专门的工具编辑，有时会显得更方便。

8.1.5 OpenGL 的颜色模式

走出黑白的世界，接着学习的是颜色的选择。OpenGL 支持两种颜色模式：一种是 RGBA，一种是颜色索引模式。无论哪种颜色模式，计算机都必须为每一个像素保存一些数据。不同的是，RGBA 模式中，数据直接就代表了颜色；而颜色索引模式中，数据代表的是一个索引，要得到真正的颜色，还必须去查索引表。

1. RGBA 颜色

RGBA 模式中，每一个像素会保存以下数据：R 值（红色分量）、G 值（绿色分量）、B 值（蓝色分量）和 A 值（alpha 分量）。其中红、绿、蓝三种颜色相组合，就可以得到所需要的各种颜色，而 alpha 不直接影响颜色，它将留待以后介绍。在 RGBA 模式下选择颜色是十分简单的事情，只需要一个函数就可以完成。glColor*系列函数可以用于设置颜色，其中 3 个参数的版本可以指定 R、G、B 的值，而 A 值采用默认；4 个参数的版本可以分别指定 R、G、B、A 的值。

例如：
```
void glColor3f(GLfloat red, GLfloat green, GLfloat blue);
void glColor4f(GLfloat red, GLfloat green, GLfloat blue, GLfloat alpha);
```

将浮点数作为参数，其中 0.0 表示不使用该种颜色，而 1.0 表示将该种颜色用到最多。例如，"glColor3f(1.0f, 0.0f, 0.0f);"表示不使用绿、蓝色，而将红色使用最多，于是得到最纯净的红色。"glColor3f(0.0f, 1.0f, 1.0f);"表示使用绿、蓝色到最多，而不使用红色。混合的效果就是浅蓝色。"glColor3f(0.5f, 0.5f, 0.5f);"表示各种颜色使用一半，效果为灰色。注意：浮点数可以精确到小数点后若干位，这并不表示计算机就可以显示如此多种颜色。实际上，计算机可以显示的颜色种数将由硬件决定。如果 OpenGL 找不到精确的颜色，会进行类似"四舍五入"的处理。可以通过改变下面代码中 glColor3f 的参数值，绘制不同颜色的矩形。

```
void myDisplay(void)
{
  glClear(GL_COLOR_BUFFER_BIT);
  glColor3f(0.0f, 1.0f, 1.0f);
```

```
glRectf(-0.5f, -0.5f, 0.5f, 0.5f);
glFlush();
}
```

注意：glColor 系列函数，在参数类型不同时，表示"最大"颜色的值也不同。采用 f 和 d 做后缀的函数，以 1.0 表示最大的使用。采用 b 做后缀的函数，以 127 表示最大的使用。采用 ub 做后缀的函数，以 255 表示最大的使用。采用 s 做后缀的函数，以 32 767 表示最大的使用。采用 us 做后缀的函数，以 65 535 表示最大的使用。这些规则看似麻烦，但熟悉后实际使用中不会有什么障碍。

2. 索引颜色

在索引颜色模式中，OpenGL 需要一个颜色表。这个表就相当于画家的调色板，虽然可以调出很多种颜色，但同时存在于调色板上的颜色种数将不会超过调色板的格数。试将颜色表的每一项想象成调色板上的一个格子，它保存了一种颜色。在使用索引颜色模式画图时，说"把第 i 种颜色设置为某某"，其实就相当于将调色板的第 i 格调为某某颜色。"需要第 k 种颜色来画图"，那么就用画笔去蘸一下第 k 格调色板。颜色表的大小是很有限的，一般在 256~4 096，且总是 2 的整数次幂。在使用索引颜色方式进行绘图时，总是先设置颜色表，然后选择颜色。

使用 glIndex*系列函数可以在颜色表中选择颜色。其中最常用的可能是 glIndexi，它的参数是一个整形：

void glIndexi(GLint c);

OpenGL 并直接没有提供设置颜色表的方法，因此设置颜色表需要使用操作系统的支持。所用的 Windows 和其他大多数图形操作系统都具有这个功能，但所使用的函数却不相同。正如没有讲述如何自己写代码在 Windows 下建立一个窗口一样，这里也不会讲述如何在 Windows 下设置颜色表。GLUT 工具包提供了设置颜色表的函数 glutSetColor，但测试始终有问题。现在为了体验索引颜色，介绍另一个 OpenGL 工具包——aux。这个工具包是 Visual Studio 自带的，不必另外安装。这里仅仅是体验一下，不必深入学习。

```
#include <windows.h>
#include <GL/gl.h>
#include <GL/glaux.h>
#pragma comment (lib, "opengl32.lib")
#pragma comment (lib, "glaux.lib")
#include <math.h>
const GLdouble Pi = 3.1415926536;
void myDisplay(void)
{
    int i;      for(i=0; i<8; ++i)
    auxSetOneColor(i, (float)(i&0x04), (float)(i&0x02), (float)(i&0x01)); glShadeModel(GL_FLAT);
    glClear(GL_COLOR_BUFFER_BIT);
    glBegin(GL_TRIANGLE_FAN);
```

```
glVertex2f(0.0f, 0.0f);
for(i=0; i<=8; ++i)
{
glIndexi(i);
glVertex2f(cos(i*Pi/4), sin(i*Pi/4));
}
glEnd();
glFlush();
}
int main(void)
{
    auxInitDisplayMode(AUX_SINGLE|AUX_INDEX);
    auxInitPosition(0, 0, 400, 400);
    auxInitWindow(L"");
    myDisplay();
        Sleep(10 * 1000);
    return 0;
}
```

其他部分都可以不管，关注 myDisplay 函数即可。首先，使用 auxSetOneColor 设置颜色表中的一格，循环八次就可以设置八格。glShadeModel 在这里暂时不做介绍。然后在循环中用 glVertex 设置顶点，同时用 glIndexi 改变顶点代表的颜色，最终得到的效果是八个相同形状、不同颜色的三角形。索引颜色的主要优势是占用空间小（每个像素不必单独保存自己的颜色，只用很少的二进制位就可以代表其颜色在颜色表中的位置），花费系统资源少，图形运算速度快，但它编程稍稍显得不够方便，并且画面效果也会比 RGB 颜色差一些。

3. 指定清除屏幕用的颜色

"glClear(GL_COLOR_BUFFER_BIT);" 意思是把屏幕上的颜色清空。但实际上什么才叫"空"呢？在宇宙中，黑色代表了"空"；在一张白纸上，白色代表了"空"；在信封上，信封的颜色才是"空"。OpenGL 用下面的函数来定义清除屏幕后屏幕所拥有的颜色。在 RGB 模式下，使用 glClearColor 来指定"空"的颜色，它需要 4 个参数，其参数的意义跟 glColor4f 相似。在索引颜色模式下，使用 glClearIndex 来指定"空"的颜色所在的索引，它需要一个参数，其意义跟 glIndexi 相似。

```
void myDisplay(void)
{
    glClearColor(1.0f, 0.0f, 0.0f, 0.0f);
    glClear(GL_COLOR_BUFFER_BIT);
    glFlush();
}
```

4. 指定着色模型

OpenGL 允许为同一多边形的不同顶点指定不同的颜色。

例如：

```
#include <math.h>
const GLdouble Pi = 3.1415926536;
void myDisplay(void)
{
    int i;      // glShadeModel(GL_FLAT);
    glClear(GL_COLOR_BUFFER_BIT);
    glBegin(GL_TRIANGLE_FAN);
    glColor3f(1.0f, 1.0f, 1.0f);
    glVertex2f(0.0f, 0.0f);
    for(i=0; i<=8; ++i)
    {
        glColor3f(i&0x04, i&0x02, i&0x01);
        glVertex2f(cos(i*Pi/4), sin(i*Pi/4));
    }
glEnd();
glFlush();
}
```

在默认情况下，OpenGL 会计算两个顶点之间的其他点，并为它们填上"合适"的颜色，使相邻的点的颜色值都比较接近。如果使用的是 RGB 模式，看起来就具有渐变的效果。如果是使用颜色索引模式，则其相邻点的索引值是接近的，如果将颜色表中接近的项设置成接近的颜色，则看起来也是渐变的效果。但如果颜色表中接近的项颜色差距很大，则看起来可能是很奇怪的效果。使用 glShadeModel 函数可以关闭这种计算，如果顶点的颜色不同，则将顶点之间的其他点全部设置为与某一个点相同（直线以后指定的点的颜色为准，而多边形将以任意顶点的颜色为准，由实现决定）。为了避免这个不确定性，尽量在多边形中使用同一种颜色。

glShadeModel 的使用方法：

```
    glShadeModel(GL_SMOOTH);   // 平滑方式，这也是默认
    glShadeModel(GL_FLAT);     // 单色方式
```

学习了如何设置颜色。其中采用 RGB 颜色方式是目前计算机上的常用方式。它可以设置 glClear 清除后屏幕所剩的颜色，也可以设置颜色填充方式：平滑方式或单色方式。

8.1.6　OpenGL 视图变换

在绘制几何图形时，是否会觉得绘图的范围太狭隘了呢？坐标只能从-1 到 1，还只能是 X 轴向右，Y 轴向上，Z 轴垂直屏幕。这些限制给绘图带来了很多不便。生活在一个三维的世界——如果要观察一个物体，可以：

（1）从不同的位置去观察它（视图变换）。

（2）移动或者旋转它，当然，如果它只是计算机里面的物体，还可以放大或缩小（模型变换）。

（3）如果把物体画下来，可以选择：是否需要一种"近大远小"的透视效果。另外，可能只希望看到物体的一部分，而不是全部（剪裁）（投影变换）。

（4）可能希望把整个看到的图形画下来，但它只占据纸张的一部分，而不是全部（视口变换）。这些都可以在 OpenGL 中实现。OpenGL 变换实际上是通过矩阵乘法来实现。无论是移动、旋转还是缩放大小，都是通过在当前矩阵的基础上乘以一个新的矩阵来达到目的。OpenGL 可以在最底层直接操作矩阵，不过作为初学，这样做的意义并不大。

1. 模型变换和视图变换

从"相对移动"的观点来看，改变观察点的位置与方向和改变物体本身的位置与方向具有等效性。在 OpenGL 中，实现这两种功能甚至使用的是同样的函数。由于模型和视图的变换都通过矩阵运算来实现，在进行变换前，应先设置当前操作的矩阵为"模型视图矩阵"。设置的方法是以 GL_MODELVIEW 为参数调用 glMatrixMode 函数，例如，"glMatrixMode(GL_MODELVIEW);"。通常，需要在进行变换前把当前矩阵设置为单位矩阵。这只需要一行代码"glLoadIdentity();"，然后就可以进行模型变换和视图变换了。进行模型和视图变换，主要涉及 3 个函数：glTranslate*，把当前矩阵和一个表示移动物体的矩阵相乘，3 个参数分别表示了在 3 个坐标上的位移值；glRotate*，把当前矩阵和一个表示旋转物体的矩阵相乘，物体将绕着(0,0,0)到(x,y,z)的直线以逆时针旋转，参数 angle 表示旋转的角度；glScale*，把当前矩阵和一个表示缩放物体的矩阵相乘，x, y, z 分别表示在该方向上的缩放比例。注意都是说"与××相乘"，而不是直接说"这个函数就是旋转"或者"这个函数就是移动"。假设当前矩阵为单位矩阵，然后先乘以一个表示旋转的矩阵 R，再乘以一个表示移动的矩阵 T，最后得到的矩阵再乘上每一个顶点的坐标矩阵 v。所以，经过变换得到的顶点坐标就是((RT)v)。由于矩阵乘法的结合率，((RT)v) = (R(Tv))，换句话说，实际上是先进行移动，然后进行旋转，即实际变换的顺序与代码中写的顺序是相反的。由于"先移动后旋转"和"先旋转后移动"得到的结果很可能不同，初学的时候需要特别注意这一点。OpenGL 之所以这样设计，是为了得到更高的效率。但在绘制复杂的三维图形时，如果每次都去考虑如何把变换倒过来，也是很痛苦的事情。这里介绍另一种思路，可以让代码看起来更自然（写出的代码其实完全一样，只是考虑问题时用的方法不同了）。想象坐标并不是固定不变的。旋转的时候，坐标系统随着物体旋转。移动的时候，坐标系统随着物体移动。如此一来，就不需要考虑代码的顺序反转的问题了。以上都是针对改变物体的位置和方向来介绍的。如果要改变观察点的位置，除了配合使用 glRotate*和 glTranslate*函数外，还可以使用这个函数：gluLookAt。它的参数比较多，前 3 个参数表示了观察点的位置，中间 3 个参数表示了观察目标的位置，最后 3 个参数代表从(0,0,0)到 (x,y,z)的直线，它表示了观察者认为的"上"方向。

2. 投影变换

投影变换就是定义一个可视空间，可视空间以外的物体不会被绘制到屏幕上。注意，从现在起坐标可以不再是-1.0 到 1.0 了。OpenGL 支持两种类型的投影变换，即透视投影和正投影。投影也是使用矩阵来实现的。如果需要操作投影矩阵，需要以 GL_PROJECTION 为参数调用 glMatrixMode 函数 "glMatrixMode(GL_PROJECTION);"。通常，需要在进行变换前把当前矩阵设置为单位矩阵。"glLoadIdentity();"透视投影所产生的结果类似于照片，有近大远小的效果，如在火车头内向前拍摄一个铁轨的照片，两条铁轨似乎在远处相交了。使用 glFrustum 函数可以将当前的可视空间设置为透视投影空间，也可以使用更常用的 gluPerspective 函数。正投影相当于在无限远处观察得到的结果，它只是一种理想状态。但对于计算机来说，使用正投影有可能获得更好的运行速度。使用 glOrtho 函数可以将当前的可视空间设置为正投影空间。如果绘制的图形空间本身就是二维的，可以使用 gluOrtho2D。其使用类似于 glOrgho。

3. 视口变换

当一切工作已经就绪，只需要把像素绘制到屏幕上。这时候还剩最后一个问题：应该把像素绘制到窗口的哪个区域呢？通常情况下，默认是完整地填充整个窗口，但完全可以只填充一半（即把整个图像填充到一半的窗口内）。使用 glViewport 来定义视口，其中前两个参数定义了视口的左下角［(0,0) 表示最左下方］，后两个参数分别是宽度和高度。

4. 操作矩阵

可以把堆栈想象成一叠盘子。开始的时候一个盘子也没有，可以一个一个往上放，也可以一个一个取下来。每次取下的都是最后一次被放上去的盘子。通常，在计算机实现堆栈时，堆栈的容量是有限的，如果盘子过多，就会出错。当然，如果没有盘子了，再要求取一个盘子，也会出错。在进行矩阵操作时，有可能需要先保存某个矩阵，过一段时间再恢复它。当需要保存时，调用 glPushMatrix 函数，它相当于把矩阵（相当于盘子）放到堆栈上。当需要恢复最近一次的保存时，调用 glPopMatrix 函数，它相当于把矩阵从堆栈上取下。OpenGL 规定堆栈的容量至少可以容纳 32 个矩阵，某些 OpenGL 实现中堆栈的容量实际上超过了 32 个。因此，不必过于担心矩阵的容量问题。通常，用这种先保存后恢复的措施，比先变换再逆变换要更方便，更快速。注意：模型视图矩阵和投影矩阵都有相应的堆栈。可以使用 glMatrixMode 来指定当前操作的究竟是模型视图矩阵还是投影矩阵。

8.2 OpenGL 在 Vega Prime 应用中绘制图形

8.2.1 理解 Vega Prime 与 OpenGL 混合编程

要在 Vega Prime 中使用 OpenGL，必须要对 Vega Prime 和 OpenGL 混合编程有比较清楚的认识。

理解起来，其实 Vega Prime 和 OpenGL 的混合编程是很简单的，问题的关键是：
（1）对 Vega Prime 的层次结构和底层 OpenGL 接口的熟悉程度如何？
（2）对 OpenGL 的流程和机制的掌握程度如何？
（3）是不是真的需要在 Vega Prime 中使用 OpenGL？

现在，这 3 个问题已经不是问题了。Vega Prime 和 OpenGL 的混合编程只是一个经验的问题。第一个要解决的就是接口，也就是说 OpenGL 代码在哪里执行的问题。

第二个要解决的问题是两者之间的差异如何统一，也就是如何让 OpenGL 画的东西能在 Vega Prime 中正常实现。

第三个要解决的问题是如何保留和恢复 Vega Prime 的状态。

对于第一个问题，首先要明白，在 Vega Prime 中需要主线程开启的同时，还开启了其他一些线程，如 cull，draw 等。OpenGL 的代码只能在 draw 这个线程里执行，那么问题又来了，draw 线程的接口在哪里呢？这是问题的本质。draw 线程的接口只有一个，因为 Vega Prime 所有的事件跟 draw 有关的就是：EVENT_PRE_DRAW 和 EVENT_POST_DRAW，那就是 vpChannel 的 EVENT_POST_DRAW，在 vpChannel::subscriber 里写 OpenGL 代码就行了，这是第一个问题。

对于第二个问题，首先搞清楚 Vega Prime 和 OpenGL 的差异，观察方式不同，Vega Prime 默认视线是沿着 y 轴的，OpenGL 是沿 z 轴。还有就是，用 Vega Prime 中的视点（也就是 vpObserver 的 6 个自由度）的位置来更新 OpenGL 的视点和用 Vega Prime 的视锥体（也就是平头截体）更新 OpenGL 的视锥体，这两个信息可以通过 vpChannel（从 vrChannel 继承来的）的几个函数得到，具体代码在下一节演示。

对于第三个问题，可以直接用 vrDrawContext->pushElment 和 popElemt() 来解决。

8.2.2 定义订阅者类

OpenGL 的代码只能在 draw 这个线程里执行，draw 线程的接口只有一个，因为 Vega Prime 所有的事件跟 draw 有关的就是：EVENT_PRE_DRAW 和 EVENT_POST_DRAW，那就是 vpChannel 的 EVENT_POST_DRAW，在 vpChannel::subscriber 里写 OpenGL 代码。

所以，从 vpChannel::subscriber 类里继承订阅类类 Mysubscriber，进行 OpenGL 绘制。

关于头文件，除了引入 Vega Prime 相关的头文件外，还需要使用 OpenGL 相关的内容。具体代码如下：

```
#include <vsgu.h>                     //相关头文件
#include <vp.h>
#include <vpApp.h>
#include <vsChannel.h>
#include <GL/gl.h>
#include <GL/glu.h>
#pragma comment( lib, "opengl32.lib" )    //引入 OpenGL 库
#pragma comment( lib, "glu32.lib" )
```

关于定义的订阅类 Mysubscriber，它继承于 vpChannel::subscriber 类。除了重载 notify 函数之外，还定义了一个变量 m_objectPosition 来储存矩阵。其定义如下代码：

vsFieldFrameData< vuMatrix< double > >   m_objectPosition

并在构造函数中进行了初始化。

另外有两个 notify 函数，一个是：

virtual void notify(vsChannel::Event, const vsChannel *,vsTraversalCull *)
对于这个与裁剪相关的函数，并没有过多的功能代码。
另一个是：
void notify(vsChannel::Event event, const vsChannel *channel, vrDrawContext *context)
对于这个与绘制相关的函数则是应该重点研究的。
首先，对现有元素进行了压栈操作：

```
//压栈操作
vrElement::const_iterator_context it, ite = vrElement::end_context();
    for (it=vrElement::begin_context();it!=ite;++it) {
        if (*it == vrTransform::ElementProjection::Id)
            context->pushElement(*it, false);
        else context->pushElement(*it, true);
    }
```

接着，获取视点矩阵：

```
//获取视点矩阵
vuMatrix<double> modelViewMat = channel->getViewMatrixInverse();
modelViewMat *= channel->getOffsetMatrix();
```

然后，读取数据：

```
//读取数据
bool bMod;
const vuMatrix<double>* mat =   m_objectPosition.getReadBuffer(&bMod,
                                    vsThread::resolveFrameNumber());
if (mat == NULL)
{ return; }
```

接着，进行了视角矩阵转换：

```
        //视角矩阵转换
        modelViewMat.preMultiply(*mat);
        vrTransform::ElementModelView modelViewElement;
        vuMatrixTruncate(&modelViewElement.m_matrix, modelViewMat);
        context->setElement(vrTransform::ElementModelView::Id,&modelViewElement);
```

然后，利用 OpenGL 代码进行了绘制，包括线段、半球、圆柱体：

```
        //绘制线段
        vuVec3<float> origin(-2.0f, 0.0f, 3.0f);
        vuVec3<float>    axisX(2.0f, 0.0f, 3.0f);
        glLineWidth(3.0f);
        glColor3f(1.0f, 0.0f, 0.0f);
        glBegin(GL_LINES);
        glVertex3f(origin[0], origin[1], origin[2]);
        glVertex3f(axisX[0], axisX[1], axisX[2]);
```

```
                glEnd();
                glFlush();
```
最后，进行了出栈操作：
```
            //出栈操作
                context->popElements(false);
```
具体详细代码如图 8.2.1 所示。

```
#include <vsgu.h>
#include <vp.h>
#include <vpApp.h>
#include <vsChannel.h>
#include <GL/glaux.h>
#include <GL/gl.h>
#pragma comment( lib, "opengl32.lib" )
#pragma comment( lib, "glu32.lib" )
#include <GL/gl.h>
#include <GL/glu.h>

#pragma once

class Mysubscriber:public vsChannel::Subscriber
{
public:
    vsFieldFrameData< vuMatrix< double > >    m_objectPosition;
    Mysubscriber(void)
    {
        vuMatrix<double> objectPositionMatrix;
        m_objectPosition.init(&objectPositionMatrix);
    }
    ~Mysubscriber(void)
    {

    }

void notify(vsChannel::Event event, const vsChannel *channel, vrDrawContext *context)
{
        //压栈操作
        vrElement::const_iterator_context it, ite = vrElement::end_context();
        for (it=vrElement::begin_context();it!=ite;++it) {
            if (*it == vrTransform::ElementProjection::Id)
                context->pushElement(*it, false);
            else context->pushElement(*it, true);
        }

        //获取视点矩阵
        vuMatrix<double> modelViewMat = channel->getViewMatrixInverse();
```

```cpp
        modelViewMat *= channel->getOffsetMatrix();

    //读取数据
    bool bMod;
    const vuMatrix<double>* mat =   m_objectPosition.getReadBuffer(&bMod,
        vsThread::resolveFrameNumber());

    if (mat == NULL) {
         return;            }
    //视角矩阵转换
    modelViewMat.preMultiply(*mat);
    vrTransform::ElementModelView modelViewElement;
    vuMatrixTruncate(&modelViewElement.m_matrix, modelViewMat);
    context->setElement(vrTransform::ElementModelView::Id,
        &modelViewElement);

    //绘制线段
    vuVec3<float> origin(-2.0f, 0.0f, 3.0f);
     vuVec3<float>     axisX(2.0f, 0.0f, 3.0f);

    glLineWidth(3.0f);
    glColor3f(1.0f, 0.0f, 0.0f);

        glBegin(GL_LINES);
        glVertex3f(origin[0], origin[1], origin[2]);
        glVertex3f(axisX[0], axisX[1], axisX[2]);
        glEnd();
        glFlush();

    //绘制半球体
    /*
    GLUquadricObj *quadObj1;
    quadObj1 = gluNewQuadric();

    glPushMatrix();
    gluQuadricDrawStyle(quadObj1,GLU_SILHOUETTE);
    glTranslatef(-5.0,-1.0,0.0);
    gluSphere(quadObj1,3.0,20.0,20.0);
    glPopMatrix();

     gluDeleteQuadric(quadObj1);

    glFlush();
    */

    //绘制圆柱体
     /*
```

```
            GLUquadricObj *quadObj1;
            quadObj1 = gluNewQuadric();
            glPushMatrix();
            gluQuadricDrawStyle(quadObj1,GLU_FILL);
            gluQuadricNormals(quadObj1,GL_FLAT);
            gluQuadricOrientation(quadObj1,GLU_INSIDE);
            gluQuadricTexture(quadObj1,GL_TRUE);

            glColor3f(1.0,0.0,0.0);
            glRotatef(0,0.0,0.0,90.0);
            glTranslatef(-6.0, 0.0, 0.0);
            gluCylinder(quadObj1,1.0,1.0,4.00,200.0,800.0);
            glPopMatrix();

            gluDeleteQuadric(quadObj1);
            glFlush();
            */
            //出栈操作
            context->popElements(false);
      }
//
      virtual void notify(vsChannel::Event, const vsChannel *,vsTraversalCull *)
      {
      }
};
```

图 8.2.1 定义订阅类

8.2.3 Vega Prime 中使用 OpenGL

在 Vega Prime 的主线程中加入自定义的订阅者类，就能够进行 OpenGL 相关的绘制操作。在进入帧循环之前获取当前场景的通道，并为该通道添加绘制事件，同时初始化相关矩阵。其代码如下：

```
//获取当前场景的通道，并添加绘制事件到通道中
Mysubscriber * MyOpenGlDraw=new Mysubscriber();
vpChannel *chan = *vpChannel::begin();
chan->addSubscriber(vsChannel::EVENT_POST_DRAW, MyOpenGlDraw);
vuMatrix<double> objectPositionMatrix;
MyOpenGlDraw->m_objectPosition.init(&objectPositionMatrix);
```

在 Vega Prime 帧循环过程中，需要获取当前帧相关数据，操作后，把数据写回，其代码如下：

```
//获取当前帧相关数据，操作后，写回数据。
uint frame = vsThread::resolveFrameNumber();
vuMatrix<double>* mat= MyOpenGlDraw ->m_objectPosition.getWriteBuffer(frame);
```

```
          if (mat == NULL)
            {
            return 0;
            }
          *mat = PublicMember::CTS_pObject_observer->getMatrixAffine().getMatrix();
          MyOpenGlDraw->m_objectPosition.writeComplete(frame);
```
最后，如果 Vega Prime 帧循环需要退出，应该先移除添加的订阅者，然后清空订阅者对象指针，其代码如下：

```
          //移除订阅者并清空订阅者对象指针
          chan->removeSubscriber(vsChannel::EVENT_POST_DRAW, MyOpenGlDraw,false);
          MyOpenGlDraw =NULL;
```

Vega Prime 和 OpenGL 的混合编程的完整代码已经设计完毕，在主线程中的详细代码如图 8.2.2 所示，其关键代码已经用黑体标识。

```
#include ".\Mysubscriber.h"
//VP 运行主线程。
UINT PublicMember::CTS_RunBasicThread(LPVOID)
{
    //初始化
    vp::initialize(__argc,__argv);

    //定义场景
    PublicMember::CTS_Define();

    //绘制场景
    vpKernel::instance()->configure();
    //
    //设置观察者
    PublicMember::CTS_pObject_observer=vpObject::find("Hummer");
    PublicMember::CTS_pObject_observer->ref();

    //设置窗体
    vpWindow * vpWin= * vpWindow::begin();
    vpWin->setParent(PublicMember::CTS_RunningWindow);
    vpWin->setBorderEnable(false);
    vpWin->setFullScreenEnable(true);

    //设置键盘
    vpWin->setInputEnable(true);
    vpWin->setKeyboardFunc((vrWindow::KeyboardFunc)PublicMember::CTS_Keyboard,NULL);
    vpWin->open();
    ::SetFocus(vpWin->getWindow());

    //获取当前场景的通道，并添加绘制事件到通道中
    Mysubscriber * MyOpenGlDraw=new Mysubscriber();
```

```
vpChannel *chan = *vpChannel::begin();
chan->addSubscriber(vsChannel::EVENT_POST_DRAW, MyOpenGlDraw);
vuMatrix<double> objectPositionMatrix;
MyOpenGlDraw->m_objectPosition.init(&objectPositionMatrix);

//帧循环
while(vpKernel::instance()->beginFrame()!=0)
{
    vpKernel::instance()->endFrame();

    //获取当前帧相关数据，操作后，写回数据
    uint frame = vsThread::resolveFrameNumber();
    vuMatrix<double>* mat= MyOpenGlDraw ->m_objectPosition.getWriteBuffer(frame);

        if (mat == NULL) {    return 0;        }
        *mat = PublicMember::CTS_pObject_observer->getMatrixAffine().getMatrix();
        MyOpenGlDraw->m_objectPosition.writeComplete(frame);

    //
    if(!PublicMember::CTS_continueRunVP)
    {
        vpKernel::instance()->endFrame();

        //移除订阅者并清空订阅者对象指针
        chan->removeSubscriber(vsChannel::EVENT_POST_DRAW, MyOpenGlDraw,false);
        MyOpenGlDraw =NULL;

        vpKernel::instance()->unconfigure();
        vp::shutdown();
        return 0;
    }
}
return 0;
}
```

图 8.2.2 使用订阅类进行绘图

在 Mysubscriber 中，如果启用绘制线段的代码，将绘制出如图 8.2.3 所示的线段，其绘制代码如下：

```
//绘制线段
vuVec3<float> origin(-2.0f, 0.0f, 3.0f);
vuVec3<float>      axisX(2.0f, 0.0f, 3.0f);
glLineWidth(3.0f);                       //设置线段宽度为 3 个像素
glColor3f(1.0f, 0.0f, 0.0f);             //设置颜色为红色

glBegin(GL_LINES);
```

glVertex3f(origin[0], origin[1], origin[2]);
glVertex3f(axisX[0], axisX[1], axisX[2]);
glEnd();
glFlush();

图 8.2.3　OpenGL 绘制的线段

在 Mysubscriber 中，如果启用绘制半球体的代码，将绘制出如图 8.2.4 所示的半球体，其绘制代码如下：

//绘制半球体
GLUquadricObj *quadObj1;
quadObj1 = gluNewQuadric();
glPushMatrix();
gluQuadricDrawStyle(quadObj1,GLU_SILHOUETTE);
glTranslatef(-5.0,-1.0,0.0);
gluSphere(quadObj1,3.0,20.0,20.0);
glPopMatrix();
gluDeleteQuadric(quadObj1);
glFlush();

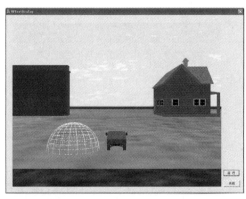

图 8.2.4　OpenGL 绘制的半球体

在 Mysubscriber 中，如果启用绘制圆柱体的代码，将绘制出如图 8.2.5 所示的圆柱体，其绘制代码如下：

//绘制圆柱体
GLUquadricObj *quadObj1;
quadObj1 = gluNewQuadric();
glPushMatrix();
gluQuadricDrawStyle(quadObj1,GLU_FILL);
gluQuadricNormals(quadObj1,GL_FLAT);
gluQuadricOrientation(quadObj1,GLU_INSIDE);
gluQuadricTexture(quadObj1,GL_TRUE);

glColor3f(1.0,0.0,0.0);
glRotatef(0,0.0,0.0,90.0);
glTranslatef(-6.0, 0.0, 0.0);
gluCylinder(quadObj1,1.0,1.0,4.00,200.0,800.0);
glPopMatrix();

gluDeleteQuadric(quadObj1);
glFlush();

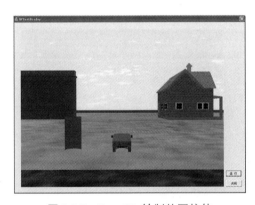

图 8.2.5 OpenGL 绘制的圆柱体

第 9 章 Vega Prime 中的实用功能实现

在实际的项目应用中，VP 提供的基本功能往往不能满足具体的需求。本章将根据实际项目需求，对 VP 的基本功能做进一步探讨，能够很好地满足实际需求。本章主要内容包括 Vega Prime 的重叠效果、自定义碰撞检测(OOBB)、窗口鼠标控制、对象的半透明处理和纹理运动仿真。整个过程充分展示了利用 Vega Prime 的强大功能，深入挖掘开发，在应用程序中可以更精确地满足实际需求。

【本章重点】

- Vega Prime 中的重叠效果；
- 自定义碰撞检测 OOBB；
- 窗口鼠标点选设计；
- 场景通道文字和图形显示；
- 纹理运动仿真；
- 仿真屏幕图片抓取；
- 仿真通道视频录制；
- 聚光灯光源使用；
- 渲染策略使用；
- 模板效果使用；
- 仿真辅助线程设计。

9.1 Vega Prime 中的重叠效果

重叠效果类继承图如图 9.1.1 所示。

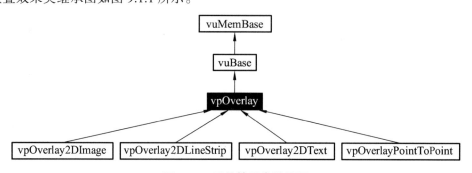

图 9.1.1 重叠效果类继承图

其中包括 4 种效果：点到点连线效果 vpOverlayPointToPoint,某个通道中一个对象和另外一个对象之间的连线；多个顶点连线效果 vpOverlay2DlineStrip,某个通道中用户自定义的多个顶点依次连线；二维文本效果 vpOverlay2DText,某个通道中二维文本输出；二维图片效果 vpOverlay2Dimage,某个通道中二维图片输出。

在 PublicMember::CTS_Define 函数中定义启用重叠效果，一定要在程序开始进行初始化，其代码如下：

//初始化重叠模块
vpModule::initializeModule("vpOverlay");

另外，还有 3 段通用代码：

设置重叠效果在有效和无效之间切换：

vpOverlay::const_iterator it = vpOverlay::begin();
vpOverlay::const_iterator ite = vpOverlay::end();
for(; it != ite; ++it)
    (*it)->setEnable(! (*it)->getEnable());

添加重叠效果到已定义的通道：

vpOverlay::const_iterator it = vpOverlay::begin();
vpOverlay::const_iterator ite = vpOverlay::end();
for(; it != ite; ++it)
    (*it)->addChannel(m_pChannel2);

从某个通道移除重叠效果：

vpOverlay::const_iterator it = vpOverlay::begin();
vpOverlay::const_iterator ite = vpOverlay::end();
for(; it != ite; ++it)
    (*it)->removeChannel(m_pChannel2);

9.1.1 点到点重叠效果

点到点连线效果 vpOverlayPointToPoint,某个通道中一个对象和另外一个对象之间的连线。这个效果的定义非常简单，首先引入头文件：

#include "vpOverlayPointToPoint.h"

其次，依次设置名字，设置使能值，设置线条颜色，设置是否显示距离值，设置是否闪烁。定义完成后，需要进行配置，包括添加通道、设置头对象和设置尾对象。

详细代码如图 9.1.2 所示。

```
//头文件
#include "vpOverlayPointToPoint.h"

//定义重叠效果
vpOverlayPointToPoint * pPoint = new vpOverlayPointToPoint();
pPoint->setName( "myOverlayPointToPoint" );
pPoint->setEnable( true );
```

```
  pPoint->setColor( 1.000000f,   0.000000f,  0.000000f,  1.000000f );
  pPoint->setDisplayDistanceEnable( true );
  pPoint->setBlinkingEnable( false );

  PublicMember::CTS_s_pInstancesToUnref->push_back( pPoint );

  //配置重叠效果
  pPoint->addChannel( pChannel_myChannel );
  pPoint->setHead(pObject_Hummer );
  pPoint->setTail(pObject_farmhouse   );
```

图 9.1.2 建立重叠效果

运行程序，其效果如图 9.1.3 所示，在汽车和房子之间多了一条红线，并且显示了两者之间的距离。

图 9.1.3 点到点重叠效果

9.1.2 闭环重叠效果

多个顶点连线效果 vpOverlay2DlineStrip，某个通道中用户自定义的多个顶点依次连线。这个效果的定义也很简单，首先引入头文件：

　　　#include "vpOverlay2DLineStrip.h"

其次，依次设置名字，设置使能值，设置线条颜色，定义顶点，设置是否闪烁。

定义完成后，需要进行配置，包括设置跟随目标和添加通道。

详细代码如图 9.1.2 所示。

```
//引入头文件
#include "vpOverlay2DLineStrip.h"

//定义重叠效果
vpOverlay2DLineStrip* pOverlay2DLineStrip = new vpOverlay2DLineStrip();
pOverlay2DLineStrip ->setName( "myOverlay2DLineStrip1" );
pOverlay2DLineStrip ->setEnable( true );
pOverlay2DLineStrip ->setColor( 1.000000f,   0.000000f,  0.000000f,  1.000000f );

//定义了五个顶点
pOverlay2DLineStrip ->addVertex( -0.100000f,   -0.100000f );
```

```
pOverlay2DLineStrip ->addVertex( -0.200000f,   0.200000f );
pOverlay2DLineStrip ->addVertex(  0.100000f,   0.100000f );
pOverlay2DLineStrip ->addVertex(  0.100000f,  -0.100000f );
pOverlay2DLineStrip ->addVertex( -0.100000f,  -0.100000f );
pOverlay2DLineStrip ->setBlinkingEnable( false );
PublicMember::CTS_s_pInstancesToUnref->push_back( pOverlay2DLineStrip );

//配置重叠效果
 pOverlay2DLineStrip ->setTrackingTarget( pObject_Hummer );
 pOverlay2DLineStrip ->addChannel( pChannel_myChannel );
```

图 9.1.4　建立闭环重叠效果

运行程序,其效果如图 9.1.5 所示,在汽车后面画了一个不规则的四边形,这个四边形将跟随汽车一起运动。

图 9.1.5　闭环重叠效果

9.1.3　二维字体重叠效果

二维文本效果 vpOverlay2DText,某个通道中二维文本输出。这个效果的定义也很简单,首先引入头文件:

　　#include "vpOverlay2DText.h"

其次,依次设置名字,设置使能值,设置文字内容,设置文字位置原点,设置颜色,设置字体,设置高度,设置是否斜体,设置是否闪烁。

定义完成后,需要进行配置,包括设置跟随目标和添加通道。

详细代码如图 9.1.6 所示。

```
//引入头文件
#include "vpOverlay2DText.h"

//定义重叠效果
vpOverlay2DText* pOverlay2DText = new vpOverlay2DText();
pOverlay2DText->setName( "myOverlay2DText2" );
pOverlay2DText->setEnable( true );
pOverlay2DText->setString( "This 2D text is tracking!" );
pOverlay2DText->setOrigin( 0.0000000f,   0.000000f );
```

```
pOverlay2DText->setColor( 1.000000f,  0.0f,  0.000000f,  1.000000f );
pOverlay2DText->setFontType( vpOverlay2DText::FONT_TYPE_ARIAL );
pOverlay2DText->setFontTypeHeight( 30 );
pOverlay2DText->setFontTypeItalic( false );
pOverlay2DText->setBlinkingEnable( false );

PublicMember::CTS_s_pInstancesToUnref->push_back( pOverlay2DText);

//配置重叠效果
pOverlay2DText->setTrackingTarget( pObject_Hummer );
pOverlay2DText->addChannel( pChannel_myChannel );
```

图 9.1.6　建立二维文字重叠效果

运行程序，其效果如图 9.1.7 所示，在汽车后面显示一个字符串，这个字符串将跟随汽车一起运动。

图 9.1.7　二维文字重叠效果

9.1.4　二维图片重叠效果

二维图片效果 vpOverlay2DImage,在某个通道中进行二维图片输出。这个效果的定义使用与前 3 个效果类似，只是需要一张图片。需要找到一个图片文件，在这里，把 C:\Program Files\MultiGen-Paradigm\resources\data\textures 目录下的 mpi_logo.rgba 文件复制到程序的当前目录中，其他步骤不变。

首先引入头文件：

```
#include "vpOverlay2DImage.h"
```

其次，依次设置名字，设置使能值，添加图片文件名称，设置显示时间，设置原点位置，设置缩放比例，设置淡入持续时间，设置淡出持续时间，设置是否闪烁。

定义完成后，需要进行配置，包括设置跟随目标和添加通道。

详细代码如图 9.1.8 所示。

```
//引入头文件
#include "vpOverlay2DImage.h"

//定义重叠效果
```

```
vpOverlay2DImage* pOverlay2DImage = new vpOverlay2DImage();
pOverlay2DImage ->setName( "myOverlay2DImage2" );
pOverlay2DImage ->setEnable( true );
pOverlay2DImage ->addImageFile( "mpi_logo.rgba" );
pOverlay2DImage ->setDisplayTime( 0.000000 );
pOverlay2DImage ->setOrigin( 0.000000f,  0.1000000f );
pOverlay2DImage ->setScale( 0.2500000f,  0.2500000f );
pOverlay2DImage ->setFadeInDuration( 0.000000 );
pOverlay2DImage ->setFadeOutDuration( 0.000000 );
pOverlay2DImage ->setBlinkingEnable( false );

PublicMember::CTS_s_pInstancesToUnref->push_back( pOverlay2DImage );

//配置重叠效果
pOverlay2DImage ->setTrackingTarget( pObject_Hummer );
pOverlay2DImage ->addChannel( pChannel_myChannel );
```

图 9.1.8　建立二维图片重叠

运行程序，其效果如图 9.1.9 所示，在汽车后面显示一个字符串，这个字符串将跟随汽车一起运动。

图 9.1.9　多通道效果

9.2　自定义碰撞检测类

9.2.1　Vega Prime 的碰撞检测

在很多可视化仿真程序中，碰撞检测是一种基本的要求。例如，飞行仿真中，飞行高度就是地面与飞机之间的垂直距离。Vega Prime 中的碰撞检测是基于线段的，基类 vpIsector 就负责维护和管理碰撞检测线段，同时也提供了一整套数据结构和方法来查询碰撞检测结果。为了找到哪一条线段产生了碰撞，场景图形不得不按照节点依次遍历查找，负责这个遍历功能的是 vsTraversalIsect 类，而不同的节点要求不同的碰撞检测程序。vpIsector 类提供了一种更高层次的 API 来配置碰撞检测和查询碰撞结果，但它是一个抽象类，只提供了是否开启碰撞检测、设置碰撞检测目标、设置碰撞检测位置和查询碰撞检测结果，它不提供配置检测线段。这样，就为开发者提供了开发自己的碰撞检测类的接口。

在 Vega Prime 帧循环过程中，应用程序将定位碰撞、触发碰撞遍历、查询碰撞检测结果。但是，在大型图形场景中，碰撞遍历可能会花费大量的时间，这有可能会影响系统性能。这样，一般不采用手动遍历的方法，而是采用碰撞检测服务类 vpIsectorService 的方法，让它负责更新碰撞检测。碰撞检测服务类被精心设计，它维护一个碰撞检测列表，这样可以实现优化遍历碰撞检测列表。

Vega Prime 提供了几个基本的碰撞检测类：

（1）VpIsectorHAT，Height Above Terrain (HAT)，它只维护一条线段，这条线段是物体与地面的垂直距离。

Segment 0: (x,y,z+zMax) to (x,y,z+zMin)可以通过 getHAT()来获取碰撞时的碰撞点、倾斜度和扭曲度。

（2）VpIsectorLOS，Line-Of-Sight (LOS)基于单线段的碰撞检测，这条线段沿着 Y 轴正向。

Segment 0: (0,minRange,0) to (0,maxRange,0)可以通过 getRange()查询碰撞距离。

（3）VpIsectorTripod，三角形碰撞检测，主要用于有方向和凹凸不平的地形。

设置三条线段：

    Segment 0: (-width/2.0,-length/2.0,zMax) to (-width/2.0,-length/2.0,zMin)

    Segment 1: (+width/2.0,-length/2.0,zMax) to (+width/2.0,-length/2.0,zMin)

    Segment 2: (0.0,+length/2.0,zMax) to (0.0,+length/2.0,zMin)

可以通过 getTripodZPR()来获取碰撞检测的相关信息。

（4）VpIsectorBump，采用 6 条线段作为碰撞检测线段，分别沿 XYZ 的正负方向。

    Segment 0: (0,0,0) to (-width, 0, 0)

    Segment 1: (0,0,0) to (+width, 0, 0)

    Segment 2: (0,0,0) to (0,-length, 0)

    Segment 3: (0,0,0) to (0,+length, 0)

    Segment 4: (0,0,0) to (0, 0,-height)

    Segment 5: (0,0,0) to (0, 0,+height)

可以通过 getBump()方法查询 6 条线段的长度。

（5）VpIsectorZ，用一条碰撞检测线段，计算碰撞点在 Z 轴上的值。

Segment 0: (x,y,z+zMax) to (x,y,z+zMin)可以通过 getZ()方法获取碰撞点及其倾斜度和扭曲度。

（6）VpIsectorZPR，基于单线段的碰撞检测。

Segment 0: (x,y,z+zMax) to (x,y,z+zMin)可以通过 getZPR()方法来获取碰撞检测的相关信息。

（7）VpIsectorXYZPR，基于单线段的碰撞检测。

Segment 0: (x,y,z+zMax) to (x,y,z+zMin)可以通过 getXYZPR()方法来获取碰撞检测的相关信息。

从这些已有的碰撞检测来看，都是基于简单的几条线段，很多时候无法满足实际需要。通过仔细分析碰撞检测的基类 vpIsector 得知，只要能获取足够多的顶点，由这些顶点连接成线段就可以组成类似的包围盒碰撞检测。整体思路非常简单：第一步就是获取物体表面的顶点；第二步继承碰撞基类 vpIsector，把顶点依次相连，第一个顶点连接第二个顶点，第二个顶点连接第三个顶点，依次往下，如果有 N 个顶点，就会有 N-1 条线段，这样基本可以把物体覆盖完整，形成自己的包围盒碰撞检测。

9.2.2 自定义查找物体顶点类

获取物体的顶点是自定义碰撞检测的第一步，依据物体自身的点进行连线，组成碰撞检测线段。

在这个物体顶点查找类 FindVertice 中，定义了 3 个成员变量和 3 个成员函数。同时必须包含 3 个头文件：

#include "vsGeometry.h"　：有关几何体的操作类；
#include "vuAllocArray.h"：有关数字分配操作的类；
#include "vuVec3.h"　　　：有关三元素向量的操作。

另外，定义了一个顶点数目的最大值限制。Vega Prime 中的检测都是基于线段的，而系统已有的碰撞检测只有几条线段，如果碰撞检测线段过于庞大，会严重影响应用程序的性能。这里，设置了一个上限 8 192，超过这个值就会提醒用户简化物体。

3 个成员变量分别为：

int　　　VerticeCount　　　　　　：物体顶点总数；
int　　　VerticeCurrentCount　　　：当前顶点总数；
vuVec3<float> * VerticesAll　　　　：顶点列表。

3 个成员函数分别为：

bool findVerticesNumber(vsNode * node)：递归统计顶点个数；
bool GetMemory()　　　　　　　　　：依据顶点个数，分配顶点列表；
bool findVertices(vsNode* node)　　　：递归获取顶点的值。

详细代码如图 9.2.1 和图 9.2.2 所示。

```
#pragma once
#include "vsGeometry.h"
#include "vuAllocArray.h"
#include "vuVec3.h"

#define PointMaxCount 8192

class FindVertice
{
public:
//
    int     VerticeCount;
    int     VerticeCurrentCount;
vuVec3<float> * VerticesAll;

    FindVertice(void);
    ~FindVertice(void);
```

```cpp
bool findVerticesNumber(vsNode * node);
bool GetMemory();
bool findVertices(vsNode* node);
};
```

图 9.2.1　获取顶点坐标类 FindVertice 的头文件

获取顶点坐标类 FindVertice 的 CPP 文件，具体代码如图 9.2.2 所示。

```cpp
#include "stdafx.h"
#include ".\FindVertice.h"

#pragma once

    //构造函数，初始化三个成员变量
    FindVertice ::FindVertice(void)
    {
        VerticesAll= vuAllocArray<vuVec3<float> >::malloc(1);
        VerticeCount=0;
        VerticeCurrentCount=0;
    }
    //析构函数
    FindVertice ::~FindVertice(void)
    {
    }
    //递归统计物体顶点数目
bool FindVertice ::findVerticesNumber(vsNode * node)
{
        size_t numChildren = node->size_child();
        for(unsigned int i = 0; i < numChildren; i++)
        {
           vsNode *childNode = *node->get_iterator_child(i);
           findVerticesNumber(childNode);
        }
        if(node->isOfClassType(vsGeometry::getStaticClassType()))
        {
           vsGeometry* geode = static_cast<vsGeometry*>(node);
           vrGeometry* geometry = geode->getGeometry();
           VerticeCount+=geometry->getNumVertices();
           return true;
        }
        return false;
}
//  动态分配数组
bool FindVertice ::GetMemory()
{
    VerticesAll= vuAllocArray<vuVec3<float> >::malloc(VerticeCount);
    if(VerticesAll==NULL)
```

```cpp
        return false;
    else
  return true;
}


//递归查找物体的顶点，并放入数组中
bool FindVertice ::findVertices(vsNode* node)
{
        size_t numChildren = node->size_child();
        for(unsigned int i = 0; i < numChildren; i++)
        {
            vsNode *childNode = *node->get_iterator_child(i);
            findVertices(childNode);
        }

        if(node->isOfClassType(vsGeometry::getStaticClassType()))
        {
          vsGeometry* geode = static_cast<vsGeometry*>(node);
          vrGeometry* geometry = geode->getGeometry();
          vuVec3<float>* vertices = geometry->getVertices();

          int number=geometry->getNumVertices();
          if( VerticeCurrentCount+number>PointMaxCount)
          {
              AfxMessageBox("物体顶点超过了 8192，请简化相关物体！");
              return false;
          }

          unsigned int i = 0;
         float x,y,z;
          for(i=0; i <number; i++,vertices++ )
            {
            vertices->get(&x,&y,&z);
            VerticesAll[VerticeCurrentCount+i].set(x,y,z);
            }
          VerticeCurrentCount+=number;
          return true;
        }
        return false;
}
```

图 9.2.2　获取顶点坐标类 FindVertice 的 CPP 文件

9.2.3　自定义碰撞检测类

自定义碰撞检测类 vpIsectorUserDefine，继承自碰撞检测基类 vpIsector。该类包含 2 个成员变量和 3 个成员函数。其中，2 个成员变量，VerticeCount 记录物体的顶点总数；VerticesAll

储存物体顶点数据。两个成员函数，SetPoint(int PointCount)负责实现动态分配顶点数组，设置碰撞检测线段索引和剪切点；updateSegments()负责设置具体的线段值，做法是把顶点列表里面的点依次相连，第一个点连接第二个点，第二个点连接第三个点，依次类推，有 N 个点则会形成 $N-1$ 条线段。

2 个成员变量是：

int   VerticeCount   ：物体顶点总数；

vuVec3<float> * VerticesAll：顶点列表。

3 个成员函数分别为：

void SetPoint(int PointCount)：实现动态分配顶点数组，并设置碰撞线段和剪切点，

virtual void updateSegments()：更新碰撞检测线段。

详细代码如图 9.2.3 和图 9.2.4 所示。

```cpp
#pragma once
#include "vpIsector.h"
#include "vuAllocArray.h"
#include "vuVec3.h"

class vpIsectorUserDefine:public vpIsector
{
public:

//物体顶点总数
int    VerticeCount;
//顶点列表
vuVec3<float> * VerticesAll;

//构造函数
vpIsectorUserDefine(void);
//析构函数
~vpIsectorUserDefine(void)

//实现动态分配顶点数组，并设置碰撞线段和剪切点
void SetPoint(int PointCount);

// vpIsector 类的虚构函数
virtual vpIsector *makeCopy(const vsgu::Options &options) const;

protected:

//更新碰撞检测线段
virtual void updateSegments();
private:
    VUBASE_HEADER_INCLUDES_COMPOSITE(vpIsectorUserDefine)
};
```

图 9.2.3　自定义碰撞检测类的头文件

```cpp
#include "StdAfx.h"
#include ".\vpisectoruserdefine.h"

VUBASE_SOURCE_INCLUDES_COMPOSITE(vpIsectorUserDefine, vpIsector,
    new vpIsectorUserDefine, true, true)

vpIsectorUserDefine::vpIsectorUserDefine(void)
{
       //   registerInstance(s_classType);
}

vpIsectorUserDefine::~vpIsectorUserDefine(void)
{
 // 从列表中移除类实例
    unregisterInstance(s_classType);
}

   // 实现复制函数

vpIsector * vpIsectorUserDefine::makeCopy(const vsgu::Options &options) const
{
    return vsgu::makeCopy(this, options);
}
   // 实现动态分配顶点数组,并设置碰撞线段和剪切点
void vpIsectorUserDefine::SetPoint(int PointCount)
   {
        VerticeCount=PointCount;

        //分配顶点数组
        VerticesAll= vuAllocArray<vuVec3<float> >::malloc(PointCount);

//产生碰撞检测线段索引
        vuSegment<double> segment;
        for (int i=0;i<PointCount;i++)
             m_vsIsector->push_back_segment(segment);

        // 设置剪切结束点
        m_vsIsector->setClip(vsIsector::CLIP_END);
   }

   //更新线段的值
void vpIsectorUserDefine::updateSegments()
{
    float x,y,z;
    int i=0;

    VerticesAll[0].get(&x,&y,&z);
```

```
    vuVec3<double> lfb(x,y,z);
    m_matrix.transformPoint(&lfb);

 for(i=1;i<VerticeCount;i++)
 {
     VerticesAll[i].get(&x,&y,&z);

     vuVec3<double> lfb2(x,y,z);
     m_matrix.transformPoint(&lfb2);
     m_vsIsector->setSegment(lfb, lfb2, i);

     lfb=lfb2;
 }
}
```

图 9.2.4　获取顶点坐标类的 CPP 文件

9.2.4　自定义碰撞检测类的使用

自定义碰撞检测类的使用，主要包括两个过程：第一步是获取物体的顶点数据，由自定义的查找物体顶点类 FindVertice 来完成；第二步是设置碰撞检测几何体，由自定义的碰撞检测类 vpIsectorUserDefine 完成。

使用第 6.1.2 节所示的主线程，进行自定碰撞检测类的使用，使房屋和汽车的碰撞检测不使用系统定义的碰撞检测类 vpIsectorBump，而是使用自定义的碰撞检测类 vpIsectorUserDefine。

在主线程 UINT PublicMember::CTS_RunBasicThread(LPVOID) 中，在进行场景配置 vpKernel::instance()->configure() 之后，首先获取产生碰撞的两个对象房屋 farmhouse 和汽车 hummer。

接着，定义了查找物体顶点对象 find，获取汽车 hummer 的顶点数目，然后根据实际顶点数目分配了顶点内存空间，利用 find 对象找到了汽车 hummer 的所有顶点值，并存入列表。其关键代码如下：

```
vpKernel::instance()->configure();
//----------------------------------------------------------------
//重新定义汽车与房屋的碰撞
//首先获取房屋和汽车对象，分别为 farmhouse 和 hummer
vpObject * farmhouse =vpObject::find( "farmhouse" );
farmhouse->ref();
vpObject * hummer =vpObject::find( "Hummer" );
hummer ->ref();
//定义查找对象 find ，获取汽车的顶点
FindVertice    find;
//获取汽车的顶点数目
if(hummer!= NULL)
```

```
            {
                vsNode *node = hummer->getRootNode();
                if(node)
                {
                    find.findVerticesNumber(node);
                }
            }
//依据获取的顶点数目,分配顶点内存空间
find.GetMemory ();
//获取汽车的每个顶点的坐标值
if(hummer != NULL)
    {
        vsNode *node =hummer ->getRootNode();
        if(node)
            {
                find.findVertices(node);
            }
    }
```

接着,定义了自定义的碰撞检测对象 m_isector,根据 find 的顶点数目,分配了顶点内存空间。然后把 find 的顶点值复制到碰撞检测对象 m_isector 的顶点列表中,设置了碰撞检测对象 m_isector 的使能属性和渲染使能属性,接着设置了碰撞的目标是房屋 farmhous,设置了碰撞的参考位置是汽车 hummer。最后把碰撞检测对象 m_isector 压入对象列表 PublicMember::CTS_s_pInstancesToUnref 中,同时设置为碰撞服务程序 pIsectorServiceInline_ myIsectorService 的孩子。具体代码如下:

```
//定义自定义的碰撞检测对象 m_isector
vpIsectorUserDefine *m_isector;
m_isector = new vpIsectorUserDefine();
//依据顶点数目,为碰撞对象分配内存空间
m_isector->SetPoint(find.VerticeCount);
//复制顶点数据
float x,y,z;
for(int i=0;i<m_isector->VerticeCount;i++)
    {
        find.VerticesAll[i].get(&x,&y,&z);
        m_isector->VerticesAll[i].set(x,y,z);
    }
//设置碰撞对象的相关属性
m_isector->setEnable(true);
```

```
m_isector->setRenderEnable(true);

//设置碰撞检测的目标为房屋，参考位置为汽车
m_isector->setTarget(farmhouse );
m_isector->setPositionReference( hummer);
m_isector->ref();
//把碰撞对象压入队列，并添加为碰撞服务程序的孩子
PublicMember::CTS_s_pInstancesToUnref->push_back( m_isector );
pIsectorServiceInline_myIsectorService->addIsector(m_isector);
//--------------------------------------------------------------------------------
```

设置完成后，在帧循环过程中检测是否有碰撞产生，有碰撞产生则弹出消息框提示发生了碰撞。代码如下：

```
if( m_isector->getHit()==true)
{
    AfxMessageBox("碰撞产生!");
}
```

完整的主线程代码如图 9.2.5 所示。

```
#include ".\vpisectoruserdefine.h"
#include ".\FindVertice.h"

UINT PublicMember::CTS_RunBasicThread(LPVOID)
{
    //初始化
    vp::initialize(__argc,__argv);

    //定义场景
    PublicMember::CTS_Define();

    //配置场景
    vpKernel::instance()->configure();

//--------------------------------------------------------------------
//重新定义汽车与房屋的碰撞
    //首先获取房屋和汽车对象，分别为 farmhouse 和 hummer
    vpObject * farmhouse =vpObject::find( "farmhouse" );
    farmhouse->ref();

    vpObject * hummer =vpObject::find( "Hummer" );
hummer ->ref();

    //定义查找对象 find ，获取汽车的顶点
    FindVertice  find;
```

```cpp
//获取汽车的顶点数目
  if(hummer!= NULL)
      {
         vsNode *node = hummer->getRootNode();
         if(node)
           {
             find.findVerticesNumber(node);
           }
      }

//依据获取的顶点数目，分配顶点内存空间
find.GetMemory ();

//获取汽车的每个顶点的坐标值
  if(hummer != NULL)
      {
         vsNode *node =hummer ->getRootNode();
         if(node)
            {
              find.findVertices(node);
            }
      }
//定义自定义的碰撞检测对象 m_isector
  vpIsectorUserDefine *m_isector;
  m_isector = new vpIsectorUserDefine();

//依据顶点数目，为碰撞对象分配内存空间
  m_isector->SetPoint(find.VerticeCount);

//复制顶点数据
  float x,y,z;
  for(int i=0;i<m_isector->VerticeCount;i++)
  {
      find.VerticesAll[i].get(&x,&y,&z);
      m_isector->VerticesAll[i].set(x,y,z);
  }
//设置碰撞对象的相关属性
  m_isector->setEnable(true);
  m_isector->setRenderEnable(true);

//设置碰撞检测的目标为房屋，参考位置为汽车
  m_isector->setTarget(farmhouse );
  m_isector->setPositionReference( hummer);

  m_isector->ref();

//把碰撞对象压入队列，并添加为碰撞服务程序的孩子
  PublicMember::CTS_s_pInstancesToUnref->push_back( m_isector );
  pIsectorServiceInline_myIsectorService->addIsector(m_isector);
```

```
//--------------------------------------------------------------------
    //设置窗体
    vpWindow * vpWin= * vpWindow::begin();
    vpWin->setParent(PublicMember::CTS_RunningWindow);
    vpWin->setBorderEnable(false);
    vpWin->setFullScreenEnable(true);

    //设置键盘
    vpWin->setInputEnable(true);
    vpWin->setKeyboardFunc((vrWindow::KeyboardFunc)PublicMember::CTS_Keyboard,NULL);

    vpWin->open();
    ::SetFocus(vpWin->getWindow());

    //帧循环
    while(vpKernel::instance()->beginFrame()!=0)
    {
        vpKernel::instance()->endFrame();

        if( m_isector->getHit()==true)
        {
            AfxMessageBox("碰撞产生！");
        }
        if(!PublicMember::CTS_continueRunVP)
        {
            vpKernel::instance()->endFrame();
            vpKernel::instance()->unconfigure();
            vp::shutdown();
            return 0;
        }
    }
    return 0;
}
```

图 9.2.5　在主线程中使用自定义碰撞检测类

运行程序 VPTestDialogOOBBsector，将出现如图 9.2.6 所示的碰撞包围盒，就克服了 Vega Prime 只能有单一线段产生的碰撞检测，利用物体自身的顶点产生了类似于包围盒的碰撞检测。

为了看得更加清楚，在碰撞检测对象 m_isector 从查找对象 find 复制顶点数据时，让所有顶点左移 4 个单位，其代码如下：

//复制顶点数据
float x,y,z;
for(int i=0;i<m_isector->VerticeCount;i++)

图 9.2.6　自定义碰撞对象产生的碰撞包围盒

```
    {
        find.VerticesAll[i].get(&x,&y,&z);
        m_isector->VerticesAll[i].set(x-4,y,z);
//原来为 m_isector->VerticesAll[i].set(x,y,z);
    }
```

这样再运行程序,就会产生如图 9.2.7 所示的情况,碰撞几何体位于汽车左面,其形状完全依据汽车本身的轮廓。与 Vega Prime 自带的碰撞检测类相比,碰撞检测有了很大的提高。

使用自定义的碰撞检测类有两个问题需要注意:

图 9.2.7　自定义碰撞对象产生的碰撞包围盒

第一,被查找顶点的物体在加载时,需要设置 vpObject 加载 setLoaderOption 选项,其中以下两个为必选项:LOADER_OPTION_PRESERVE_GENERIC_NAMES,LOADER_OPTION_ PRESERVE_GENERIC_NODES。在 LynX Prime 中,加载物体的高级属性里面,勾选上这两个选项即可。

在 Vega Prime 的代码加载中,需要使用下面两句代码实现:

```
pObject_Hummer->setLoaderOption( vsNodeLoader
            ::Data::LOADER_OPTION_PRESERVE_GENERIC_NAMES,  true );
pObject_Hummer->setLoaderOption( vsNodeLoader
            ::Data::LOADER_OPTION_PRESERVE_GENERIC_NODES,  true );
```

第二,被查找的对象不能有孩子结点,否则,查找到的顶点是这个物体对象最小的孩子。为了避免这个问题,需要在这个物体对象拥有孩子之前查找顶点。另外,被查找对象必须是某个物体对象的孩子,并且必须是通过物体对象指针成为其他物体的孩子,不能是通过被查找物体上的 DOF 成为其他物体的孩子。

9.3　窗口鼠标控制的完整实现

鼠标操作会给窗口程序带来很大方便。对于虚拟现实环境,能够利用鼠标对场景进行操作也会带来极大方便。

9.3.1　窗口鼠标函数的认识

Vega Prime 并没有完整地实现鼠标操作。但是它提供了良好的实现接口。首先,对于鼠标的认识需要普及一下。对于鼠标的按键,分为左键、中键和右键,这 3 个键的使用是通过判断该键是否被按下来决定的。对于鼠标的运动状态:中键是通过判断前后滚动的方向来决定的;左键和右键是通过平面坐标来判断的,其中就包含 X 值和 Y 值。对于左右键,X 值逐步变大,表示鼠标从左向右滑动;反之亦然,X 值逐步变小,表示鼠标从右向左滑动;Y 值逐步变大,表示鼠标从下向上滑动;反之亦然,Y 值逐步变小,表示鼠标从上向下滑动。中键前后滑动包含两种状态；左键按下，X 值变化有两种状态；左键按下，Y 值变化有两种状态，右键按下，X 值变化有两种状态；右键按下，Y 值变化有两种状态。总共有 8 种状态，基本能满足使用需求。

Vega Prime 的鼠标类包含设计与配置两个部分。设置部分,主要是利用重载通知函数 notify 来操作具体的对象。配置部分，主要是把设计的窗口鼠标类绑定到指定的 VP 窗口上。

9.3.2 窗口鼠标类的设计实现

对于窗口鼠标类的设计，采用了多继承的方式，继承的父类为下列 4 个：vpInputSource Boolean::Subscriber，vpInputSourceFloat::Subscriber，vpInputSourceInteger:: Subscriber，vsChannel::Subscriber。

重载实现的方法为以下两个： void notify(vpInputSourceBoolean::Event event, vpInputSourceBoolean *source)，该方法主要用于判断鼠标是否有键被按下；void notify (vpInputSourceInteger:: Event, vpInputSourceInteger *source)，该方法主要用于获取鼠标坐标值 x 和 y 的大小或鼠标中键滑动方向的变化。

另外,需要设计相关的变量协助处理。首先,设计了两个布尔变量:"bool LeftButtonDown;" "bool RightButtonDown;" 用于判断鼠标按下的是左键还是右键，初始值都为 false。其次，设计了两个整数变量："int ValX;""int ValY;"分别用于记录鼠标的坐标 x 值和 y 值，依据坐标值的大小变化，判断鼠标滑动的方向，初始值为 0。最后，设计了一个浮点变量："float myChannelAngle;"用于记录通道视角大小，初始值为 16.0，变化量是 0.5。另外，还涉及两个操作对象的变量，一个对象是 PublicMember::pObject_m1_tank，另一个对象时通道对象 PublicMember::myChannel。这两个对象都被设计为 PublicMember 的静态成员，相当于作为全局变量使用，具体代码如图 9.3.1 所示。

```
#pragma once
#include ".\publicmember.h"
#include "resource.h"

#include <vpInputMouse.h>

class WxpMouseInputScence: public vpInputSourceBoolean::Subscriber,
            public vpInputSourceFloat::Subscriber,
            public vpInputSourceInteger::Subscriber,
            public vsChannel::Subscriber
{
public:
    WxpMouseInputScence(void)
        {
LeftButtonDown=false;
        RightButtonDown=false;
ValX=0;
        ValY=0;
        myChannelAngle=16.5;
```

```cpp
}
    ~WxpMouseInputScence(void){};

private:
    bool LeftButtonDown;
    bool RightButtonDown;
int ValX;
    int ValY;
    float myChannelAngle;

    void notify(vsChannel::Event event, const vsChannel *channel, vrDrawContext *context) {     }

      virtual void notify(vsChannel::Event, const vsChannel *,vsTraversalCull *) {   }
    /**
     * notify method to catch boolean source subscriber events
        判断左右键中,具体是哪一个键被按下
     */
    void notify(vpInputSourceBoolean::Event event, vpInputSourceBoolean *source)
    {
        CString name=source->getName();
        int val=source->getValue();
            //判断左键是否被按下
        if(name=="vpInputMouse::SOURCE_BOOLEAN_BUTTON_LEFT")
        {
            if(val==1)
                LeftButtonDown=true;

            if(val==0)
                LeftButtonDown=false;
        }

         //判断右键是否被按下
        if(name=="vpInputMouse::SOURCE_BOOLEAN_BUTTON_RIGHT")
        {
            if(val==1)
                RightButtonDown=true;
            if(val==0)
                RightButtonDown=false;
        }
    }

    /**
     * notify method to catch float source subscriber events
     */
    void notify(vpInputSourceFloat::Event, vpInputSourceFloat *source){     }
    /**
     * notify method to catch integer source subscriber events
```

```cpp
    */
    void notify(vpInputSourceInteger::Event, vpInputSourceInteger *source)
{
    //获取键值名称,包含左键、右键、中键
    CString name=source->getName();
    int val=source->getValue();

        //按下左键控制 y z 面的位置
    if ( (name=="vpInputMouse::SOURCE_INTEGER_POSITION_X") &&    LeftButtonDown )
    {    //按下左键从左向右滑动
        if(val>ValX)
        {
            PublicMember::pObject_m1_tank->setTranslateX(-0.001,true);
        }

        //按下左键从右向左滑动
        if(val<ValX)
        {
            PublicMember::pObject_m1_tank->setTranslateX(0.001,true);
        }
        ValX=val;
    }

    if ( (name=="vpInputMouse::SOURCE_INTEGER_POSITION_Y") &&    LeftButtonDown )
    {    //按下左键从下向上滑动
        if(val>ValY)
        {
            PublicMember:: pObject_m1_tank->setTranslateZ(-0.001,true);
        }
        //按下左键从上向下滑动
        if(val<ValY)
        {
            PublicMember:: pObject_m1_tank->setTranslateZ(0.001,true);
        }
        ValY=val;
    }

        //按下右键控制 H P 方向的转动
    if ( (name=="vpInputMouse::SOURCE_INTEGER_POSITION_X") &&    RightButtonDown )
    {
            //按下右键从左向右滑动
        if(val>ValX)
        {
            PublicMember:: pObject_m1_tank->setRotateR(-1.0,true);
        }
            //按下右键从右向左滑动
        if(val<ValX)
```

```
                {
                        PublicMember:: pObject_m1_tank->setRotateR(1.0,true);
                }
                ValX=val;
        }

        //按下右键
        if ( (name=="vpInputMouse::SOURCE_INTEGER_POSITION_Y") &&   RightButtonDown )
        {
                //按下右键从下向上滑动
                if(val>ValY)
                {
                        PublicMember:: pObject_m1_tank->setRotateP(-1.0,true);
                }
                //按下右键从上向下滑动
                if(val<ValY)
                {
                        PublicMember:: pObject_m1_tank->setRotateP(1.0,true);
                }
                ValY=val;
        }
//中键滑动控制远近
  if  (name=="vpInputMouse::SOURCE_INTEGER_WHEEL_DIRECTION")
{
                //中键向前滑动，通道角度逐步变大，场景物体变小
                if(val==1)
                {
                        myChannelAngle=GasRelay_myChannelAngle+0.5;
PublicMember::myChannel->setFOVSymmetric( GasRelay_myChannelAngle ,   -1.000000f );
                }
                //中键向后滑动，通道角度逐步变小，场景物体变大
                if(val==-1)
                {
                        myChannelAngle=GasRelay_myChannelAngle-0.5;
PublicMember::myChannel->setFOVSymmetric( GasRelay_myChannelAngle ,   -1.000000f );
                }
}
 }
 //end of class
};
```

图 9.3.1 窗口鼠标类

9.3.3 窗口鼠标类的配置

窗口鼠标类的设计完成后，就是把自己设计的类用于指定的窗口，整个配置都在 VP 主线程中完成。首先实例化对象 "WxpMouseInputScence * wang=new WxpMouseInputScence();"，

然后才是具体的配置操作，具体代码如图 9.3.2 所示。

```cpp
#pragma once
#include ".\publicmember.h"
#include "resource.h"
#include <vpInputMouse.h>
#include ".\wxpmouseinputscence.h"
UINT PublicMember::CTS_RunBasicThread(LPVOID)
{
    //初始化
    vp::initialize(__argc,__argv);
    //定义场景
    PublicMember::CTS_Define();

    //绘制场景
    vpKernel::instance()->configure();
    //
    //设置观察者
    PublicMember::CTS_pObject_observer=vpObject::find("Hummer");
    PublicMember::CTS_pObject_observer->ref();

    //设置窗体
    vpWindow * vpWin= * vpWindow::begin();
    vpWin->setParent(PublicMember::CTS_RunningWindow);
    vpWin->setBorderEnable(false);
    vpWin->setFullScreenEnable(true);

    //设置键盘
    vpWin->setInputEnable(true);
    vpWin->setKeyboardFunc((vrWindow::KeyboardFunc)PublicMember::CTS_Keyboard,NULL);

    //设置鼠标===================================================
            WxpMouseInputScence * wang=new WxpMouseInputScence();
            vpWindow* window = vpWindow::empty() ? NULL : *vpWindow::begin();
            vpChannel* channel = vpChannel::empty() ? NULL : *vpChannel::begin();
            PublicMember::myChannel=channel;         //获取通道

            // 创建鼠标
            vpInputMouse *m_mouse;
            m_mouse = new vpInputMouse();
            m_mouse->setWindow(window);
            m_mouse->setChannel(channel);
            m_mouse->ref();

            // 创建订阅者，捕捉鼠标输入值的变化
            vpInputMouse::const_iterator_source_boolean bit;
            for (bit=m_mouse->begin_source_boolean();
                 bit!=m_mouse->end_source_boolean();++bit) {
```

```
            (*bit)->addSubscriber(
                vpInputSourceBoolean::EVENT_VALUE_CHANGED, wang);
        }

        vpInputMouse::const_iterator_source_integer iit;
        for (iit=m_mouse->begin_source_integer();
            iit!=m_mouse->end_source_integer();++iit) {
            (*iit)->addSubscriber(
                vpInputSourceInteger::EVENT_VALUE_CHANGED, wang);
        }
    vpWin->open();
    ::SetFocus(vpWin->getWindow());
//==============================鼠标配置结束==============================
vpWin->open();
::SetFocus(vpWin->getWindow());

//帧循环
while(vpKernel::instance()->beginFrame()!=0)
{
    vpKernel::instance()->endFrame();
    if(!PublicMember::CTS_continueRunVP)
    {
     vpKernel::instance()->endFrame();
     vpKernel::instance()->unconfigure();
     vp::shutdown();
        return 0;
     }
}
return 0;
}
```

图 9.3.2　配置使用窗口鼠标类

运行示例 VPTestDialogMouse，效果如图 9.3.3 所示。分别按下鼠标左键进行滑动，可以移动坦克位置；按下鼠标右键进行滑动，可以转动坦克；按下鼠标中键，前后滑动，可以拉近或推远场景，实现物体的放大缩小。

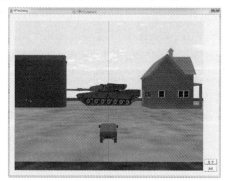
（a）正常场景

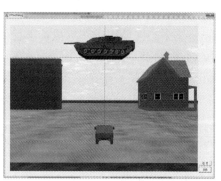
（b）左键上下滑动鼠标

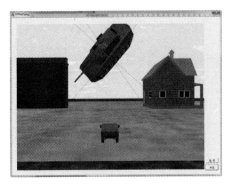

（c）右键旋转　　　　　　　　　　　　（d）中键滑动

图 9.3.3　窗口鼠标操作效果

9.4　场景通道屏幕文字和图形显示

物体在仿真场景中的位置坐标使用的都是全局坐标，包含（x，y，z），而场景通道屏幕坐标是二维的（x，y，0）。利用好这个映射转换，无须借助于第三方控件，就可以在场景通道屏幕上显示相关信息。

9.4.1　通道屏幕文字显示类的设计实现

全局坐标与场景通道屏幕坐标的转换类可以单独设计实现，也可以与窗口鼠标类合并实现。因为通道屏幕显示的数据往往与鼠标操作相关，本书采用了后者，与窗口鼠标类结合实现。因此，通道屏幕坐标转换类在完全利用窗口鼠标类的基础上扩展了部分功能而实现。

对于坐标转换类的设计，采用了多继承的方式，继承的父类为下列 4 个：vpInputSourceBoolean::Subscriber，vpInputSourceFloat::Subscriber，vpInputSourceInteger::Subscriber，vsChannel::Subscriber。

重载实现的方法有以下 3 个，其中前两个与鼠标类完全相同：

void notify(vpInputSourceBoolean::Event event, vpInputSourceBoolean *source)，该方法主要用于判断鼠标是否有键被按下。void notify(vpInputSourceInteger::Event, vpInputSourceInteger *source)，该方法主要用于获取鼠标坐标值 x 和 y 的大小或鼠标中键滑动方向的变化。void notify(vsChannel::Event event, const vsChannel *channel, vrDrawContext *context)，该方法是通道屏幕坐标转换类唯一增加的方法，该方法包含事件 event、通道 channel 和绘制上下文 context，可以完全实现通道屏幕的显示。在该方法中，主要涉及坐标转换和字体设置显示问题。

坐标转换，主要是利用通道内的目标对象 PublicMember::pObject_m1_tank 的三维坐标，使用 vuMatrixf 对象 world2screen 转换为通道屏幕坐标。转换后的通道屏幕坐标，以通道屏幕左下角为坐标原点，左下角坐标值为（0,0），左上角坐标为（0,1），右下角坐标为（1,0），右上角坐标为（1,1）。

字体设置显示，主要利用 vrFont2D 对象 m_font 实现。

字体设置的初始化代码：

　　m_font=new vrFont2D ("system", 25, 25, false);

第一个参数为字体类型，可以为系统支持的字体；第二个参数为字的高度，第三个参数为字的宽度；第四个参数控制字是否倾斜。遗憾的是，目前暂不支持中文。

字体显示的代码：

　　m_font->displayStringAt(context, m_text1.c_str(), color, 0.80f, 0.09f);

第一个参数为通道绘制上下文，第二个参数为字符串，第三个参数为 RGB 值的字体颜色，第四个参数和第五个参数分别为横坐标 x 和纵坐标 y。需要注意的是，坐标值的范围为[0~1]。

另外，为了使显示位置跟随目标对象一起移动，需要进行更多的转换：

　　m_font->displayStringAt(context, m_text2.c_str(), color, fx, fy+0.14);

其中的 fx 和 fy 就是依据目标对象的实际三维坐标转换而来的。

具体代码如图 9.4.1 所示。

```
void notify(vsChannel::Event event, const vsChannel *channel, vrDrawContext *context)
{
// get the world to screen matrix and viewport

    vrFont2D * m_font;
    vuString m_text1;
    vuString m_text2;
    vpObject *m_object=PublicMember:: pObject_m1_tank;
    int ox, oy, sx, sy;
    vuMatrixf world2screen;

    //控制字体大小           字体  ，高度，宽度，倾斜
    m_font=new vrFont2D ("system", 25,    25,   false);

    // set the default text
    m_text1.sprintf("object pos (0.0, 0.0, 0.0)");
    m_text2.sprintf("This text starts at the 2D point %d",::GetTickCount());

    channel->getVrChannel()->getViewport(&ox, &oy, &sx, &sy);
    world2screen = channel->getVrChannel()->getWorldToScreenMatrix();

    // Create a point in world space to transform
    //
    // Note: it would be better to use getAbsolutePosition() instead of
    // getTranslate() if you need to work with additional transforms in
    // the scene graph. Also, getting a position in the draw thread
    // may not be accurate if the object is moving.

    vuVec4f point;
point.set(m_object->getTranslateX(),m_object->getTranslateY(),m_ object->getTranslateZ(),1.0f);
```

```cpp
// update the text with the object position
m_text1.sprintf("object pos (%.1f, %.1f, %.1f)",point[0], point[1], point[2]);

// transform the 4d point into clip space
world2screen.transform(&point);

// divide by w
float fx = point[0]/point[3]; // x between [-1,1]
float fy = point[1]/point[3]; // y between [-1,1]
float fz = point[2]/point[3]; // z between [-1,1]

// normalize to [0,1] range
fx = (fx+1.0f)/2.0f;
fy = (fy+1.0f)/2.0f;
fz = (fz+1.0f)/2.0f;

// find the 2d screen location
int x_coord = (int)(ox + fx*sx + 0.5f);
int y_coord = (int)(oy + fy*sy + 0.5f);

// it is recommended that all modifications to state be done via
// vrDrawContext.   Although we're not preventing you from accessing
// OpenGL or Direct3D directly, it is much safer to go through state
// manager as it will guarantee that the software representation of
// the hardware graphics state maintained by vrDrawContext stays in sync
// w/ the hardware.   If this gets out of sync for some reason, visual
// anomalies can occur which can be very difficult to track down.   If
// you do need to drop down to the graphics library layer, you need to
// make sure that the graphics library hardware is in the same state
// when you leave the function as when you enter it.

// reset all of the graphics state elements to their default values.
// Note that this also resets the model view matrix to identity, so
// all vertex information, etc. will be specified relative to that

context->pushElements(true);

// need to disable the depth buffer since it's enabled by default
vrDepthTest::Element depthTestElement;
depthTestElement.m_enable = false;
context->setElement(vrDepthTest::Element::Id, &depthTestElement);

// set up an orthographic projection.   In this case we'll just map the
// channel viewport to 0 to 1 both horizontally and vertically.
vrTransform::ElementProjection projectionElement;

projectionElement.makeOrthographic(0.0f, 1.0f, 0.0f, 1.0f, -1.0f, 1.0f);
context->setElement(vrTransform::ElementProjection::Id,&projectionElement);
```

```
    // draw the text   字体颜色控制
    vuVec4<float> color(1.0f, 0.0f, 0.0f, 0.1f);
    vuVec4<float> color2(0.0f, 1.0f, 0.0f, 1.0f);

    //屏幕位置：x 位置，从左到右[0~1];y 位置，从下到上[0~1]。
    m_font->displayStringAt(context, m_text1.c_str(), color,   0.80f, 0.09f);

    // draw the following text only if the object is in the frustum
    if( (fz > 0.0f) && (fz < 1.0f) &&
                    (fx > 0.0f) && (fx < 1.0f) &&
                    (fy > 0.0f) && (fy < 1.0f))
                                                            //位置偏移   对象信息
       m_font->displayStringAt(context, m_text2.c_str(), color, fx, fy+0.14);

    // restore all of the graphics state elements to their previous value
    context->popElements();

}
```

图 9.4.1　通道屏幕坐标转换的函数实现

与窗口鼠标类相同，需要设计相关的变量协助处理。首先，设计了两个布尔变量："bool LeftButtonDown;""bool RightButtonDown;"。它们用于判断鼠标按下的是左键还是右键，初始值都为 false。其次，设计了两个整数变量："int ValX;""int ValY;"。它们分别用于记录鼠标的坐标 x 值和 y 值，依据坐标值的大小变化，判断鼠标滑动的方向，初始值为 0。最后，设计了一个浮点变量："float myChannelAngle;"，用于记录通道视角大小，初始值为 16.0，变化量是 0.5。另外，还涉及两个操作对象的变量，一个对象是 PublicMember::pObject_m1_tank，另一个对象是通道对象 PublicMember::myChannel。这两个对象都被设计为 PublicMember 的静态成员，相当于作为全局变量使用，具体代码如图 9.4.2 所示。

```
#pragma once
#include ".\publicmember.h"
#include "resource.h"

#include <vpInputMouse.h>

class WxpMouseInputScence: public vpInputSourceBoolean::Subscriber,
                public vpInputSourceFloat::Subscriber,
                public vpInputSourceInteger::Subscriber,
                public vsChannel::Subscriber
{
public:
    WxpMouseInputScence(void)
        {
            LeftButtonDown=false;
```

```cpp
                RightButtonDown=false;
                 ValX=0;
                 ValY=0;
                 myChannelAngle=16.5;
            }
    ~WxpMouseInputScence(void){};

private:
    bool LeftButtonDown;
    bool RightButtonDown;
    int ValX;
    int ValY;
    float myChannelAngle;

    void notify(vsChannel::Event event, const vsChannel *channel, vrDrawContext *context)
    {
// get the world to screen matrix and viewport

        vrFont2D * m_font;
        vuString m_text1;
        vuString m_text2;
        vpObject *m_object=PublicMember:: pObject_m1_tank;
        int ox, oy, sx, sy;
        vuMatrixf world2screen;

        //控制字体大小            字体    ,高度,宽度,倾斜
        m_font=new vrFont2D ("system", 25,    25,   false);

        // set the default text
        m_text1.sprintf("object pos (0.0, 0.0, 0.0)");
        m_text2.sprintf("This text starts at the 2D point %d",::GetTickCount());

        channel->getVrChannel()->getViewport(&ox, &oy, &sx, &sy);
        world2screen = channel->getVrChannel()->getWorldToScreenMatrix();

        // Create a point in world space to transform
        //
        // Note: it would be better to use getAbsolutePosition() instead of
        // getTranslate() if you need to work with additional transforms in
        // the scene graph. Also, getting a position in the draw thread
        // may not be accurate if the object is moving.

        vuVec4f point;
        point.set(m_object->getTranslateX(),m_object->getTranslateY(),m_object->getTranslateZ(),1.0f);

        // update the text with the object position
        m_text1.sprintf("object pos (%.1f, %.1f, %.1f)",point[0], point[1], point[2]);
```

```cpp
// transform the 4d point into clip space
world2screen.transform(&point);

// divide by w
float fx = point[0]/point[3]; // x between [-1,1]
float fy = point[1]/point[3]; // y between [-1,1]
float fz = point[2]/point[3]; // z between [-1,1]

// normalize to [0,1] range
fx = (fx+1.0f)/2.0f;
fy = (fy+1.0f)/2.0f;
fz = (fz+1.0f)/2.0f;

// find the 2d screen location
int x_coord = (int)(ox + fx*sx + 0.5f);
int y_coord = (int)(oy + fy*sy + 0.5f);

// it is recommended that all modifications to state be done via
// vrDrawContext.   Although we're not preventing you from accessing
// OpenGL or Direct3D directly, it is much safer to go through state
// manager as it will guarantee that the software representation of
// the hardware graphics state maintained by vrDrawContext stays in sync
// w/ the hardware.   If this gets out of sync for some reason, visual
// anomalies can occur which can be very difficult to track down.   If
// you do need to drop down to the graphics library layer, you need to
// make sure that the graphics library hardware is in the same state
// when you leave the function as when you enter it.

// reset all of the graphics state elements to their default values.
// Note that this also resets the model view matrix to identity, so
// all vertex information, etc. will be specified relative to that

context->pushElements(true);

// need to disable the depth buffer since it's enabled by default
vrDepthTest::Element depthTestElement;
depthTestElement.m_enable = false;
context->setElement(vrDepthTest::Element::Id, &depthTestElement);

// set up an orthographic projection.   In this case we'll just map the
// channel viewport to 0 to 1 both horizontally and vertically.
vrTransform::ElementProjection projectionElement;
projectionElement.makeOrthographic(0.0f, 1.0f, 0.0f, 1.0f, -1.0f, 1.0f);
context->setElement(vrTransform::ElementProjection::Id,&projectionElement);

// draw the text    字体颜色控制
vuVec4<float> color(1.0f, 0.0f, 0.0f, 0.1f);
vuVec4<float> color2(0.0f, 1.0f, 0.0f, 1.0f);
```

```cpp
            //屏幕位置：x 位置，从左到右[0~1];y 位置，从下到上[0~1]。
            m_font->displayStringAt(context, m_text1.c_str(), color,   0.80f, 0.09f);

        // draw the following text only if the object is in the frustum
        if( (fz > 0.0f) && (fz < 1.0f) &&
                    (fx > 0.0f) && (fx < 1.0f) &&
                    (fy > 0.0f) && (fy < 1.0f) )
                                                           //位置偏移，对象信息
            m_font->displayStringAt(context, m_text2.c_str(), color, fx, fy+0.14);

        // restore all of the graphics state elements to their previous value
        context->popElements();
    }

    virtual void notify(vsChannel::Event, const vsChannel *,vsTraversalCull *) {  }
    /**
     * notify method to catch boolean source subscriber events
       判断左右键中，具体是哪一个键被按下
     */
    void notify(vpInputSourceBoolean::Event event, vpInputSourceBoolean *source)
    {
        CString name=source->getName();
        int val=source->getValue();
            //判断左键是否被按下
        if(name=="vpInputMouse::SOURCE_BOOLEAN_BUTTON_LEFT")
        {
            if(val==1)
                LeftButtonDown=true;

            if(val==0)
                LeftButtonDown=false;
        }

            //判断右键是否被按下
        if(name=="vpInputMouse::SOURCE_BOOLEAN_BUTTON_RIGHT")
        {
            if(val==1)
                RightButtonDown=true;
            if(val==0)
                RightButtonDown=false;
        }
    }

    /**
     * notify method to catch float source subscriber events
     */
```

```cpp
                    void notify(vpInputSourceFloat::Event, vpInputSourceFloat *source){   }
    /**
     * notify method to catch integer source subscriber events
     */
    void notify(vpInputSourceInteger::Event, vpInputSourceInteger *source)
{
        //获取键值名称  包含左键  右键  中键
        CString name=source->getName();
        int val=source->getValue();

                //按下左键控制 y z面的位置
        if ( (name=="vpInputMouse::SOURCE_INTEGER_POSITION_X") &&   LeftButtonDown )
        {       //按下左键从左向右滑动
            if(val>ValX)
            {
                PublicMember::pObject_m1_tank->setTranslateX(-0.001,true);
            }

            //按下左键从右向左滑动
            if(val<ValX)
            {
                PublicMember::pObject_m1_tank->setTranslateX(0.001,true);
            }
            ValX=val;
        }


        if ( (name=="vpInputMouse::SOURCE_INTEGER_POSITION_Y") &&   LeftButtonDown )
        {   //按下左键从下向上滑动
            if(val>ValY)
            {
                PublicMember:: pObject_m1_tank->setTranslateZ(-0.001,true);
            }
//按下左键从上向下滑动
            if(val<ValY)
            {
                PublicMember:: pObject_m1_tank->setTranslateZ(0.001,true);
            }
            ValY=val;
        }


        //按下右键控制 H P 方向的转动
        if ( (name=="vpInputMouse::SOURCE_INTEGER_POSITION_X") &&   RightButtonDown )
        {
                //按下右键从左向右滑动
            if(val>ValX)
```

```
            {
                PublicMember:: pObject_m1_tank->setRotateR(-1.0,true);
            }
            //按下右键从右向左滑动
            if(val<ValX)
            {
                PublicMember:: pObject_m1_tank->setRotateR(1.0,true);
            }
            ValX=val;
        }

        //按下右键
        if ( (name=="vpInputMouse::SOURCE_INTEGER_POSITION_Y") &&   RightButtonDown )
        {
            //按下右键从下向上滑动
            if(val>ValY)
            {
                PublicMember:: pObject_m1_tank->setRotateP(-1.0,true);
            }
            //按下右键从上向下滑动
            if(val<ValY)
            {
                PublicMember:: pObject_m1_tank->setRotateP(1.0,true);
            }
            ValY=val;
        }
    //中键滑动控制远近
    if   (name=="vpInputMouse::SOURCE_INTEGER_WHEEL_DIRECTION")
{
            //中键向前滑动，通道角度逐步变大，场景物体变小
            if(val==1)
            {
                    myChannelAngle=GasRelay_myChannelAngle+0.5;
PublicMember::myChannel->setFOVSymmetric( GasRelay_myChannelAngle ,   -1.000000f );
            }
            //中键向后滑动，通道角度逐步变小，场景物体变大
            if(val==-1)
            {
                    myChannelAngle=GasRelay_myChannelAngle-0.5;
PublicMember::myChannel->setFOVSymmetric( GasRelay_myChannelAngle ,   -1.000000f );
            }
}
  }
    //end of class
};
```

图 9.4.2   通道屏幕坐标转换类

9.4.2 通道屏幕文字显示类的配置调用

通道屏幕坐标转换类的设计完成后，需要把自己设计的类用于指定的通道，整个配置都在 VP 主线程中完成。首先实例化对象"WxpMouseInputScence * wang=new WxpMouseInputScence();"，然后是具体的配置操作，大多数代码与鼠标窗口类相同，唯一不同的代码是需要为通道添加订阅类，设置触发事件 vsChannel::EVENT_POST_DRAW，代码如下：

    channel->addSubscriber(vsChannel::EVENT_POST_DRAW, wang);

具体代码如图 9.4.3 所示。

```
#pragma once
#include ".\publicmember.h"
#include "resource.h"
#include <vpInputMouse.h>
#include ".\wxpmouseinputscence.h"

UINT PublicMember::CTS_RunBasicThread(LPVOID)
{
    //初始化
    vp::initialize(__argc,__argv);
    //定义场景
    PublicMember::CTS_Define();

    //绘制场景
    vpKernel::instance()->configure();
    //
    //设置观察者
    PublicMember::CTS_pObject_observer=vpObject::find("Hummer");
    PublicMember::CTS_pObject_observer->ref();

    //设置窗体
    vpWindow * vpWin= * vpWindow::begin();
    vpWin->setParent(PublicMember::CTS_RunningWindow);
    vpWin->setBorderEnable(false);
    vpWin->setFullScreenEnable(true);

    //设置键盘
    vpWin->setInputEnable(true);
    vpWin->setKeyboardFunc((vrWindow::KeyboardFunc)PublicMember::CTS_Keyboard,NULL);

    //设置鼠标=========================================================
    WxpMouseInputScence * wang=new WxpMouseInputScence();
    vpWindow* window = vpWindow::empty() ? NULL : *vpWindow::begin();
    vpChannel* channel = vpChannel::empty() ? NULL : *vpChannel::begin();
    PublicMember::myChannel=channel;        //获取通道

    //添加通道订阅类，设置通道触发事件
```

```
channel->addSubscriber(vsChannel::EVENT_POST_DRAW, wang);

// 创建鼠标
 vpInputMouse *m_mouse;
 m_mouse = new vpInputMouse();
 m_mouse->setWindow(window);
 m_mouse->setChannel(channel);
 m_mouse->ref();

// 创建订阅者，捕捉鼠标输入值的变化
 vpInputMouse::const_iterator_source_boolean bit;
 for (bit=m_mouse->begin_source_boolean();
    bit!=m_mouse->end_source_boolean();++bit) {
    (*bit)->addSubscriber(
       vpInputSourceBoolean::EVENT_VALUE_CHANGED, wang);
     }

 vpInputMouse::const_iterator_source_integer iit;
 for (iit=m_mouse->begin_source_integer();
    iit!=m_mouse->end_source_integer();++iit) {
    (*iit)->addSubscriber(
       vpInputSourceInteger::EVENT_VALUE_CHANGED, wang);
     }
     vpWin->open();
     ::SetFocus(vpWin->getWindow());
//==============================鼠标配置结束==============================
vpWin->open();
::SetFocus(vpWin->getWindow());

//帧循环
while(vpKernel::instance()->beginFrame()!=0)
{
    vpKernel::instance()->endFrame();
    if(!PublicMember::CTS_continueRunVP)
    {
     vpKernel::instance()->endFrame();
     vpKernel::instance()->unconfigure();
     vp::shutdown();
         return 0;
     }
 }
 return 0;
}
```

图 9.4.3 配置使用通道屏幕坐标转换类

运行示例 VPTestDialogMouse，效果如图 9.4.4 所示。按下鼠标左键进行滑动，可以移动坦克位置，通道显示文字的位置随之变化，通道右下角将显示对象的位置变化值。

第 9 章　Vega Prime 中的实用功能实现

 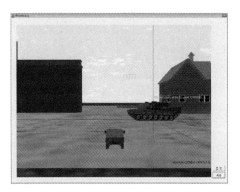

图 9.4.4　通道屏幕文字显示

9.4.3　通道屏幕绘图功能设计实现

通道屏幕坐标转换类中的关键函数：

　　void notify(vsChannel::Event event, const vsChannel *channel, vrDrawContext *context)

最后一个参数 vrDrawContext *context 为绘制上下文，可以进行图形绘制。其中的 fx 和 fy 就是依据目标对象的实际三维坐标转换而来，Width 和 Height 分别为场景窗口的实际宽度和高度，ValX 和 ValY 为窗口鼠标函数获取的当前鼠标在通道窗口的位置。

绘制线段时，以窗口鼠标函数为基础，绘制了四条线段：一条是从通道左下角到鼠标点击位置绘制一条蓝色线段；一条是从通道右下角到鼠标点击位置绘制一条红色线段；一条是绘制宽度为 15 的蓝色线段为背景；一条是绘制宽度为 15 的红色线段为对象生命值。后两条线段的以目标对象 PublicMember:: CTS_pObject_observer 的动态位置(fx,fy)为参考点绘制。

整个绘制过程中，完全使用 OpenGL 的线段绘制函数。需要绘制复杂图形的，可自行查阅相关函数。

glLineWidth(15.0)用来控制线段的宽度，glColor4f(0.0f, 0.0f, 1.0f, 1.0f)用来控制线段的颜色。

另外，通道屏幕的高度和宽度都进行了归一化处理，范围为[0~1]，以通道屏幕的左下角为坐标原点，以通道屏幕的右上角为[1,1]坐标点。

具体代码如图 9.4.5 所示。

```
void notify(vsChannel::Event event, const vsChannel *channel, vrDrawContext *context)
{
    //绘图------------------------------------------------
    vpObject *m_object=PublicMember:: CTS_pObject_observer;
    int ox, oy, sx, sy;
    vuMatrixf world2screen;
    channel->getVrChannel()->getViewport(&ox, &oy, &sx, &sy);
    world2screen = channel->getVrChannel()->getWorldToScreenMatrix();

    vuVec4f point;
    point.set(m_object->getTranslateX(),m_object->getTranslateY(),m_object->getTranslateZ(),1.0f);
    // transform the 4d point into clip space
    world2screen.transform(&point);
```

```cpp
// divide by w
float fx = point[0]/point[3]; // x between [-1,1]
float fy = point[1]/point[3]; // y between [-1,1]
float fz = point[2]/point[3]; // z between [-1,1]

// normalize to [0,1] range
fx = (fx+1.0f)/2.0f;
fy = (fy+1.0f)/2.0f;
fz = (fz+1.0f)/2.0f;
//获取场景实际窗口的长度  屏幕左下角为坐标原点
vpWindow * vpWin= * vpWindow::begin();
int Width,Height;

vpWin->getSize(&Width,&Height);
 // store the current graphics state and reset all of the graphics
 // state elements to their default values
 context->pushElements( true);

 // disable the depth test and set up the projection and model view
 // matrices
 glDisable(GL_DEPTH_TEST);
 glMatrixMode(GL_PROJECTION);
 glPushMatrix();
 glLoadIdentity();
 gluOrtho2D(0.0, 1.0, 0.0, 1.0);
 glMatrixMode(GL_MODELVIEW);
 glPushMatrix();
 glLoadIdentity();

 //屏幕线段绘制
 //从通道左下角到鼠标点击位置，绘制一条蓝色线段
 glColor4f(0.0f, 0.0f, 1.0f, 1.0f);
 glBegin(GL_LINES);
 glVertex3f(0.00f, 0.00f, 0.0f);
 glVertex3f((float)ValX/Width,(float)ValY/Height, 0.0f);
 glEnd();

 //从通道右下角到鼠标点击位置，绘制一条红色线段
 glColor4f(1.0f, 0.0f, 0.0f, 1.0f);
 glBegin(GL_LINES);
 glVertex3f(1.00f, 0.00f, 0.0f);
 glVertex3f((float)ValX/Width,(float)ValY/Height, 0.0f);
 glEnd();

 // 根据物体对象的移动，重新绘制线段，以对象的生命值为目标
  //绘制宽度为15的蓝色线段为背景，绘制宽度为15的红色线段为对象生命值
  //(fx,fy)为物体中心

  glLineWidth(15.0);
```

```
        glColor4f(0.0f, 0.0f, 1.0f, 1.0f);

        glBegin(GL_LINES);
        glVertex3f(fx, fy+0.14, 0.0f);
        glVertex3f(fx, fy+0.14+0.1,0.0f);    //以 0.1 为满值
        glEnd();

        glColor4f(1.0f, 0.0f, 0.0f, 1.0f);
        glBegin(GL_LINES);
        float val=(ValY % 101 )/1000.0;    //红色值变化
        glVertex3f(fx, fy+0.14, 0.0f);
        glVertex3f(fx, fy+0.14+val,0.0f); /
        glEnd();
        // restore all of the graphics state elements to their previous value.
        // Since we've been making changes to OpenGL directly, the software
        // representation of the graphics hardware maintained by vrDrawContext is
        // now out of sync.   By passing 'false' as the 2nd argument to popAll,
        // we'll disable lazy applies and thereby force a resynchronization
        // of the graphics hardware w/ vrDrawContext.
        context->popElements( false);
}
```

图 9.4.5　通道屏幕坐标转换的绘图功能

图形绘制函数编写完毕后，也需要在 Vega Prime 的主线程中配置使用。

运行例子 VPTestDialogMouseScreenDisplay，效果如图 9.4.6 所示。按下鼠标左键进行上下滑动，汽车头顶的生命值即红色线段值会随之改变。

图 9.4.6　通道屏幕图形绘制

当然，图形绘制和文字显示，可以结合使用，效果更好。

9.5　鼠标点选通道对象功能设计

鼠标点选通道对象的设计相对较为复杂，主要涉及三个方面：一是以视线碰撞检测

vpIsectorLOS 为核心的 Picker 功能类的设计，该功能主要以当前通道场景为检测对象，判断出鼠标点击处的通道屏幕二维坐标通过坐标映射是否存在三维空间中存在挑选的目标对象，若存在则返回目标对象指针；二是在窗口鼠标类的鼠标单击事件中调用对象挑选类 Picker，对于选中的对象进行不同的渲染策略，以便突出显示；三是在 VP 通用类 PublicMember 中定义对象挑选类 Picker 的静态指针，并在 VP 主线程中进行具体实例化，以便在窗口鼠标类中可以使用。

9.5.1 通道挑选 Picker 类的设计实现

鼠标点选通道对象的设计思路是，利用视线碰撞检测 vpIsectorLOS 检测鼠标点击位置是否与通道内三维空间物体发生碰撞，如果发生了碰撞，以碰撞点为起点，寻找父节点，找到鼠标点选的对象，然后使用不同的渲染策略，对选中的对象进行不同的渲染。该功能的核心是视线碰撞检测 vpIsectorLOS，该碰撞检测以当前通道场景为检测对象，以碰撞检测体的位置和方向为基准，沿着 Y 坐标正向设置检测线段，在碰撞检测中该线段进行了 H 和 P 方向 360°旋转检测。

Picker 类包含两个私有成员，第一个是视线碰撞检测对象 m_isector，另一个是自定义的节点类型对象 m_mode。这里，对于鼠标点选的目标类型 Mode 分为四种，分别为物体对象 MODE_OBJECT、几何体对象 MODE_GEOMETRY、DOF 对象 MODE_DOF 和 LOD 对象 MODE_LOD。物体对象 MODE_OBJECT 包含整个的目标对象，是本书设计中默认使用的值；几何体对象 MODE_GEOMETRY 可以为物体对象的某一部分的几何体。

构造函数 Picker 包含两个参数：一个是视线碰撞检测对象，另一个为视线碰撞检测掩码。在该函数中，对节点类型 m_mode 初始化为 MODE_OBJECT；对视线碰撞检测对象 m_isector 进行实例化和设置。

Picker 类的核心功能函数为：

vsNode *pick(vpChannel *channel, float mx, float my);

该函数较为复杂，第一个参数为当前场景通道，mx 和 my 分别为鼠标点击处的通道屏幕平面坐标的 x 值和 y 值。

pick 函数首先解决的问题是投影问题。获取当前视觉矩阵，如果为正射投影 orthographic projection，则直接计算视线碰撞检测的相关参数；如果为透视投影 perspective projection，则需要对鼠标坐标进行矩阵转换，由世界坐标转换计算得到视线碰撞检测的相关参数。

得到视线碰撞检测对象 m_isector 的相关参数后，就可以利用以下代码设置碰撞参数：

m_isector->setTranslate(x, y, z);

m_isector->setRotate(h, p, 0.0);

m_isector->setSegmentRange(range);

然后利用以下代码实施碰撞检测，查看碰撞是否发生：

vsNode *node = NULL;

m_isector->update();

如果碰撞发生，就根据设置的节点类型返回 vsNode 类型的目标节点对象 node。

具体代码如图 9.5.1 所示。

```cpp
#include <vuAllocTracer.h>
#include <vuMath.h>
#include <vrLight.h>
#include <vrMode.h>
#include <vrRenderStrategy.h>
#include <vrDrawContext.h>
#include <vsDOF.h>
#include <vsLOD.h>
#include <vpIsectorLOS.h>
#include <vpObject.h>
#include <vpObserver.h>

#pragma once
// create a simple picker class to perform intersection tests w/ the scene
// graph for a given mouse position
class Picker {
public:
    Picker(vpScene *scene, uint mask = 0xFFFFFFFF)
    {           // set the parameters to their default values
        m_mode = MODE_OBJECT;
        m_isector = new vpIsectorLOS();
        m_isector->setTarget(scene);
        m_isector->setIsectMask(mask);
    }
Picker::~Picker(void){}
    // picking modes to select either objects or nodes
    enum Mode {
        MODE_OBJECT,    // select all geometry under the closest parent of the
                        // picked geometry that is an object
        MODE_GEOMETRY,  // select only the picked geometry.  Note that this may
                        // be in a different LOD than the one you're seeing.
        MODE_DOF,       // select all geometry under the closest parent of the
                        // picked geometry that is a dof
        MODE_LOD        // select all geometry under the closest parent of the
                        // picked geometry that is a lod
    };
    // set / get the picking mode
    void setMode(Mode mode) { m_mode = mode; }
    Mode getMode() const { return m_mode; }
    // intersect w/ the scene at the given mouse position
    vsNode *pick(vpChannel *channel, float mx, float my)
    {
        float n, f;
        double x, y, z, h, p, r = 0.0, range;
        const vuMatrix<double> &viewMat = channel->getViewMatrix();
        channel->getNearFar(&n, &f);
        // if we've got an orthographic projection, then things are easy.
```

```cpp
// In this case the mouse position gives us the xy location within
// the orthographic frustum and our isector will run from the near
// to the far clipping plane.
if (channel->getProjection() == vrChannel::PROJECTION_ORTHOGRAPHIC) {

    float l, r, b, t;
    channel->getFrustum(&l, &r, &b, &t);
    x = l + ((mx + 1.0f) * 0.5f) * (r - l);
    y = b + ((my + 1.0f) * 0.5f) * (t - b);
    z = viewMat[3][2] + n;
    h = 0.0;
    p = -90.0;
    range = f - n;
}
// if we've got a perspective projection, then things are a little
// more difficult
else {
    // transform the mouse position, which is in normalized channel
    // screen space, back into world space.  In order to accomplish
    // this, we need to "undo" the projection and graphics library
    // specific offset transformations by transforming the point
    // through the inverse of their respective matrices.   This gives
    // us the mouse position in the eye's coordinate system.   Then a
    // final transformation through the view matrix will put us into
    // world coordinates.   Note that in the initial transformation
    // that the Z coordinate doesn't matter since the inverse of the
    // projection transformation "ignores" it.
    vuVec3<float> mouse(mx, my, -1);
    vuVec3<double> vec;
    vuMatrix<float> projInv;
    projInv.invert(channel->getVrChannel()->getProjectionMatrix());
    projInv.transformPoint(&mouse);
    channel->getVrChannel()->getOffsetMatrixInverse(
        ).transformPoint(&mouse);
    channel->getVrChannel()->getViewMatrix().transformPoint(&mouse);
    // create a vector from the eye point through the mouse position
    x = viewMat[3][0];
    y = viewMat[3][1];
    z = viewMat[3][2];

    vec[0] = mouse[0] - x;
    vec[1] = mouse[1] - y;
    vec[2] = mouse[2] - z;

    // calculate the heading and pitch of this vector
    h = vuRad2Deg(-vuArcTan(vec[0], vec[1]));
    p = vuRad2Deg(vuArcTan(vec[2],
```

```cpp
                    vuSqrt(vuSq(vec[0]) + vuSq(vec[1])))); 

            range = 2 * f;
        }
        // update the isector to align w/ the vector
        m_isector->setTranslate(x, y, z);
        m_isector->setRotate(h, p, 0.0);
        m_isector->setSegmentRange(range);

        // perform the intersection test and see if we hit anything
        vsNode *node = NULL;
        m_isector->update();
        if (m_isector->getHit()) {
            switch(m_mode) {

            case MODE_OBJECT: // pick the object
                node = m_isector->getHitObject();
                break;

            case MODE_GEOMETRY: // pick the geometry
                node = m_isector->getHitNode();
                break;

            case MODE_DOF: // pick the dof
                node = getParent(m_isector->getHitNode(), vsDOF::getStaticClassType());
                if(node == NULL)
                    printf("Geometry is not parented by a DOF node\n");
                break;

            case MODE_LOD: // pick the lod
                // if we don't have a parent whose a lod, return the hit geometry
                node = getParent(m_isector->getHitNode(), vsLOD::getStaticClassType());
                if(node == NULL)
                    printf("Geometry is not parented by a LOD node\n");
                break;

            }
        }
        return node;
    }
private:
    //Walk up the parents till a vsDOF node is found
    vsNode *getParent(vsNode *child, vuClassType *classType)
    {
        if (child != NULL) {
            vsNode::const_iterator_parent it, ite = child->end_parent();
            for(it=child->begin_parent();it!=ite;++it) {
                if((*it)->isOfClassType(classType))
```

```
                    return *it;
                else return getParent(*it, classType);
        }
    }
    return NULL;
}
// select either picking of objects or nodes
Mode m_mode;
// an isector for use w/ intersection tests
vuField<vpIsectorLOS *>m_isector;
};
```

图 9.5.1 通道挑选 Picker 类

9.5.2 窗口鼠标类的辅助设计实现

通道对象的挑选主要通过鼠标点击场景通道的屏幕来完成，这是因为挑选类的功能使用是在窗口鼠标类中完成的。其中，这里会使用到 VP 公共类里的一个静态成员 PublicMember::CTS_pObject_observer，用于记录选中的目标对象。

在该窗口鼠标类中，设置了一个私有变量 m_boundsStrategy，该对象为挑选对象的渲染策略，该渲染策略的类型为 vrRenderStrategyBounds，可以绘制一个矩形的包围盒。该变量在窗口类的构造函数中进行了实例化和初始化。

  m_boundsStrategy = new vrRenderStrategyBounds();
  m_boundsStrategy->setRenderGeometryEnable(true);
  m_boundsStrategy->setLineWidth (1);
  m_boundsStrategy->setColor (1.0,1.0,0.0,1.0);
  m_boundsStrategy->setWireframeEnable(true);
  m_boundsStrategy->ref();

可以依次控制线框可见性、线框线条粗细度、线框线条颜色和线框类型。

对于窗口鼠标类对于挑选类的功能调用，主要在下面一个方法中实现：

  void notify(vpInputSourceBoolean::Event event, vpInputSourceBoolean *source)

该方法包含 vpInputSourceBoolean::Event 事件和 vpInputSourceBoolean，该方法首先获取鼠标点击通道屏幕处的坐标，然后使用 Picker 类的 pick 方法返回目标对象。对于返回的目标对象，需要区别对待，对于已经选中并进行了高亮渲染的对象，首先应取消高亮渲染；对于新选中的第一个对象，才使用高亮渲染策略。当然，也可以实现逐次选中多个物体实现高亮渲染。核心代码如下：

  vsNode *node =PublicMember::m_picker->pick(*vpChannel::begin(),
     sourceX->getValue(), sourceY->getValue());

   // 对于已经有选中的对象，首先取消高亮渲染
   if (selectedNode != NULL)

```
        {
            setRenderStrategy(selectedNode, NULL);
            PublicMember::CTS_pObject_observer=(vpObject *)selectedNode;
        }

            // 对于最新的选中对象,实现高亮渲染
            if ((node != NULL) && (node != selectedNode))
              {
                setRenderStrategy(node, m_boundsStrategy);
                selectedNode = node;
              }
            else
              selectedNode = NULL;
```

窗口鼠标类的辅助设计的具体代码如图 9.5.2 所示。

```cpp
#pragma once
#include ".\publicmember.h"
#include "resource.h"

#include <vrMode.h>
#include <vrFontFactory.h>
#include <vpInputMouse.h>

class WxpMouseInputScence: public vpInputSourceBoolean::Subscriber,
            public vpInputSourceFloat::Subscriber,
            public vpInputSourceInteger::Subscriber,
            public vsChannel::Subscriber
{
public:
  WxpMouseInputScence(void)
{
    m_boundsStrategy = new vrRenderStrategyBounds();
   m_boundsStrategy->setRenderGeometryEnable(true);
   m_boundsStrategy->setLineWidth (1);
   m_boundsStrategy->setColor (1.0,1.0,0.0,1.0);
   m_boundsStrategy->setWireframeEnable(true);
   m_boundsStrategy->ref();

//控制包围框的属性
   m_boundsStrategy->setColor(1.0,0,0.0,1);           //颜色
   m_boundsStrategy->setRenderGeometryEnable(true);   //控制物体可见性
   m_boundsStrategy->setLineWidth(1);                 //线条宽度
}
   ~WxpMouseInputScence(void){};
```

```cpp
private:
    //渲染策略，渲染一个矩形框，包围点选的目标对象
    vrRenderStrategyBounds* m_boundsStrategy;

    vsTraversal::Result travFuncGeometry(vsNode *node, vrRenderStrategy *strategy)
    {
        vrGeometry *geometry = static_cast<vsGeometry*>(node)->getGeometry();
        geometry->setRenderStrategy(strategy);
        return vsTraversal::RESULT_CONTINUE;
    }

    void setRenderStrategy(vsNode *root, vrRenderStrategy *strategy)
    {
        vsTraversalUser<vrRenderStrategy *, vsTraversalLookUpNodeId> trav(strategy);
        trav.addPreVisit(vsGeometry::getStaticNodeId(), this, travFuncGeometry);
        trav.visit(root);
    }

    void notify(vsChannel::Event event, const vsChannel *channel, vrDrawContext *context) { }

    virtual void notify(vsChannel::Event, const vsChannel *,vsTraversalCull *) { }

    //判断鼠标单击事件是否发生，发生则触发点选功能------------------------
    // notify method to catch boolean source subscriber events
    void notify(vpInputSourceBoolean::Event event, vpInputSourceBoolean *source)
    {
        //选择物体
        static vsNode *selectedNode = NULL;
        vpInputMouse *mouse = *vpInputMouse::begin();
        vpInputSourceFloat *sourceX = mouse->getSourceFloat(
            vpInputMouse::SOURCE_FLOAT_POSITION_X);
        vpInputSourceFloat *sourceY = mouse->getSourceFloat(
            vpInputMouse::SOURCE_FLOAT_POSITION_Y);
        vsNode *node =PublicMember::m_picker->pick(*vpChannel::begin(),
            sourceX->getValue(), sourceY->getValue());

        // 对于已经有选中的对象，首先取消高亮渲染
        if (selectedNode != NULL)
        {
            setRenderStrategy(selectedNode, NULL);
            PublicMember::CTS_pObject_observer=(vpObject *)selectedNode;
        }

        // 对于最新的选中对象，实现高亮渲染
        if ((node != NULL) && (node != selectedNode))
        {
```

```
                    setRenderStrategy(node, m_boundsStrategy);
                    selectedNode = node;
            }
            else
selectedNode = NULL;
    }
    // notify method to catch float source subscriber events
    void notify(vpInputSourceFloat::Event, vpInputSourceFloat *source)
    {    }
    // notify method to catch integer source subscriber events
    void notify(vpInputSourceInteger::Event, vpInputSourceInteger *source)
    {    }
  //end of class
};
```

图 9.5.2　窗口鼠标类的辅助设计

9.5.3　通道挑选 Picker 类的配置使用

通道挑选类 Picker 的使用必须在 VP 主线程中进行必要的配置才能使用。配置包含两个过程，第一是鼠标窗口函数的配置，第二个是 Picker 类的配置。当然，首先需要在 VP 的公共类 PublicMember 类里面定义一个 Picker 类的静态变量：

　　　static Picker *m_picker;

同时进行实例化：

　　　Picker *PublicMember::m_picker=NULL;

第一个配置，完成窗口鼠标类的配置。首先把设计的窗口鼠标类进行实例化，并添加为鼠标值变化事件的订阅者，捕捉 vpInputSourceBoolean::EVENT_VALUE_CHANGED 事件的发生。具体代码如下：

```
//设置鼠标函数
    WxpMouseInputScence * wang=new WxpMouseInputScence();

    vpWindow* window = vpWindow::empty() ? NULL : *vpWindow::begin();
    vpChannel* channel = vpChannel::empty() ? NULL : *vpChannel::begin();
    PublicMember::myChannel=channel;
//创建鼠标
    vpInputMouse *m_mouse;
    m_mouse = new vpInputMouse();
    m_mouse->setWindow(window);
    m_mouse->setChannel(channel);
    m_mouse->ref();

//添加订阅者，捕捉鼠标输入值的变化
```

```
vpInputMouse::const_iterator_source_boolean bit;
for (bit=m_mouse->begin_source_boolean();
    bit!=m_mouse->end_source_boolean();++bit) {
    (*bit)->addSubscriber(
        vpInputSourceBoolean::EVENT_VALUE_CHANGED, wang);
}
```

第二个配置，完成通道挑选类的配置。首先把设计的 Picker 类进行实例化，然后获取鼠标左值变化输入源，并为之添加事件订阅者，捕捉 vpInputSourceBoolean::EVENT_FALLING_EDGE 事件的发生。具体代码如下：

```
//配置鼠标选择物体
m_picker = new Picker(*vpScene::begin());
vpInputMouse *mouse = new vpInputMouse();
mouse->setWindow(*vpWindow::begin());
mouse->setChannel(*vpChannel::begin());
vpInputSourceBoolean *leftMouseButton = mouse->getSourceBoolean (vpInputMouse::
                        SOURCE_BOOLEAN_BUTTON_LEFT);
leftMouseButton->addSubscriber(vpInputSourceBoolean::EVENT_FALLING_EDGE,
                        wang);
```

完整的通道挑选 Picker 类的配置使用具体代码如图 9.5.3 所示。

```
UINT PublicMember::CTS_RunBasicThread(LPVOID)
{
    //初始化
    vp::initialize(__argc,__argv);
    //定义场景
    PublicMember::CTS_Define();
    //绘制场景
    vpKernel::instance()->configure();
    //设置观察者
    PublicMember::CTS_pObject_observer=vpObject::find("Hummer");
    PublicMember::CTS_pObject_observer->ref();

    //设置窗体
    vpWindow * vpWin= * vpWindow::begin();
    vpWin->setParent(PublicMember::CTS_RunningWindow);
    vpWin->setBorderEnable(false);
    vpWin->setFullScreenEnable(true);
    //设置键盘
    vpWin->setInputEnable(true);
    vpWin->setKeyboardFunc((vrWindow::KeyboardFunc)PublicMember::CTS_Keyboard,NULL);
        //----------------鼠标点选功能设置--------------------
        //设置鼠标函数
        WxpMouseInputScence * wang=new WxpMouseInputScence();
```

```
vpWindow* window = vpWindow::empty() ? NULL : *vpWindow::begin();
vpChannel* channel = vpChannel::empty() ? NULL : *vpChannel::begin();
PublicMember::myChannel=channel;
// 创建鼠标
vpInputMouse *m_mouse;
m_mouse = new vpInputMouse();
m_mouse->setWindow(window);
m_mouse->setChannel(channel);
m_mouse->ref();

// 创建订阅者，捕捉鼠标输入值的变化
vpInputMouse::const_iterator_source_boolean bit;
for (bit=m_mouse->begin_source_boolean();
     bit!=m_mouse->end_source_boolean();++bit) {
    (*bit)->addSubscriber(
        vpInputSourceBoolean::EVENT_VALUE_CHANGED, wang);
}
//===============
//配置鼠标选择物体
   m_picker = new Picker(*vpScene::begin());

   vpInputMouse *mouse = new vpInputMouse();
   mouse->setWindow(*vpWindow::begin());
   mouse->setChannel(*vpChannel::begin());
   vpInputSourceBoolean*leftMouseButton=
       mouse->getSourceBoolean(vpInputMouse::SOURCE_BOOLEAN_BUTTON_LEFT);
   leftMouseButton->addSubscriber(vpInputSourceBoolean::EVENT_FALLING_EDGE, wang);

//---------------------------------------
 vpWin->open();
 ::SetFocus(vpWin->getWindow());
 //帧循环
while(vpKernel::instance()->beginFrame()!=0)
{
    vpKernel::instance()->endFrame();
    if(!PublicMember::CTS_continueRunVP)
    {
       vpKernel::instance()->endFrame();
       vpKernel::instance()->unconfigure();
       vp::shutdown();
       return 0;
    }
}
  return 0;
}
```

图 9.5.3　通道挑选 Picker 类的配置使用

运行代码，其效果如图 9.5.4 所示。点选汽车，其效果如图 9.5.4（a）所示；点选房屋，其效果如图 9.5.4（3）所示。

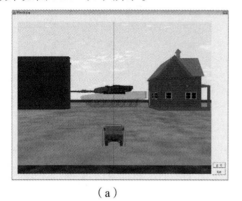
（a）

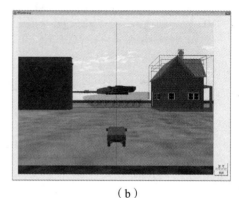
（b）

图 9.5.4　通道对象鼠标点选效果

对于挑选类的使用，可以进行进一步设计开发。例如，可以对渲染策略进行改进，可以结合通道屏幕显示功能对选中的对象显示相关信息，可以稍加改进实现多个对象的选择，等等。

当选取模型为 DOF 时，即"m_mode=MODE_DOF;"时，显示对象的指针不能为 DOF 的指针，需要重现确定为一个 vpObject，否则计算显示坐标时将出错，即：

void notify(vsChannel::Event event, const vsChannel *channel, vrDrawContext *context)
{
　　vuString m_text2;
　　vuString m_text3;

　　vpObject *m_object=PublicMember::CTS_pObject_other;
　　//vpObject *m_object=PublicMember::CTS_pObject_observer;
}

m_object 必须为 vpObject 类型，不能为 vsDOF 类型，处理方式是选取另一个固定对象物体，计算屏幕坐标，显示在固定位置上。

9.6　仿真通道的屏幕图片抓取

对于仿真通道的屏幕场景，很多时候需要进行抓取，储存为常规图片。本节将通过窗口鼠标类的事件，鼠标右键单击通道窗口，保存一张图片。

9.6.1　通道屏幕图片抓取功能的设计

对于通道屏幕图片的抓取，需要用到通道数据的读取，然后保存为 jpg 图片。整个功能设计全部在窗口鼠标类中完成。

设计中，定义了两个常量：MAX_WINDOW_WIDTH 为通道屏幕宽度，默认值为 1 920；MAX_WINDOW_HEIGHT 为通道屏幕高度，默认值为 1 080。另外定义了一个布尔变量 RightButtonDown，用于控制是否进行图片保存，鼠标右键单击后该变量改变为真值，允许保存图片，保存完成后该值自动修改为假值，取消保存图片。该值的控制在下列函数中实现：

void notify(vpInputSourceBoolean::Event event, vpInputSourceBoolean *source);

通道屏幕图片的抓取功能，其功能的实现完全在下面整个函数中完成：

void notify(vsChannel::Event event, const vsChannel *channel, vrDrawContext *context)

该函数中，首先定义了一个 vuImageFactory 类型的变量 m_factory，然后根据通道的高度和宽度，获取通道的图像数据，设置像素类型。

//读取通道像素数据

glReadPixels(ox, oy, sx, sy, GL_RGB, GL_UNSIGNED_BYTE, m_data);

// 转换为图像数据

vuImageUserBuffer   image(m_data);

// 设置图像的宽度高度

image.setDimensions(sx, sy);

//设置像素类型   image.setPixelType(vuImageBase::PixelType (vuImageBase::TYPE_UNSIGNED_BYTE,8,8,8,0))

为了避免图片名称重复，使用帧序列号来作为图片名称，保存为 jpg 图片。最后，把图片写入当前目录。

m_factory->write(filename.c_str(), &image);

具体代码如图 9.6.1 所示。

```
#pragma once
#include ".\publicmember.h"
#include "resource.h"

#include <vuImage.h>
#include <vuImageFactory.h>
#include <vuAllocTracer.h>
#include <GL/gl.h>

#define MAX_WINDOW_WIDTH    1920
#define MAX_WINDOW_HEIGHT 1080

class WxpMouseInputScence: public vpInputSourceBoolean::Subscriber,
            public vpInputSourceFloat::Subscriber,
            public vpInputSourceInteger::Subscriber,
            public vsChannel::Subscriber
{
public:
    WxpMouseInputScence(void){ RightButtonDown=false;}
    ~WxpMouseInputScence(void){ }
```

```cpp
private:
    bool RightButtonDown;
    void notify(vsChannel::Event event, const vsChannel *channel, vrDrawContext *context)
    {
        //场景通道捕捉--------------------------------------------------
        unsigned char *m_data;
        vuField<vuImageFactory*> m_factory= new vuImageFactory;   //必须实例化
        if(RightButtonDown)
        {
        // allocate space large enough for any frame capture (in case
        // the window is maximized/resized before capture)
          m_data = vuAllocArray<uchar >::malloc(MAX_WINDOW_WIDTH*MAX_WINDOW_HEIGHT*3);
        // get the viewport
        int ox, oy, sx, sy;
        channel->getVrChannel()->getViewport(&ox, &oy, &sx, &sy);

            // error checking
            if(sx > MAX_WINDOW_WIDTH || sy > MAX_WINDOW_HEIGHT)
            {
                vuNotify::print(vuNotify::LEVEL_WARN, NULL,
                    "Window is larger than %d %d, image will be cropped",
                    MAX_WINDOW_WIDTH, MAX_WINDOW_HEIGHT);
                sx = MAX_WINDOW_WIDTH;
                sy = MAX_WINDOW_HEIGHT;
            }
            // 读取通道像素数据
            glReadPixels(ox, oy, sx, sy, GL_RGB, GL_UNSIGNED_BYTE, m_data);
            // 转换为图像数据
            vuImageUserBuffer   image(m_data);
            // 设置图像宽度和高度
            image.setDimensions( sx, sy );
            // 设置图像像素类型  image.setPixelType(vuImageBase::PixelType (vuImageBase:: TYPE_
                                                       UNSIGNED_BYTE,8,8,8,0));

             // the image filename will be frame########.jpg
            vuString filename;
            filename.sprintf("frame%08d.jpg", vsThread::resolveFrameNumber());
            m_factory->write( filename.c_str(), &image );
            // reset the flag
            RightButtonDown = false;
        }
    }
    virtual void notify(vsChannel::Event, const vsChannel *,vsTraversalCull *)   { }
      /* notify method to catch boolean source subscriber events       */
    void notify(vpInputSourceBoolean::Event event, vpInputSourceBoolean *source)
        {
            CString name=source->getName();
```

```
            int val=source->getValue();
            if(name=="vpInputMouse::SOURCE_BOOLEAN_BUTTON_RIGHT")
            {
                if(val==1)
                    RightButtonDown=true;
                if(val==0)
                    RightButtonDown=false;
            }
        }
    /* notify method to catch float source subscriber events    */
    void notify(vpInputSourceFloat::Event, vpInputSourceFloat *source){ }
    /** notify method to catch integer source subscriber events */
    void notify(vpInputSourceInteger::Event, vpInputSourceInteger *source){ }
    //end of class
};
```

图 9.6.1　通道图片抓取功能设计

9.6.2　通道屏幕图片抓取功能的配置使用

通道屏幕图片抓取功能的配置与窗口鼠标类的配置基本相同，就是把自己设计的类用于指定的窗口，整个配置都在 VP 主线程中完成。首先实例化对象"WxpMouseInputScence * wang=new WxpMouseInputScence();"，然后才是具体的配置操作，主要是为当前通道添加订阅者，捕捉 vsChannel::EVENT_POST_DRAW 事件。

　　　　channel->addSubscriber(vsChannel::EVENT_POST_DRAW, wang);

具体代码如图 9.6.2 所示。

```
UINT PublicMember::CTS_RunBasicThread(LPVOID)
{
    //初始化
    vp::initialize(__argc,__argv);
    //定义场景
    PublicMember::CTS_Define();
    //绘制场景
    vpKernel::instance()->configure();
    //设置观察者
    PublicMember::CTS_pObject_observer=vpObject::find("Hummer");
    PublicMember::CTS_pObject_observer->ref();

    //设置窗体
    vpWindow * vpWin= * vpWindow::begin();
    vpWin->setParent(PublicMember::CTS_RunningWindow);
    vpWin->setBorderEnable(false);
    vpWin->setFullScreenEnable(true);

    //设置键盘
```

```cpp
vpWin->setInputEnable(true);
vpWin->setKeyboardFunc((vrWindow::KeyboardFunc)PublicMember::CTS_Keyboard,NULL);

//设置鼠标================================================
    WxpMouseInputScence * wang=new WxpMouseInputScence();
    vpWindow* window = vpWindow::empty() ? NULL : *vpWindow::begin();
    vpChannel* channel = vpChannel::empty() ? NULL : *vpChannel::begin();
    PublicMember::myChannel=channel;

    channel->addSubscriber(vsChannel::EVENT_POST_DRAW, wang);
    // 创建鼠标
    vpInputMouse *m_mouse;
    m_mouse = new vpInputMouse();
    m_mouse->setWindow(window);
    m_mouse->setChannel(channel);
    m_mouse->ref();
    // 创建订阅者，捕捉鼠标输入值的变化
    vpInputMouse::const_iterator_source_boolean bit;
    for (bit=m_mouse->begin_source_boolean();
         bit!=m_mouse->end_source_boolean();++bit) {
        (*bit)->addSubscriber(
            vpInputSourceBoolean::EVENT_VALUE_CHANGED, wang);
    }
    vpInputMouse::const_iterator_source_float fit;
    for (fit=m_mouse->begin_source_float();
         fit!=m_mouse->end_source_float();++fit) {
        (*fit)->addSubscriber(
            vpInputSourceFloat::EVENT_VALUE_CHANGED, wang);
    }
    vpInputMouse::const_iterator_source_integer iit;
    for (iit=m_mouse->begin_source_integer();
         iit!=m_mouse->end_source_integer();++iit) {
        (*iit)->addSubscriber(
            vpInputSourceInteger::EVENT_VALUE_CHANGED, wang);
    }
//================================================
    vpWin->open();
    ::SetFocus(vpWin->getWindow());
//帧循环
while(vpKernel::instance()->beginFrame()!=0)
{
    vpKernel::instance()->endFrame();
    if(!PublicMember::CTS_continueRunVP)
    {
        vpKernel::instance()->endFrame();
        vpKernel::instance()->unconfigure();
        vp::shutdown();
```

```
                return 0;
            }
    }
    return 0;
}
```

图 9.6.2　通道屏幕图片抓取配置

运行程序，其效果如图 9.6.3 所示，图 9.6.3（a）所示为整个仿真窗口图片，图 9.6.3（b）所示为通道图片。当然，如果能逐帧保存图片，就能存储为连续的视频。

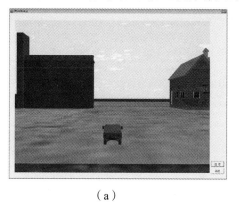 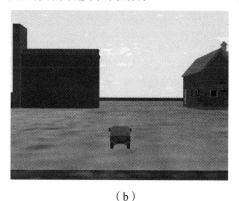

　　　　　　（a）　　　　　　　　　　　　　　　　（b）

图 9.6.3　通道屏幕图片抓取效果

9.7　仿真通道的视频录制

对于仿真通道的屏幕场景，某些时候需要进行场景通道视频录制。本节设计了一个场景通道的视频录制类，通过键盘函数按键启用或停止视频录制。

9.7.1　通道屏幕视频录制类的设计

通道屏幕视频录制主要是利用 Windows 系统的 vfw32.lib 库，继承通道订阅类的功能，在 vpChannel::EVENT_POST_DRAW 事件中完成帧数据的提取。

通道屏幕视频录制类的设计，第一步需要使用系统的较多功能，需要引入以下几个头文件：

```
#include <windows.h>
#include <windowsx.h>
#include <memory.h>
#include <wingdi.h>
#include <mmsystem.h>
#include <vfw.h>
#include <GL/gl.h>
```

并导入 vfw32 库：
// Automatically link in vfw32.lib
#pragma comment(lib, "vfw32.lib")
#pragma message("Will automatically link with " "vfw32.lib")
第二步，需要定义较多的私有变量，用于处理视频的相关信息：

int	m_status;
int	m_frameCnt;
int	m_width;
int	m_height;
void*	m_imageBuffer;
vuLock	m_deleteLock;
vuString	m_curFrameFileName;
PAVIFILE	m_pAVIFile;
PAVISTREAM	m_ps;
PAVISTREAM	m_psCompressed;
PAVISTREAM	m_psText;
BITMAPINFOHEADER	m_bi;

依次包含状态、帧数目、视频宽度、视频高度、图像存储指针、操作锁、视频文件名、视频流指针、压缩视频流指针、视频流文字、BITMAP 信息头。这些信息在类的构造函数中都进行了初始化，某些指针变量在析构函数中进行了处理。

第三步，在配置函数中配置相关功能：

int configure(const char* fileName, int fpsRate, bool compressionDialog, int winWidth, int winHeight);

所带 5 个函数依次为视频文件名、帧率、压缩与否对话框、宽度和高度，函数功能主要是调用 AVI 相关函数完成配置，AVIFileCreateStream 函数创建 AVI 视频流，AVIMakeCompressedStream 使用 AVI 视频压缩流，AVIStreamSetFormat 设置视频流格式。

第四步，视频帧的抓取函数"captureFrame();"主要调用了 glReadPixels 读取帧数据，使用 AVIStreamWrite 把帧数据写入视频文件。

第五步，视频写入操作主要在下面这个事件响应函数中完成,其功能主要是锁定帧数据，抓取写入视频流，解锁，其代码如下：

void notify(vsChannel::Event event, const vsChannel *channel, vrDrawContext *context){
  m_deleteLock.lock();
  captureFrame();
  m_deleteLock.unlock();
}

具体代码如图 9.7.1 所示。

```
#ifndef VSMOVIERECORDER_H
#define VSMOVIERECORDER_H
```

```cpp
#include "vsChannel.h"

#if VSG_OS == VSG_WINDOWS
#include <windows.h>
#include <windowsx.h>
#include <memory.h>
#include <wingdi.h>
#include <mmsystem.h>
#include <vfw.h>
#include <GL/gl.h>

// Automatically link in vfw32.lib
#pragma comment(lib, "vfw32.lib")
#pragma message( "Will automatically link with " "vfw32.lib" )
class vsMovieRecorder : public vsChannel::Subscriber
{   public:
    /** Default constructor*/
    vsMovieRecorder() : m_pAVIFile(NULL),
                        m_ps(NULL),
                        m_psCompressed(NULL),
                        m_psText(NULL),
                        m_status(AVIERR_OK),
                        m_width(0),
                        m_height(0),
                        m_frameCnt(0),
                        m_imageBuffer(NULL)
    {
        memset(&m_bi,0,sizeof(BITMAPINFOHEADER));
        AVIFileInit();
    }
    /*** Default destructor */
    ~vsMovieRecorder()
    {
        m_deleteLock.lock();
        closeFile();
        if(m_imageBuffer != NULL)
            vuFreeRef(m_imageBuffer);
        AVIFileExit();
        m_deleteLock.unlock();
    }
    /*** Configure a stream for recording purposes.   This method will try to
     * set up a recording stream at the specified filename with the specified
     * frame rate and the specified dimensions.
     * @return vsgu::FAILURE in the event of a failure condition.
vsgu::SUCCESS otherwise. */
    int configure(const char* fileName, int fpsRate, bool compressionDialog, int winWidth, int winHeight){
        if(openFile(fileName) == false)
```

```cpp
    {
            vuNotify::print( vuNotify::LEVEL_WARN, NULL,"unable to open file %s", fileName);
            return false;
    }
    m_frameCnt = 0;
    if(!m_pAVIFile) {
            vuNotify::print( vuNotify::LEVEL_WARN, NULL,"need to open an AVI file first");
            return false;
    }

    AVISTREAMINFO strhdr;
    _fmemset(&strhdr, 0, sizeof(strhdr));
    strhdr.fccType                  = streamtypeVIDEO;// stream type
    strhdr.fccHandler               = 0;
    strhdr.dwScale                  = 1;
    strhdr.dwRate                   = fpsRate;        // rate fps
    strhdr.dwSuggestedBufferSize    = 0;

    m_height = winHeight;
    m_width = winWidth;

    if( (m_width % 160) || (m_height % 160))
    {
            vuNotify::print( vuNotify::LEVEL_WARN, NULL, "window dimensions are not divisible by 160, some codecs may not like this!");
    }
    m_imageBuffer = vuMallocRef(m_width*m_height*3);
    SetRect(&strhdr.rcFrame, 0, 0, m_width, m_height); // rectangle for stream
    // And create the stream;
    m_status = AVIFileCreateStream(m_pAVIFile,      // file pointer
        &m_ps,                  // returned stream pointer
        &strhdr);               // stream header

    if (m_status != AVIERR_OK)
        return false;
    assert(m_ps);
    //set options and compression
    AVICOMPRESSOPTIONS opts;
    AVICOMPRESSOPTIONS FAR * aopts[1] = {&opts};
    _fmemset(&opts, 0, sizeof(opts));

    if(compressionDialog)
        if (!AVISaveOptions(NULL, 0, 1, &m_ps, (LPAVICOMPRESSOPTIONS FAR *) &aopts))
            return false;

    m_status = AVIMakeCompressedStream(&m_psCompressed, m_ps, &opts, NULL);
    if (m_status != AVIERR_OK) { return false; }
```

```cpp
        m_bi.biSize = sizeof(BITMAPINFOHEADER);
        m_bi.biWidth = m_width;
        m_bi.biHeight = m_height;
        m_bi.biPlanes = 1;
        m_bi.biBitCount = 24;
        m_bi.biCompression = BI_RGB;
        m_bi.biSizeImage = 0;
        m_bi.biXPelsPerMeter = 0;
        m_bi.biYPelsPerMeter = 0;
        m_bi.biClrUsed = 0;
        m_bi.biClrImportant = 0;

        m_status = AVIStreamSetFormat(m_psCompressed, 0,
            &m_bi,                          // stream format
            m_bi.biSize                     // format size
            + m_bi.biClrUsed * sizeof(RGBQUAD)
            );
        if (m_status != AVIERR_OK) {
            return vsgu::FAILURE;
        }
        return vsgu::SUCCESS;
    }

    /** Retrieves the current status of the AVI library.  */
    inline int getStatus(void) const { return m_status;}
private:

    /** Open an AVI stream for recording purposes. */
    int     openFile(const char* fileName, unsigned int mode = OF_WRITE | OF_CREATE){
        if(m_pAVIFile) {
            vuNotify::print( vuNotify::LEVEL_WARN, NULL," You already have an AVI file opened. Close it first");
            return vsgu::FAILURE;
        }

        m_status = AVIFileOpen(   &m_pAVIFile,         // returned file pointer
            fileName,      // file name
            mode,          // mode to open file with
            NULL);         // use handler determined
        // from file extension....
        if (m_status != AVIERR_OK)
            return vsgu::FAILURE;
        return vsgu::SUCCESS;
    }

    /*** Close the AVI stream.  */
    int     closeFile(){
```

```cpp
        if(!m_pAVIFile) return vsgu::FAILURE;

        //close all opened streams
        if (m_ps) {
            AVIStreamClose(m_ps);
            m_ps = NULL;
        }

        if (m_psCompressed) {
            AVIStreamClose(m_psCompressed);
            m_psCompressed = NULL;
        }

        if (m_psText) {
            AVIStreamClose(m_psText);
            m_psText = NULL;
        }

        if (m_pAVIFile) {
            AVIFileClose(m_pAVIFile);
            m_pAVIFile = NULL;
        }

        return vsgu::SUCCESS;
}

/** Captures a single frame of video to the stream   */
int     captureFrame(){
        if( ! (m_pAVIFile && m_psCompressed && m_imageBuffer)) {
        vuNotify::print( vuNotify::LEVEL_WARN, NULL,"need to create a stream first");
            return vsgu::FAILURE;
        }

        // capture the front buffer
        glReadPixels(0, 0, m_width, m_height, GL_BGR_EXT, GL_UNSIGNED_BYTE, m_imageBuffer);

        m_status = AVIStreamWrite(m_psCompressed,           // stream pointer
            m_frameCnt,                                      // time of this frame
            1,                                               // number to write
            m_imageBuffer,                                   // pointer to data
            m_width*m_height*3,                              // size of this frame
            AVIIF_KEYFRAME,                                  // flags....
            NULL,
            NULL);

        if (m_status != AVIERR_OK)
            return vsgu::FAILURE;
```

```
            m_frameCnt++;
            return vsgu::SUCCESS;
    }

    void        resetFrameCount(int frm = 0) { m_frameCnt = frm;}

    /**The actual subscriber notification handler    */
void notify(vsChannel::Event, const vsChannel *channel, vrDrawContext *context)
{
            m_deleteLock.lock();
            captureFrame();
            m_deleteLock.unlock();
    }

    virtual void notify(vsChannel::Event, const vsChannel *, vsTraversalCull *) {};

    int                     m_status; //e.g. AVIERR_OK
    int                     m_frameCnt;
    int                     m_width;
    int                     m_height;
    void*                   m_imageBuffer;
    vuLock                  m_deleteLock;
    vuString                m_curFrameFileName;
    PAVIFILE                m_pAVIFile;
    PAVISTREAM              m_ps;
    PAVISTREAM              m_psCompressed;
    PAVISTREAM              m_psText;
    BITMAPINFOHEADER        m_bi;
};

#else // VSG_OS == VSG_WINDOWS

// The subscriber is windows-only
static int THIS_SUBSCRIBER_WILL_NOT_WORK_ON_NON_WINDOWS_OPERATING_SYSTEMS;
#endif // VSG_OS
#endif
```

图 9.7.1　通道屏幕视频录制功能设计

9.7.2　通道屏幕视频录制功能的配置使用

通道屏幕视频录制功能可以通过外部事件进行触发或结束撤销。在这里，采用键盘函数进行控制完成，启动仿真场景后，按下 F2 键开始视频录制，按下 F3 键结束视频录制。

首先，需要定义一个视频录制类为 PublicMember 类的静态成员，以方便使用：

//定义视频录制类为 PublicMember 的静态成员

```
static vsMovieRecorder * m_movieRecorder;
```

//初始化 PublicMember 的成员视频录制类为 NULL
```
vsMovieRecorder * PublicMember::m_movieRecorder=NULL;
```
其次,定义一个开始录制视频的函数为 PublicMember 类的静态成员函数:
```
void PublicMember::CTS_RecordBegin()
{
//获取场景通道
vpChannel* m_channel;
m_channel = *vpChannel::begin();
m_channel->ref();

//设置视频的宽度和高度
int width, height;
vpWindow * vpWin= * vpWindow::begin();
vpWin->getSize(&width,&height);
width=160*(width/160);
height=160*(height/160);

//添加订阅者事件,进行视频录制
m_movieRecorder = new vsMovieRecorder();
char buffer[2048];
sprintf(buffer, "m%d.avi", vsThread::resolveFrameNumber());
m_movieRecorder->configure(buffer, 60, true, width, height );
m_channel->addSubscriber(vpChannel::EVENT_POST_DRAW, m_movieRecorder);
}
```
该函数完成三个功能:第一,获取当前通道;第二,获取当前仿真通道窗口的高度和宽度;第三,实例化视频录制类并进行必要的配置,并添加当前通道的事件订阅者,触发视频录制。

然后,定义一个删除录制视频的函数为 PublicMember 类的静态成员函数:
```
void PublicMember::CTS_RecordStop()
{
    vpChannel* m_channel;
    m_channel = *vpChannel::begin();
    m_channel->ref();

    m_channel->removeSubscriber(vpChannel::EVENT_POST_DRAW,m_movieRecorder,
                                true);
        if(m_movieRecorder != NULL)
```

```
        {
            delete m_movieRecorder;
            m_movieRecorder = NULL;
        }
```

该函数主要移除通道订阅者事件,并删除视频录制类对象。

最后,是窗口键盘函数:

```
void PublicMember::CTS_Keyboard(vpWindow *window,vpWindow::Key key, int modifier, void *)
{
    switch(key)
    {
        case vpWindow::KEY_F2:
            { PublicMember::CTS_RecordBegin();    }
            break;

        case vpWindow::KEY_F3:
            { PublicMember::CTS_RecordStop(); }
            break;
             default: ;
    }
}
```

按下键盘 F2 键,弹出视频格式窗口,点击"确认"后,开始录制视频。按下键盘 F3 键,停止录制视频。

具体代码如图 9.7.2 所示。

```
//定义视频录制类为 PublicMember 的静态成员
static vsMovieRecorder * m_movieRecorder;

//初始化 PublicMember 的成员视频录制类为 NULL
vsMovieRecorder * PublicMember::m_movieRecorder=NULL;

//屏幕录制开始----------------------------------------------------------
void PublicMember::CTS_RecordBegin()
{
    //获取场景通道
    vpChannel* m_channel;
    m_channel = *vpChannel::begin();
    m_channel->ref();

    //设置视频的宽度和高度
    int width, height;
    vpWindow * vpWin= * vpWindow::begin();
    vpWin->getSize(&width,&height);
    width=160*(width/160);
    height=160*(height/160);
```

```cpp
    //添加订阅者事件，进行视频录制
    m_movieRecorder = new vsMovieRecorder();
    char buffer[2048];
    sprintf(buffer, "m%d.avi", vsThread::resolveFrameNumber());
    m_movieRecorder->configure(buffer, 60, true, width, height );
    m_channel->addSubscriber(vpChannel::EVENT_POST_DRAW, m_movieRecorder);
}
//屏幕录制结束--------------------------------------------------------
void PublicMember::CTS_RecordStop()
{
    vpChannel* m_channel;
    m_channel = *vpChannel::begin();
    m_channel->ref();

     m_channel->removeSubscriber(vpChannel::EVENT_POST_DRAW, m_movieRecorder, true);
     if(m_movieRecorder != NULL)
     {
            delete m_movieRecorder;
            m_movieRecorder = NULL;
     }
}
//键盘函数控制视频录制------------------------------------------------
void   PublicMember::CTS_Keyboard(vpWindow *window,vpWindow::Key key, int modifier,void *)
{
   switch(key)
   {
        //控制观察位置
        case vpWindow::KEY_LEFT:
            {PublicMember::CTS_pObject_observer->setTranslateY(1.0,true);}
            break;
            //
        case vpWindow::KEY_RIGHT:
            {PublicMember::CTS_pObject_observer->setTranslateY(-1.0,true); }
            break;

        case vpWindow::KEY_F2:
            { PublicMember::CTS_RecordBegin();}
             break;

        case vpWindow::KEY_F3:
            { PublicMember::CTS_RecordStop(); }
             break;
              default: ;
   }
}
```

图 9.7.2　通道屏幕视频录制功能的配置使用

运行程序，其效果如图 9.7.3 所示。图 9.7.3（a）为按下 F2 键开始录制图片，图 9.7.3（b）为录制视频播放界面。

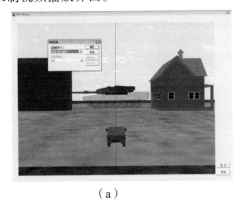 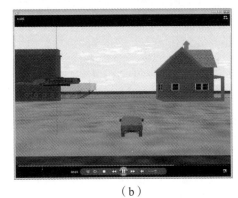

（a） （b）

图 9.7.3　通道屏幕视频录制效果

9.8　虚拟仿真中的半透明处理和纹理运动仿真

物体的半透明控制可以通过 creator 操作完成。它实现仿真中的透明度处理，用于实现需要半透明仿真的物体对象。纹理运动的仿真主要是通过操作物体的纹理坐标实现，用于仿真场景中的车轮运动或液体流动等。

9.8.1　虚拟仿真中的半透明处理

半透明处理，完全可以由 creator 处理完成。整体上分为两步，第一步是设置物体对象的透明度，第二步是重新计算阴影。

打开目录里面的 data\m1_tank 目录，打开 m1_tank.flt 文件，在 Creator 里面，选中根节点 body，发图 9.8.1 所示。

第一步，设置物体对象透明度。

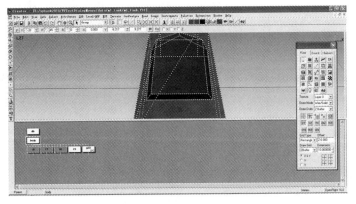

图 9.8.1　设置坦克的透明度

双击根节点,将弹出操作窗口,如图 9.8.2 所示。

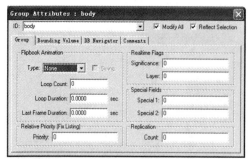
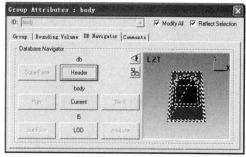
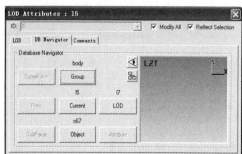
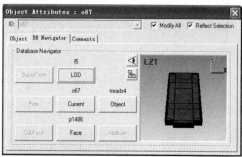
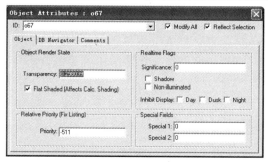

图 9.8.2　设置坦克的透明度

物体对象的透明度范围从 0 到 1,0 为完全透明,1 为不透明。可根据需要设置自己的效果,这里设置为 0.3,系统自动转换为 0.298039。

第二步,计算物体对象的阴影。这一步很重要,如果缺少这一步,透明将无法实现。选择 Creator 菜单中的"Attributes"→"Calculate Shading",如图 9.8.3 所示。

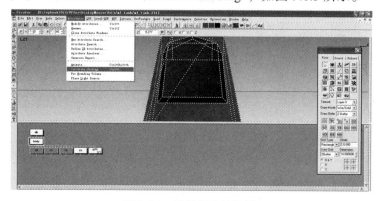

图 9.8.3　计算坦克的阴影

这时会弹出阴影计算小窗口，如图 9.8.4 所示。选择一种阴影计算方式（Shading Model），然后点击 Ok 或 Apply 按钮，完成阴影计算。

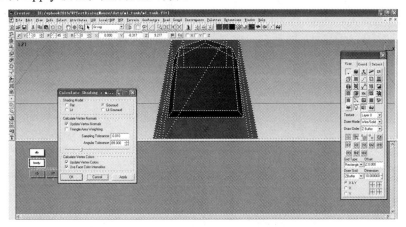

图 9.4.4　设置坦克的阴影计算参数

运行示例 VPTestDialogMouse，效果如图 9.4.5 所示。坦克的下半部分实现了透明，至于最后的效果，可以根据需要设置不同的透明值。

在实际项目中的透明效果如图 9.4.6 所示。

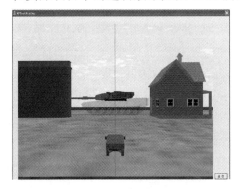 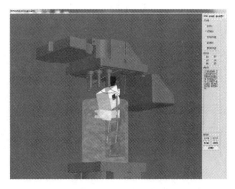

　　图 9.4.5　坦克透明实例　　　　　　图 9.8.6　实际项目中的纹理透明实例

9.8.2　虚拟仿真中的纹理运动仿真

在虚拟仿真中，纹理的运动仿真可以实现车轮、传送带、液体流动等仿真。纹理运动仿真主要分为两个过程：第一个过程是设计纹理渲染类；第二个过程是进行纹理渲染类的配置使用。

纹理渲染类的设计的代码如图 9.8.7 所示。纹理渲染类 TextureAnimation 继承自系统的渲染类 vrRenderStrategy，该类主要是重载了绘制函数：

　　　　virtual void draw(vrDrawContext *context, const vrDrawFunc::Data &data)

该函数获取纹理元素的坐标后，对坐标进行了修改，从而实现纹理运动仿真。在纹理渲染类 TextureAnimation 中，设置了一个整型变量 m_frameNumber，用于记录仿真场景是否进行了帧更新，更新后就对纹理元素坐标进行修改。

```cpp
#include <vuAllocTracer.h>
#include <vuField.h>
#include <vrRenderStrategy.h>
#include <vrTransform.h>
#include <vpApp.h>
#include <vpMotionDrive.h>
#include <vpObject.h>

#pragma once

class TextureAnimation: public vrRenderStrategy
{
public:
    TextureAnimation(void)
{   m_frameNumber = 0;}
    //~TextureAnimation(void);
  virtual void draw(vrDrawContext *context, const vrDrawFunc::Data &data)
        {
            // do pre draw stuff...
            // store the previous value of the texture matrix
            context->pushElement(vrTransform::ElementTexture::Id);
            // update the x translation of the texture matrix based upon the speed
            // of the motion model.  The movement produced by this is very trivial,
            // so if you were doing this for real you'd probably want to add some
            // more realistic physics behind this, but this will have to do for
            // now.  Note that we only want to update the matrix   once per frame,
            // so that the render strategy can be referenced by more than one
            // geometry.   If we didn't do this we'd have to create a unique render
            // strategy per geometry.
            if (vpKernel::instance()->getFrameNumber() > m_frameNumber)
            {
                vuMatrix<float> &m = m_textureTransformElement.m_matrix[0];
                //m[3][0] += 0.02f;//加减代表方向  上下
                m[3][0]-=0.06f;   //左右
                m_frameNumber = vpKernel::instance()->getFrameNumber();
            }

            // update the texture matrix
            context->setElement(vrTransform::ElementTexture::Id,
                &m_textureTransformElement);

            // draw the geometry...
            vrDrawFunc::call(context, data);

            // do post draw stuff...
            // reset the texture matrix to its previous value
            context->popElement(vrTransform::ElementTexture::Id);
```

```
            }
private:
    // the frame number of our last update
    uint m_frameNumber;
    vrTransform::ElementTexture m_textureTransformElement;
    VUMEMBASE_HEADER_INCLUDES(TextureAnimation)
};
```

图 9.8.7　纹理渲染策略类设计

对于纹理坐标的修改代码：

  vuMatrix<float> &m = m_textureTransformElement.m_matrix[0];

    //m[3][0] += 0.02f;　　//加减代表方向上下

    m[3][0]-=0.06f;　　　//左右

首先，vrTransform::ElementTexture m_textureTransformElement;用于储存纹理元素的转换矩阵，这里获取的是"m_textureTransformElement.m_matrix[0];"，它表示获取的是第一重纹理，单个物体对象是可以配置多重纹理的。然后对纹理转换矩阵进行处理：

    //m[3][0] += 0.02f;　　//加减代表方向上下

    m[3][0]-=0.06f;　　　//左右

等号右边的值大小，可以控制纹理运动的快慢；加减可以控制纹理的运动方向。具体应用时，可以多次反复试验，以得到最好的效果。另外，可以自行设计变量和机制用于控制纹理运动的方向和快慢。

设计好纹理渲染类以后，就需要有效使用它。在主线程中，内核对象配置"vpKernel::instance()->configure();"之后，帧循环之前，就是主要的配置场所，如图 9.8.8 所示。

首先是获取需要动态仿真的对象指针：

  vpObject *tank = vpObject::find("tank");

  tank->ref();

然后是实例化纹理渲染类：

  TextureAnimation *m_strategy ;

  m_strategy=new TextureAnimation() ;

  m_strategy->ref();

最后是遍历对象的子孙结点的几何体"vrGeometry *geometry;"，对所有的几何体应用设计的纹理渲染类。

```
UINT PublicMember::CTS_RunBasicThread(LPVOID)
{
    //初始化
    vp::initialize(__argc,__argv);
    //定义场景
    PublicMember::CTS_Define();
    //绘制场景
    vpKernel::instance()->configure();
    //
```

```cpp
//设置观察者
PublicMember::CTS_pObject_observer=vpObject::find("Hummer");
PublicMember::CTS_pObject_observer->ref();

//设置窗体
vpWindow * vpWin= * vpWindow::begin();
vpWin->setParent(PublicMember::CTS_RunningWindow);
vpWin->setBorderEnable(false);
vpWin->setFullScreenEnable(true);

//设置键盘
vpWin->setInputEnable(true);
vpWin->setKeyboardFunc((vrWindow::KeyboardFunc)PublicMember::CTS_Keyboard,NULL);
vpWin->setMouseFunc((vrWindow::MouseFunc)PublicMember::CTS_MouseFunc,NULL);

vpWin->open();
::SetFocus(vpWin->getWindow());

//设置纹理动画------------------------------------------------------------
    vpObject *tank = vpObject::find("tank");
 tank->ref();
 TextureAnimation *m_strategy ;
 m_strategy=new TextureAnimation() ;
 m_strategy->ref();

 vrGeometry *geometry;
 vpObject::const_iterator_named it, ite = tank->end_named();
 vsNode::const_iterator_child cit, cite;
 for (it=tank->begin_named();it!=ite;++it)
     {
            //if (strstr((*it)->getName(), "tread") != NULL)
            //if(strcmp( (*it)->getName(),"tread") >=0)
              {
                 cite = (*it)->end_child();
                 for (cit=(*it)->begin_child();cit!=cite;++cit)
                    {
                       if ((*cit)->isExactClassType( vsGeometry::getStaticClassType()))
                         {
                            geometry = static_cast<vsGeometry *>(*cit)->getGeometry();
                            geometry->setRenderStrategy(m_strategy);
                         }
                    }
              }
     }

//纹理动画配置结束--------------------------------------------------
```

```
    //帧循环
    while(vpKernel::instance()->beginFrame()!=0)
    {
        vpKernel::instance()->endFrame();
        if(!PublicMember::CTS_continueRunVP)
        {
            vpKernel::instance()->endFrame();
            vpKernel::instance()->unconfigure();
            vp::shutdown();
            return 0;
        }
    }
    return 0;
}
```

图 9.8.8　在主线程中配置使用纹理控制类

运行示例 VPTestDialogAnimateTexture，可以得到如图 9.8.9 所示的效果，图中的坦克的履带在滚动。

图 9.8.9　纹理运动控制效果

9.9　虚拟仿真中的聚光灯光源使用

光照对于物体在三维场景中的展现具有至关重要的作用。因此，合理有效地使用仿真场景中的光照，可以有效提高仿真场景中的物体的逼真度。仿真场景中的光照包含太阳光、一般灯光和聚光灯光源。

太阳光作为场景中的主要光源，具有唯一性。它对整个场景起作用，可以通过调整当前环境（vpEnv）的时间、经纬度等参数调整光线的位置和强度，但有时难以兼顾不同位置上的物体。

　　PublicMember::CTS_vpEnv->setReferencePosition(18.000001f, 24.000001f);
　　PublicMember::CTS_vpEnv->setTimeOfDay(15.000000f);

一般光源（vpLight），只对小范围内的物体起作用，而且可控性较弱。只有聚光灯光源

（vpLightLobe）具有很强的局部性，同时可以利用纹理对灯光效果进行控制，功能非常强大。

9.9.1 聚光灯光源 vpLightLobe 的理解

Vega Prime 中的普通光照对象 vpLight 可以进行光照操作，但是可控性较弱，光照效果也不明显。

聚光灯光源 vpLightLobe 的类视图如图 9.9.1 所示。该光源采用点光源方式，在场景中的某个位置投射到场景中的物体上。该光源可以配置各种参数，包括位置、亮度、角度等，对于突出照射某个局部，具有非常明显的效果。

如图 9.9.2 所示，在黑暗中，坦克上面设置了一个聚光灯光源（vpLightLobe），照射到墙上，在汽车上设置了一个普通光源（vpLight），其效果仅仅显示为一个小白斑。

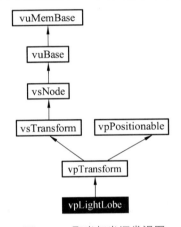
图 9.9.1 聚光灯光源类视图

图 9.9.2 黑夜中的光照效果

9.9.2 聚光灯光源 vpLightLobe 的使用

使用 vpLightLobe 的完整代码如图 9.9.3 所示。

第一步，引入必需的头文件：

  #include "vpLightLobe.h"

  #include "vpLightLobeControl.h"

第二步，启用聚光灯模块：

  vpModule::initializeModule("vpLightLobe");

第三步，考虑使用聚光灯集中控制模块：

vpLightLobeControl::instance()可以获取聚光灯控制对象,可以集中控制场景中的所有聚光灯，如控制使能开关。

  bool enable = !vpLightLobeControl::instance()->getEnable();

  vpLightLobeControl::instance()->setEnable(enable);

```
//必需的头文件
#include "vpLightLobe.h"
#include "vpLightLobeControl.h"
```

```cpp
//-----------------------------------------------------------
//启用聚光灯模块
vpModule::initializeModule( "vpLightLobe" );
//-----------------------------------------------------------
//多个灯源时，可以使用 vpLightLobeControl 集中控制，否则可以不使用
vpLightLobeControl* pLightLobeControl_myLightLobeControl = vpLightLobeControl::instance();
pLightLobeControl_myLightLobeControl->setEnable( true );
pLightLobeControl_myLightLobeControl->setEyepointFadeRange(500.000000f);
pLightLobeControl_myLightLobeControl->setRenderQuality( vpLightLobeControl::RENDER_QUALITY_
                                    GOOD );
//-----------------------------------------------------------
//整个处理过程与控制一般的对象 vpObject 类似。单个灯源定义
vpLightLobe* pLightLobe_myLightLobe1 = new vpLightLobe();
pLightLobe_myLightLobe1->setName( "myLightLobe1" );
 pLightLobe_myLightLobe1->setCullMask( 0x0FFFFFFFF );
 pLightLobe_myLightLobe1->setRenderMask( 0x0FFFFFFFF );
 pLightLobe_myLightLobe1->setIsectMask( 0x0FFFFFFFF );
 pLightLobe_myLightLobe1->setStrategyEnable( true );
 //位置、角度和缩放
 pLightLobe_myLightLobe1->setTranslate( 0.550000, 0.300000, 5.000000 );
 pLightLobe_myLightLobe1->setRotate( 0.000000, 0.000000, 0.000000 );
 pLightLobe_myLightLobe1->setScale( 1.000000, 1.000000, 1.000000 );
 pLightLobe_myLightLobe1->setStaticEnable( false );
 pLightLobe_myLightLobe1->setEnable( true );                 //使能控制
 pLightLobe_myLightLobe1->setBrightness( 20.000000f );       //亮度控制
 pLightLobe_myLightLobe1->setSpreadAngle(30.000000f);        //光照广角角度
 pLightLobe_myLightLobe1->setFadeRange( 50.000000f );        //衰减范围
 pLightLobe_myLightLobe1->setFalloff( vpLightLobe::FALLOFF_SQUARED );

 //可以使用图片文件，控制灯光效果。jpg 图片
 pLightLobe_myLightLobe1->setTextureFile( "" );
 pLightLobe_myLightLobe1->setTextureSize( 128 );

 PublicMember::CTS_s_pInstancesToUnref->push_back( pLightLobe_myLightLobe1 );
 //-----------------------------------------------------------
 //通过搭载到一般物体上，便于控制灯光的位置、角度等
 //当然，物体也需要加入场景中
 pObject_m1_tank->addChild( pLightLobe_myLightLobe1 );
 pScene_myScene->addChild(pObject_m1_tank);
```

图 9.9.3 聚光灯光源类的使用

第四步，定义使用聚光灯对象 vpLightLobe。定义使用聚光灯对象与定义使用一般的场景对象 vpObject 类似，可以控制其位置、姿态等，其中包括亮度、照射角度和灯光衰减范围等。最后，需要把 vpLightLobe 对象作为某个场景对象的孩子，vpLightLobe 对象随着场景对象的变化而变化。在图 9.9.3 中，vpLightLobe 对象作为坦克的孩子，随着坦克运动而运动。运行结果如图 9.9.4（a）所示。当然，可以定义多个聚光灯，如图 9.9.4（b）所示。

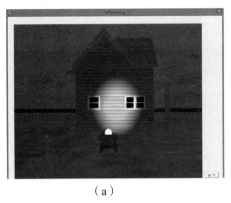
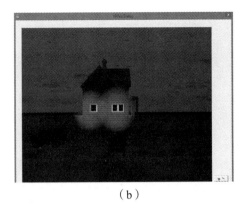

(a)   (b)

图 9.9.4 聚光灯光照效果

聚光灯对象 vpLightLobe 还可以使用纹理，对照射光线进行渲染。

  pLightLobe_myLightLobe1->setTextureFile("sun.jpg");

如图 9.9.5 所示，（a）为纹理图片，（b）为聚光灯的纹理照射效果。

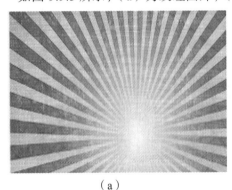
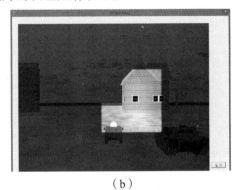

(a)   (b)

图 9.9.5 聚光灯的纹理照射效果

合理使用聚光灯对象 vpLightLobe，可以为场景创造出绚丽的灯光效果。

9.10　渲染策略的使用

渲染策略是通过一个类允许用户对具体的几何体重新定义渲染功能。渲染策略可以对每一个几何体使用，也可以对每一个通道进行使用。对几何体使用时，只对选定的几何体起作用；对通道使用时，则对通道内的所有几何体都起作用。

9.10.1　物体渲染策略的使用

vrRenderStrategy 的类视图如图 9.10.1 所示，常用的渲染策略类包含 3 个：vrRenderStrategyBounds，渲染出一个包围盒；vrRenderStrategyHighlight，渲染出一个高亮显示的几何体；vrRenderStrategyNormals，渲染出法线几何体。

图 9.10.1 渲染策略的类视图

高亮渲染策略 vrRenderStrategyHighlight 的设置，主要包含线条宽度、线条颜色和是否使用线框模式：

//高亮显示 1
vrRenderStrategyHighlight *m_strategyHighlight;
m_strategyHighlight = new vrRenderStrategyHighlight();
m_strategyHighlight->ref();
m_strategyHighlight->setLineWidth(1.0f);
m_strategyHighlight->setColor(1.0f, 0.0f, 0.0f, 1.0f);
m_strategyHighlight->setWireframeEnable(true);

包围盒渲染策略 vrRenderStrategyBounds 的设置，主要包含线条宽度、线条颜色和是否使用线框模式：

//包围盒 2
vrRenderStrategyBounds *m_strategyBounds;
m_strategyBounds = new vrRenderStrategyBounds();
m_strategyBounds->ref();
m_strategyBounds->setLineWidth(1.0f);
m_strategyBounds->setColor(1.0f, 0.0f, 0.0f, 1.0f);
m_strategyBounds->setWireframeEnable(true);

法线渲染策略 vrRenderStrategyNormals 的设置，主要包含法线长度和线条颜色：

//法线 3
vrRenderStrategyNormals *m_strategyNormals;
m_strategyNormals = new vrRenderStrategyNormals();
m_strategyNormals->setLength(0.5f);
m_strategyNormals->setColor(1.0f, 0.0f, 0.0f, 1.0f);
m_strategyNormals->ref();

具体渲染策略的使用，还需要遍历三维物体中所包含的几何体 vrGeometry，该几何体对象使用 setRenderStrategy()方法进行渲染，以下代码使对象 obj 使用高亮渲染策略进行渲染：

// 对该物体的所有几何体实施渲染策略
vrGeometry *geometry;
vpObject::const_iterator_geometry it, ite = obj->end_geometry();
for (it=obj->begin_geometry();it!=ite;++it)
    if ((*it)->isExactClassType(vsGeometry::getStaticClassType()))

```
            {
                geometry = static_cast<vsGeometry *>(*it)->getGeometry();
                geometry->setRenderStrategy(m_strategyHighlight);
            }
```

为了便于使用，把渲染策略编写成 PublicMember 类的一个静态函数：

    void setObjRenderStrategy(vpObject *obj,int m=0);

该函数包含两个参数：第一个参数是对象指针 obj；第二个参数是整数 m。m 默认值为 0，表示不使用渲染策略；m 为 1，表示使用高亮渲染策略；m 为 2，表示使用包围盒渲染策略；m 为 3，表示使用法线渲染策略。具体代码如图 9.10.2 所示。

```
//静态函数声明
static void setObjRenderStrategy(vpObject *obj,int m=0);

//函数实现
void PublicMember::setObjRenderStrategy(vpObject *obj,int m)
{
    if(!obj)
        return ;
    //高亮显示   1
    vrRenderStrategyHighlight *m_strategyHighlight;
    m_strategyHighlight = new vrRenderStrategyHighlight();
    m_strategyHighlight->ref();
    m_strategyHighlight->setLineWidth( 1.0f  );
    m_strategyHighlight->setColor( 1.0f, 0.0f, 0.0f, 1.0f );
    m_strategyHighlight->setWireframeEnable(true);
    //包围盒   2
    vrRenderStrategyBounds *m_strategyBounds;
    m_strategyBounds = new vrRenderStrategyBounds();
    m_strategyBounds->ref();
    m_strategyBounds->setLineWidth( 0.50f  );
    m_strategyBounds->setColor( 1.0f, 0.0f, 0.0f, 1.0f );
    m_strategyBounds->setWireframeEnable(true);
    //法线   3
    vrRenderStrategyNormals *m_strategyNormals;
    m_strategyNormals = new vrRenderStrategyNormals();
    m_strategyNormals->setLength( 1.0f  );
    m_strategyNormals->setColor( 1.0f, 0.0f, 0.0f, 1.0f );
    m_strategyNormals->ref();
    // 对该物体的所有几何体实施渲染策略
    vrGeometry *geometry;
    vpObject::const_iterator_geometry it, ite = obj->end_geometry();
    for (it=obj->begin_geometry();it!=ite;++it)
    {  if ((*it)->isExactClassType(vsGeometry::getStaticClassType()))
        {
            geometry = static_cast<vsGeometry *>(*it)->getGeometry();
            switch(m)
```

```
                {
                case 1:geometry->setRenderStrategy(m_strategyHighlight);break;
                case 2:geometry->setRenderStrategy(m_strategyBounds);break;
                case 3:geometry->setRenderStrategy(m_strategyNormals);break;
                default:geometry->setRenderStrategy(NULL);
                }
            }
        }
}
```

图 9.10.2　物体渲染策略方式设计

如图 9.10.3 所示，对场景中的汽车分别进行了高亮渲染、包围盒渲染和法线渲染策略，只需要在窗口相应的按钮下调用函数：

PublicMember::setObjRenderStrategy(obj,m);

与 m 值对应，可对物体按相应的渲染策略进行渲染。

（a）高亮渲染

（b）包围盒渲染

（c）法线渲染

（d）默认效果

图 9.10.3　物体的几种渲染效果

9.10.2　通道渲染的使用

通道渲染的使用主要通过通道订阅类 vsChannel::Subscriber 的 vsChannel::EVENT_PRE_DRAW 事件来完成。

在这里，设计了一个通道渲染类 myChannRender 来执行通道的渲染：

class myChannRender:public vsChannel::Subscriber

通道渲染类 myChannRender 继承于 vsChannel::Subscriber，重载了通道的绘制函数 notify，在该函数中处理 vsChannel::EVENT_PRE_DRAW 事件。该事件中，首先清理掉原有的通道色彩，根据需要设置 RGB 和 ALPHA 通道，RGB 分别代表红绿蓝 3 个颜色通道，ALPHA 通道则代表透明。

通道颜色的设置通过 glColorMask 函数完成：

  glColorMask(GL_FALSE, GL_FALSE, GL_FALSE, GL_FALSE);

glColorMask 函数通过设置颜色掩码，由 RGB 形成单色或者混合颜色。在设计里，通过一个静态整数变量 PublicMember::RenderColor 传递数值，0 代表红色，1 代表绿色，2 代表蓝色，对通道进行颜色渲染。其他情况则恢复通道的正常渲染：

  glColorMask(GL_TRUE, GL_TRUE, GL_TRUE, GL_TRUE);

通道渲染类 myChannRender 的完整代码如图 9.10.4 所示。

```cpp
#include <vpChannel.h>
#include ".\publicmember.h"
#pragma once

class myChannRender:public vsChannel::Subscriber
{
public:
    myChannRender(void){};
    ~myChannRender(void){};

    /* inherited pre / post cull notification method. We have to implement
     * this even though we don't care about these events since the method is
     * abstract, however, we can just make it a noop.        */
    virtual void notify(vsChannel::Event, const vsChannel *,vsTraversalCull *) {}

    /** inherited pre / post draw notification method       */
    virtual void notify(vsChannel::Event event, const vsChannel *channel,
        vrDrawContext *context)
    {
        // in the pre draw, set the color mask to render the channel green
        // if necessary.   A couple of points of note here.   First, since
        // glColorMask() is not a function wrapped by VSG, we don't need to
        // go through vrStateMgr.   Second, if we are going to render to only
        // the green channel, we need to clear the other channels to black
        // before hand so they don't have anything left over from the previous
        // frame in the frame buffer.   Note that this has to be done here,
        // before the color mask is set, because once its set, even the normal
        // channel clear will only effect the green channel.   Also note that
        // if this channel was always going to be rendered green that we could
        // get by w/ only performing this action once on the first frame.
        // Again, since glColorMask() isn't being called by VSG, we wouldn't
        // have to worry about our settings getting changed.
        if (event == vsChannel::EVENT_PRE_DRAW)
        {
            channel->getVrChannel()->setClearColor(0.0f, 0.0f, 0.0f, 0.0f);
            channel->getVrChannel()->apply(context);
```

```
            switch(PublicMember::RenderColor)
             {
             case 0:glColorMask(GL_TRUE,  GL_FALSE, GL_FALSE, GL_FALSE);break;//红色
             case 1:glColorMask(GL_FALSE, GL_TRUE,  GL_FALSE, GL_FALSE);break;//绿色
             case 2:glColorMask(GL_FALSE, GL_FALSE, GL_TRUE,  GL_FALSE);break;//蓝色
             default:;
             }
         }
        // in the post draw reset the color mask for normal rendering
         else glColorMask(GL_TRUE, GL_TRUE, GL_TRUE, GL_TRUE);
    }
};
```

图 9.10.4　通道渲染订阅类设计

通道渲染类 myChannRender 必须在 VP 的主线程中为通道添加订阅事件才能发挥作用，其主要代码如下：

```
// 为通道添加订阅事件
myChannRender * ChannRender=new myChannRender();
vpChannel *channel = *vpChannel::begin();
channel->addSubscriber(vsChannel::EVENT_PRE_DRAW, ChannRender);
channel->addSubscriber(vsChannel::EVENT_POST_DRAW, ChannRender);
```

首先实例化一个通道渲染对象指针，然后获取当前通道，最后分别为当前通道添加 vsChannel::EVENT_PRE_DRAW 和 vsChannel::EVENT_POST_DRAW 事件，以便使用通道渲染功能。其完整代码如图 9.10.5 所示。

```
#include ".\mychannrender.h"
 UINT PublicMember::CTS_RunBasicThread(LPVOID)
{
  vp::initialize(__argc,__argv);      //初始化
  PublicMember::CTS_Define();         //定义场景
  vpKernel::instance()->configure();  //绘制场景
   //设置观察者
  PublicMember::CTS_pObject_observer=vpObject::find("Hummer");
  PublicMember::CTS_pObject_observer->ref();

  //为通道添加订阅事件
   myChannRender * ChannRender=new myChannRender();
   vpChannel *channel = *vpChannel::begin();
   channel->addSubscriber(vsChannel::EVENT_PRE_DRAW, ChannRender);
   channel->addSubscriber(vsChannel::EVENT_POST_DRAW, ChannRender);
  //

  //设置窗体
  vpWindow * vpWin= * vpWindow::begin();
  vpWin->setParent(PublicMember::CTS_RunningWindow);
  vpWin->setBorderEnable(false);
```

```
vpWin->setFullScreenEnable(true);

//设置键盘
vpWin->setInputEnable(true);
vpWin->setKeyboardFunc((vrWindow::KeyboardFunc)PublicMember::CTS_Keyboard,NULL);

vpWin->open();
::SetFocus(vpWin->getWindow());

//帧循环
while(vpKernel::instance()->beginFrame()!=0)
{
    vpKernel::instance()->endFrame();
    if(!PublicMember::CTS_continueRunVP)
    {
      vpKernel::instance()->endFrame();
       vpKernel::instance()->unconfigure();
      vp::shutdown();
      return 0;
      }
}
return 0;
}
```

图 9.10.5　通道渲染类的使用

（a）红色通道

（b）绿色通道

（c）蓝色通道

（d）

图 9.10.6　各色通道渲染效果

在操作界面上，设置了红色通道、绿色通道和蓝色通道，以及正常渲染，如图 9.10.6 所示。只需要修改 PublicMember::RenderColor 的值即可：0 代表红色，1 代表绿色，2 代表蓝色。当然，可以通过设置对通道进行更多的颜色混合渲染。

9.11　模板效果的设计与使用

Vega Prime 提供了一个命名空间 vrStencilTest 来提供模板 Stencil 效果。模板 Stencil 效果本质上是对默认为矩形的屏幕空间进行渲染控制，渲染出用户需要的屏幕形状，如圆形（可参考本节后面的渲染效果图）模拟望远镜的放大效果，以便增加仿真的真实程度。

模板 Stencil 效果是对场景输出的控制，因此，主要借助于对场景通道的控制来完成功能，场景通道的控制主要是对场景通道订阅类 vsChannel::Subscriber 的事件 vsChannel::EVENT_PRE_DRAW 和 vsChannel::EVENT_POST_DRAW 进行操作以达到实际效果。命名空间 vrStencilTest 提供了一个结构体 vrStencilTest::Element 来储存模板效果的参数，该结构体的值通过 vrDrawContext::setElement() 函数传递给 vrDrawContext 进行绘制。场景通道订阅类 vsChannel::Subscriber 正好可以在事件中操作 vrDrawContext。

9.11.1　模板效果的基本要素

模板效果的实现，主要包含 3 个基本要素：打开场景窗口设置模板效果开关，设置通道可绘制区域，设置模板显示效果。

（1）窗口开关。

在默认设置中，vpWindow 的模板效果是关闭的：

　　pWindow_myWindow->setNumStencilBits(0);

因此，要使用模板效果，必须打开此开关：

　　pWindow_myWindow->setNumStencilBits(1);

（2）通道绘制区域。

在默认设置中，vpChannel 的绘制区域，是归一化处理后的标准矩形：

　　pChannel_myChannel->setDrawArea(0.00 , 1.00 , 0.00 , 1.00);

实际上，setDrawArea 的 4 个参数是归一化处理后相对于屏幕 4 个边线的距离：

　　int setDrawArea　　（double L, double R, double B, double T）;

4 个参数 L,R,B,T 分别对应父窗口的左边距、右边距、下边距和上边距，取值范围都是[0, 1]。另外需要注意，4 个参数的设置必须保证符合实际情况，否则绘制区域将不能正常显示。例如，绘制区域的左右边距应该保证绘制区域的左右两边不能错位，左边距过大可能使左边位于右边的右边，绘制区域将不能正常显示。

（3）设置模板显示效果。

前面两项都非常简单，这一项是模板效果的核心，相对复杂一些。模板效果借助于对场景通道的控制来完成，场景通道订阅类 vsChannel::Subscriber 通过事件 vsChannel:: EVENT_

PRE_DRAW 和 vsChannel::EVENT_POST_DRAW 对通道进行处理。

命名空间 vrStencilTest 提供的结构体 vrStencilTest::Element 用来储存模板效果的参数，vrStencilTest::Element 的继承视图如图 9.11.1 所示。

另外，还涉及简单的 OpenGL 操作。在显示区域绘制实际的模板效果需要由 OpenGL 绘制函数来完成。

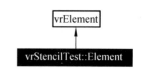

图 9.11.1　模板结构体继承视图

9.11.2　单通道模板效果设计

按照模板效果的 3 个基本要素，前 2 个要素非常简单，只要记住进行设置即可。

窗口模板效果开关采用如下代码设置即可：

　　pWindow_myWindow->setNumStencilBits(1);

通道绘制区域设置采用如下代码即可：

　　pChannel_myChannel->setDrawArea(0.00,1.00,0.00,1.00);

参数值采用了默认设置，也就是表示绘制区域是整个窗口。开发者也可以自己尝试[0, 1]的合理值，试试效果。比如：

　　pChannel_myChannel->setDrawArea(0.01,0.99,0.01,0.99);

可以设置绘制区域为窗口的中间[0.01, 0.99]的区域，四周留下黑色边框，如图 9.11.2 所示。

模板显示效果的核心设计在于通道订阅类的设计。

首先，在 PublicMember 类里面设计两个静态浮点数，默认值为 0.5，表示圆形模板的中心位于绘制区域的中心，用于控制圆形模板的中心位置：

　　//圆形模板中心坐标

　　float　　PublicMember::drawX=0.5;

　　float　　PublicMember::drawY=0.5;

其次，进行窗口订阅类的设计，其详细代码如图 9.11.3 所示。

图 9.11.2　设置通道绘制区域效果

单通道模板效果通道订阅类 myChannelStencil 继承于通道订阅类 vsChannel::Subscriber，实现原有的两个虚函数。

虚函数 virtual void notify(vsChannel::Event event, const vsChannel *channel,vrDrawContext *context)是主要的功能设计区。这里首先定义了一个模板元素 stencilTestElement，然后在 vsChannel::EVENT_PRE_DRAW 事件中处理全部事情，其他事件则不再进行特殊处理。

由于 OpenGL 没有直接绘制圆形的函数，就利用静态变量只进行一次赋值的特性，准备了一个包含 37 个元素的数组，每个元素包含 x 和 y 两个值，也就是平面上的 37 个点，这 37 个点是圆形模板上的 37 个点，从而通过绘制扇形得到圆形。

然后，设置通道的基本颜色，设置为蓝色：

　　channel->getVrChannel()->setClearColor(0.0f, 0.0f, 1.0f, 1.0f);

启用通道的绘制上下文：
  channel->getVrChannel()->apply(context);
接下来，允许重新设置图形状态，并启用其默认值：
  context->pushElements(true);
模板效果只对平面起作用，需要设置相关项目：
  glDisable(GL_DEPTH_TEST);
  glMatrixMode(GL_PROJECTION);
  glPushMatrix();
  glLoadIdentity();
  gluOrtho2D(0.0, 1.0, 0.0, 1.0);
  glMatrixMode(GL_MODELVIEW);
  glPushMatrix();
  glLoadIdentity();
清理模板缓冲区的值，并设置为 0：
  glClearStencil(0);
  glClear(GL_STENCIL_BUFFER_BIT);
设置并启用模板元素值，其他值均被模板值代替：
  stencilTestElement.m_enable = true;
  stencilTestElement.m_mode = vrStencilTest::MODE_ALWAYS;
  stencilTestElement.m_ref = 1;
  stencilTestElement.m_mask = 1;
  stencilTestElement.m_fail = vrStencilTest::OPERATION_REPLACE;
  stencilTestElement.m_dfail = vrStencilTest::OPERATION_REPLACE;
  stencilTestElement.m_dpass = vrStencilTest::OPERATION_REPLACE;
  context->setElement(vrStencilTest::Element::Id,&stencilTestElement);//启用模板值
绘制模板显示的形状，这里绘制为圆形：
  glBegin(GL_TRIANGLE_FAN);

  for (i=0;i<numVertices;i++)
  glVertex2f(PublicMember::drawX + vertex[i][0], PublicMember::drawY + vertex[i][1]);
  glEnd();
其中的 PublicMember::drawX 和 PublicMember::drawY 代表圆心位置，默认值为绘制区域中心，后面可以在键盘事件中修改它们的值，以便于控制模板特效的显示位置。
设置并启用模板元素值，在绘制的圆形区域内，场景值得以保持：
  context->popElements(false);
  stencilTestElement.m_enable = true;
  stencilTestElement.m_mode = vrStencilTest::MODE_EQUAL;
  stencilTestElement.m_ref = 1;
  stencilTestElement.m_mask = 1;

stencilTestElement.m_fail = vrStencilTest::OPERATION_KEEP;
stencilTestElement.m_dfail = vrStencilTest::OPERATION_KEEP;
stencilTestElement.m_dpass = vrStencilTest::OPERATION_KEEP;
context->setElement(vrStencilTest::Element::Id,&stencilTestElement);

至此，单通道模板效果的设计全部完成，所有设计均在 vsChannel::EVENT_PRE_DRAW 事件中完成，详细代码如图 9.11.3 所示。

```cpp
#include <vrMode.h>
#include <vpChannel.h>
#include <GL/gl.h>
#include <GL/glu.h>
#include ".\publicmember.h"
#pragma once

class myChannelStencil: public vsChannel::Subscriber
{
public:
   myChannelStencil(void);
   ~myChannelStencil(void);

   /**
    * inherited pre / post cull notification method. We have to implement
    * this even though we don't care about these events since the method is
    * abstract, however, we can just make it a noop.
    */
   virtual void notify(vsChannel::Event, const vsChannel *,vsTraversalCull *) {}

   /**inherited pre / post draw notification method.   Note here that since
    * we're using this function for both channels that we'll need to add some
    * extra logic to organize our code here.      */
   virtual void notify(vsChannel::Event event, const vsChannel *channel,vrDrawContext *context)
   {
      // for the main channel we'll do all of our stencil stuff so
       // we can get the rounded outputs

           vrStencilTest::Element stencilTestElement;

           // for the pre draw we'll write our mask to the stencil buffer
           // and enable the stencil test for the channel rendering.   Note
           // that all of the operations that are being performed below wrt
           // writing to the stencil buffer technically only need to be
           // performed whenever the window is resized, but for the sake of
           // simplicity, we'll do them every frame here.   A
           // vsWindow::EVENT_RESIZE event could be used to be more optimal,
           // or if you knew your window size was not going to change, you
           // could just do these operations the first time through.
```

```cpp
if (event == vsChannel::EVENT_PRE_DRAW) {

    // since OpenGL doesn't directly support rendering of a
    // circle, we'll create a simple triangle fan primitive to
    // approximate one the first time through.
    int i;
    static bool first = true;
    const int numVertices = 38;
    static vuVec2<float> vertex[numVertices];

    if (first) {
        float s, c, angle = 0.0f;
        vertex[0].set(0.0f, 0.0f);
        for (i=1;i<numVertices;i++,angle+=10.0f) {
            vuSinCos(360.0f - angle, &s, &c);
            vertex[i].set(0.25f * s, 0.5f * c);
        }
        vertex[37] = vertex[1];
        first = false;
    }

    // since we're in a pre draw callback here, the channel hasn't
    // actually been configured yet (i.e. the viewport, projection
    // matrix, etc. don't get specified until the channel's draw
    // function is called), so we'll need to call
    // vrChannel::apply() here to force the update.   This update
    // also clears the frame buffer, so this is a nice opportunity
    // to reset the clear color as well.   Note that just changing
    // the clear color in the ACF wouldn't work, since the clear
    // color is being overridden by the environment.   By the action
    // we're taking here we're actually overriding the clear color
    // specified by the environment.
    channel->getVrChannel()->setClearColor(0.0f, 0.0f, 1.0f, 1.0f);
    channel->getVrChannel()->apply(context);

    // store the current graphics state and reset all of the
    // graphics state elements to their default values
    context->pushElements(true);

    // disable the depth test and set up the projection and model
    // view matrices for orthographic rendering
    glDisable(GL_DEPTH_TEST);
    glMatrixMode(GL_PROJECTION);
    glPushMatrix();
    glLoadIdentity();
    gluOrtho2D(0.0, 1.0, 0.0, 1.0);
    glMatrixMode(GL_MODELVIEW);
    glPushMatrix();
```

```cpp
            glLoadIdentity();

            // clear the stencil buffer to all zeros
            glClearStencil(0);
            glClear(GL_STENCIL_BUFFER_BIT);

            // write ones to the stencil buffer where the circles that
            // represent the stencil are
            stencilTestElement.m_enable = true;
            stencilTestElement.m_mode = vrStencilTest::MODE_ALWAYS;
            stencilTestElement.m_ref = 1;
            stencilTestElement.m_mask = 1;
            stencilTestElement.m_fail = vrStencilTest::OPERATION_REPLACE;
            stencilTestElement.m_dfail = vrStencilTest::OPERATION_REPLACE;
            stencilTestElement.m_dpass = vrStencilTest::OPERATION_REPLACE;
            context->setElement(vrStencilTest::Element::Id,&stencilTestElement);

            glBegin(GL_TRIANGLE_FAN);
            for (i=0;i<numVertices;i++)
              glVertex2f(PublicMember::drawX + vertex[i][0], PublicMember::drawY + vertex[i][1]);

            glEnd();

            // restore the graphics state
            context->popElements(false);

            // now configure the stencil buffer to only accept fragments
            // whose stencil value is equal to one
            stencilTestElement.m_enable = true;
            stencilTestElement.m_mode = vrStencilTest::MODE_EQUAL;
            stencilTestElement.m_ref = 1;
            stencilTestElement.m_mask = 1;
            stencilTestElement.m_fail = vrStencilTest::OPERATION_KEEP;
            stencilTestElement.m_dfail = vrStencilTest::OPERATION_KEEP;
            stencilTestElement.m_dpass = vrStencilTest::OPERATION_KEEP;
            context->setElement(vrStencilTest::Element::Id,&stencilTestElement);
        }
        // for the post draw, just disable the stencil test
        else {context->setElement(vrStencilTest::Element::Id,&stencilTestElement);}
    };//end of function
};//end of class
```

图 9.11.3 单通道模板效果通道订阅类的设计

单通道模板效果设计完成后,还需要在主线程中进行启用,才会有实际效果。为了使用方便,需要预先定义一个通道静态变量 PublicMember::MainChannel。整个启用过程是在 VP

的主线程中配置完成的，主要是为通道事件 EVENT_PRE_DRAW 和 EVENT_POST_DRAW 添加处理程序。

首先，获取通道，并保存于通道指针中：

PublicMember::MainChannel = vpChannel::find("myChannel");
PublicMember::MainChannel->ref();

其次，实例化一个模板特效通道订阅类：

myChannelStencil *myChStencil=new myChannelStencil();

最后，为通道添加订阅者事件及其处理程序：

PublicMember::MainChannel->addSubscriber(vsChannel::EVENT_PRE_DRAW,
         myChStencil);
PublicMember::MainChannel->addSubscriber(vsChannel::EVENT_POST_DRAW,
         myChStencil);

当然，在主线程退出时，可以移除订阅者：

PublicMember::MainChannel->removeSubscriber(vsChannel::EVENT_PRE_DRAW,
         myChStencil,true);
PublicMember::MainChannel->removeSubscriber(vsChannel::EVENT_POST_DRAW,
         myChStencil,true);

详细代码如图 9.11.4 所示。

```cpp
#include "StdAfx.h"
#include ".\publicmember.h"
#include ".\mychannelstencil.h"

UINT PublicMember::CTS_RunBasicThread(LPVOID)
{
    vp::initialize(__argc,__argv);          //初始化
    PublicMember::CTS_Define();             //定义场景
    vpKernel::instance()->configure();      //绘制场景

//单通道模板效果的启用
//---------------------------------------------------------------
    PublicMember::MainChannel = vpChannel::find("myChannel");
    PublicMember::MainChannel->ref();

    myChannelStencil *myChStencil=new myChannelStencil();

    PublicMember::MainChannel->addSubscriber(vsChannel::EVENT_PRE_DRAW, myChStencil);
    PublicMember::MainChannel->addSubscriber(vsChannel::EVENT_POST_DRAW, myChStencil);

//---------------------------------------------------------------

//设置窗体
    vpWindow * vpWin= * vpWindow::begin();
```

```
vpWin->setParent(PublicMember::CTS_RunningWindow);
vpWin->setBorderEnable(false);
vpWin->setFullScreenEnable(true);

//设置键盘
vpWin->setInputEnable(true);
vpWin->setKeyboardFunc((vrWindow::KeyboardFunc)PublicMember::CTS_Keyboard,NULL);

vpWin->open();
::SetFocus(vpWin->getWindow());

//帧循环
while(vpKernel::instance()->beginFrame()!=0)
{
    vpKernel::instance()->endFrame();
    if(!PublicMember::CTS_continueRunVP)
    {
      PublicMember::MainChannel->removeSubscriber(vsChannel::EVENT_PRE_DRAW,
                                    myChStencil,true);
      PublicMember::MainChannel->removeSubscriber(vsChannel::EVENT_POST_DRAW,
                                    myChStencil,true);
      vpKernel::instance()->unconfigure();
      vp::shutdown();
            return 0;
      }
   }
   return 0;
}
```

图 9.11.2　模板通道订阅类在主线程中的使用

运行结果如图 9.11.5（a）所示。

在键盘的左移键功能中减小 PublicMember::drawX 的值，使模板圆形左移，其效果如图 9.11.3（b）所示当然，右移与此类似。

在键盘的下移键功能中减小 PublicMember::drawY 的值，使模板圆形下移，其效果如图 9.11.3（d）所示当然，上移与此类似。

（a）居中模板　　　　　　　　　　　　　　（b）中心左移

（c）场景放大　　　　　　　　　　（d）中心下移

图 9.11.5　单通道模板效果

为了更好地体现模板效果，需要重新设计键盘函数的功能。这里增加了一个静态 float 类型的变量 PublicMember::m_magnification，用于控制放大倍数。

放大与缩小效果则放在了字母 m 功能之下，PublicMember::m_magnification 默认为 1，当其值被修改时，通过修改通道的 FOV 角度值实现场景的放大与缩小：

PublicMember::MainChannel->setFOVSymmetric(45.0/PublicMember::m_magnification ,-1.0f);
PublicMember:: MainChannel ->setLODVisibilityRangeScale(1.0f/ PublicMember::m_ magnification);
键盘函数的详细代码如图 9.11.6 所示。

```
void   PublicMember::CTS_Keyboard(vpWindow *window,vpWindow::Key key, int modifier,void *)
{
  switch(key)
  {
  case vpWindow::KEY_LEFT:
       PublicMember::drawX=PublicMember::drawX-0.01; //模板中心左移
       break;
  case vpWindow::KEY_RIGHT:
       PublicMember::drawX=PublicMember::drawX+0.01; //模板中心右移
       break;
  case vpWindow::KEY_UP:
       PublicMember::drawY=PublicMember::drawY+0.01; //模板中心上移
       break;
  case vpWindow::KEY_DOWN:
       PublicMember::drawY=PublicMember::drawY-0.01;    //模板中心下移
       break;

    case vrWindow::KEY_m:                              // 减小放大倍数
       PublicMember::m_magnification=PublicMember::m_magnification-1;
       if( PublicMember::m_magnification=PublicMember::m_magnification<1)
           PublicMember::m_magnification=1;

PublicMember::MainChannel->setFOVSymmetric( 45.0/PublicMember::m_magnification ,-1.0f );
PublicMember:: MainChannel ->setLODVisibilityRangeScale(1.0f/ PublicMember::m_magnification);
       break;
```

```
case vrWindow::KEY_M:                                    // 增大放大倍数
  PublicMember::m_magnification=PublicMember::m_magnification+1;
  PublicMember::MainChannel->setFOVSymmetric( 45.0/PublicMember::m_magnification ,-1.0f );
  PublicMember:: MainChannel ->setLODVisibilityRangeScale(1.0f/PublicMember::m_magnification);
    break;

    default:  ;
  }
}
```

图 9.11.6 单模板效果的键盘函数

效果如图9.11.5(c)所示,通道的FOV角度默认为45°,修改 PublicMember::m_magnification 的值也就能修改通道的FOV角度,从而实现场景的放大与缩小。

9.11.3 双通道模板效果设计

在实际应用中,双通道模板效果更为实用,如图 9.11.7 所示。主通道总览全局,子通道关注红绿灯路口。

双通道模板效果的实现与单通道模板效果的实现步骤类似,但逻辑上包含主通道和子通道。主通道与平常使用的通道功能类似,只是增加了绘制任务,如图 9.11.7 所示,主通道绘制了中心区域的矩形绘制观察区域和左下角的放大倍数。子通道如图 9.11.7 的右下角所示,使用模板效果。

图 9.11.7 双通道模板效果

主要分为以下 3 个步骤实现:
(1)建立双通道效果,主要代码如图 9.11.8 所示。
窗口代码需要打开模板效果开关:

    pWindow_myWindow->setNumStencilBits(1);

建立主通道:

    vpChannel* pChannel_myChannel = new vpChannel();
    vpChannel_myChannel->setName("myChannel");
    pChannel_myChannel->setDrawArea(0.00 ,1.00 ,0.00 ,1.00);//全窗口绘制
    pChannel_myChannel->setFOVSymmetric(40.00f, -1.00f); //FOV 角度为 40

建立子通道:

    vpChannel* pChannel_bigger = new vpChannel();
    pChannel_bigger->setName("biggerChannel");
    pChannel_bigger->setDrawArea(0.69 ,0.99,0.01 ,0.31);//绘制于窗口右下角
    pChannel_bigger->setFOVSymmetric(20.00f, -1.00f); //FOV 角度为 20

建立主观察者:

vpObserver* pObserver_myObserver = new vpObserver();
pObserver_myObserver->setName("myObserver");

建立子观察者：

vpObserver* pObserver_bigger = new vpObserver();
pObserver_bigger->setName("biggerObserver");

对象也需要建立，建立完成后，还需要为窗口配置双通道，为两个观察者配置必要的要素。详细代码如图 9.11.8 所示。

```
vpWindow* pWindow_myWindow = new vpWindow();
pWindow_myWindow->setName( "myWindow" );
pWindow_myWindow->setNumStencilBits( 1 );
//其他代码采用默认值********************************************
PublicMember::CTS_s_pInstancesToUnref->push_back( pWindow_myWindow );

vpChannel* pChannel_myChannel = new vpChannel();
pChannel_myChannel->setName( "myChannel" );
pChannel_myChannel->setOffsetTranslate( 0.000000 ,   0.000000 ,   0.000000 );
pChannel_myChannel->setOffsetRotate( 0.000000 ,   0.000000 ,   0.000000 );
pChannel_myChannel->setDrawArea( 0.000000 ,   1.000000 ,   0.000000 ,   1.000000 );
pChannel_myChannel->setFOVSymmetric( 45.000000f ,   -1.000000f );
 //其他代码采用默认值********************************************
PublicMember::CTS_s_pInstancesToUnref->push_back( pChannel_myChannel );

vpChannel* pChannel_bigger = new vpChannel();
pChannel_bigger->setName( "biggerChannel" );
pChannel_bigger->setOffsetTranslate( 0.000000 ,   0.000000 ,   0.000000 );
pChannel_bigger->setOffsetRotate( 0.000000 ,   0.000000 ,   0.000000 );
pChannel_bigger->setDrawArea( 0.690000 ,   0.990000 ,   0.010000 ,   0.310000 );
pChannel_bigger->setFOVSymmetric( 20.000000f ,   -1.000000f );
//其他代码采用默认值********************************************
PublicMember::CTS_s_pInstancesToUnref->push_back( pChannel_bigger );

vpObserver* pObserver_myObserver = new vpObserver();
pObserver_myObserver->setName( "myObserver" );
pObserver_myObserver->setStrategyEnable( false );
pObserver_myObserver->setTranslate( 2000.000000 ,   2500.000000 ,   15.000000 );
pObserver_myObserver->setRotate( -90.000000 ,   0.000000 ,   0.000000 );
pObserver_myObserver->setLatencyCriticalEnable( false );
PublicMember::CTS_s_pInstancesToUnref->push_back( pObserver_myObserver );

vpObserver* pObserver_bigger = new vpObserver();
pObserver_bigger->setName( "biggerObserver" );
pObserver_bigger->setStrategyEnable( false );
pObserver_bigger->setTranslate( 2300.000000 ,   2500.000000 ,   5.000000 );
pObserver_bigger->setRotate( -90.000000 ,   0.000000 ,   0.000000 );
pObserver_bigger->setLatencyCriticalEnable( false );
```

```
PublicMember::CTS_s_pInstancesToUnref->push_back( pObserver_bigger );

//配置双通道
pWindow_myWindow->addChannel( pChannel_myChannel );
pWindow_myWindow->addChannel( pChannel_bigger );

//配置主观察者
pObserver_myObserver->setStrategy( pMotionDrive_myMotion );
pObserver_myObserver->addChannel( pChannel_myChannel );
pObserver_myObserver->addAttachment( pEnv_myEnv );
pObserver_myObserver->setScene( pScene_myScene );

//配置子观察者
pObserver_bigger->setStrategy( pMotionDrive_myMotion );
pObserver_bigger->addChannel( pChannel_bigger );
pObserver_bigger->addAttachment( pEnv_myEnv );
pObserver_bigger->setScene( pScene_myScene );
```

图 9.11.8 建立双通道场景

（2）模板效果类设计。

双通道模板效果类的设计与单通道模板效果类的设计基本相同，最大的不同就是要在实现函数中区分主通道和子通道，实现不同的功能。

对于主通道，需要绘制一个矩形框和显示放大倍数，因此，需要设计一个字符串成员：

```
vrString *m_string;
```

需要在构造函数中启用字体，并调节字体间距：

```
char str[256], *cp = getenv("MPI_LOCATE_VEGA_PRIME");
sprintf(str, "%s/config/vsg/vsgr", cp);
vuSearchPath *path = vrFontFactory::getDefaultSearchPath();
path->set(str);
vrFontFactory *fontFactory = new vrFontFactory();
vrFont *font = fontFactory->read("mpi_font.dat");
fontFactory->unref();

float scale = 0.02f;
vuVec3<float> spacing;
vuMatrix<float> mat;
vrGeometry *geometry;
vrFont3D *font3D = static_cast<vrFont3D *>(font);
mat.makeScale(scale, scale, scale);
for (int i=0;i<256;i++) {
    if ((geometry = font3D->getGeometry(i)) != NULL)
        geometry->transform(mat);
```

```
            font3D->getSpacing(i, &spacing);
            spacing *= scale;
            font3D->setSpacing(i, spacing);
        }
    m_string = new vrString();
    m_string->setFont(font);
    m_string->ref();
```
在析构函数中释放字符指针：
```
    m_string=NULL;
```
主通道绘制矩形和放大倍数：
```
    // store the current graphics state and reset all of the graphics
    // state elements to their default values
    context->pushElements(true);

    // disable the depth test and set up the projection and model view
    // matrices for orthographic rendering
    glDisable(GL_DEPTH_TEST);
    glMatrixMode(GL_PROJECTION);
    glPushMatrix();
    glLoadIdentity();
    gluOrtho2D(0.0, 1.0, 0.0, 1.0);
    glMatrixMode(GL_MODELVIEW);
    glPushMatrix();
    glLoadIdentity();

    // get the fields of view for the main channel.   Note that this
    // code assumes the vertical field of view was specified as -1,
    // so we'll need to determine the actual field of view by
    // computing the aspect ratio.
    int width, height;
    float hfov, vfov, aspect;
    pWindow *window = *vpWindow::begin();
    assert(window);
    window->getSize(&width, &height);
    channel->getFOVSymmetric(&hfov, &vfov);
    if (width > height)
        aspect = (float) width / (float) height;
    else aspect = (float) height / (float) width;
    vfov = hfov / aspect;
```

```
// draw a box around what the binoculars are looking at.    We'll
// compute the size of the box by computing the ratio between the
// main and binocular fields of view.
float x = 40.0f / PublicMember::m_magnification / hfov / 2.0f;
float y = 40.0f / PublicMember::m_magnification / 2.0f / vfov / 2.0f;

glColor4f(1.0f, 1.0f, 1.0f, 1.0f);
glBegin(GL_LINE_LOOP);

glVertex2f(PublicMember::drawX + -x, PublicMember::drawY +  y);
glVertex2f(PublicMember::drawX + -x, PublicMember::drawY + -y);
glVertex2f(PublicMember::drawX +  x, PublicMember::drawY + -y);
glVertex2f(PublicMember::drawX +  x, PublicMember::drawY +  y);

glEnd();

// draw the magnification level
char str[256];
sprintf(str, "%dX", PublicMember::m_magnification);
m_string->setColor(1.0f, 1.0f, 1.0f, 1.0f);
m_string->setString(str);
m_string->setPosition(0.02f, 0.02f, 0.0f);
m_string->draw(context);

// restore the graphics state
context->popElements(false);
```

其他代码则与单通道模板效果相同，设置子通道的模板效果，详细代码如图 9.11.9 所示。其中的几个变量都是定义在 PublicMember 中的静态成员：

PublicMember::drawX : float 类型，默认值为 0.5，模板圆心横坐标；

PublicMember::drawY : float 类型，默认值为 0.5，模板圆心纵坐标；

PublicMember::biggerChannel: 子通道指针；

PublicMember::m_magnification: int 类型，默认值为 2，放大倍数。

```
#include <vrFontFactory.h>
#include <vrString.h>
#include <vrMode.h>
#include <vpChannel.h>
#include <GL/gl.h>
#include <GL/glu.h>
#include ".\publicmember.h"
```

```cpp
#pragma once
class myChannelStencil: public vsChannel::Subscriber
{
public:
    // a string to render with
    vrString *m_string;

    myChannelStencil(void)
    {
        //初始化主通道绘制功能需要的字体
        // our Mainchannel load a font to render our string
        char str[256], *cp = getenv("MPI_LOCATE_VEGA_PRIME");
        sprintf(str, "%s/config/vsg/vsgr", cp);
        vuSearchPath *path = vrFontFactory::getDefaultSearchPath();
        path->set(str);
        vrFontFactory *fontFactory = new vrFontFactory();
        vrFont *font = fontFactory->read("mpi_font.dat");
        fontFactory->unref();

        // We'll need to scale the font, otherwise it will be too big.  Note
        // that when scaling the font that you must also reset the spacing to
        // avoid characters running together or being too far apart.
        float scale = 0.02f;
        vuVec3<float> spacing;
        vuMatrix<float> mat;
        vrGeometry *geometry;
        vrFont3D *font3D = static_cast<vrFont3D *>(font);
        mat.makeScale(scale, scale, scale);
        for (int i=0;i<256;i++) {
            if ((geometry = font3D->getGeometry(i)) != NULL)
                geometry->transform(mat);
            font3D->getSpacing(i, &spacing);
            spacing *= scale;
            font3D->setSpacing(i, spacing);
        }

        m_string = new vrString();
        m_string->setFont(font);
        m_string->ref();
    }

    ~myChannelStencil(void)    {   m_string=NULL;}

    /* inherited pre / post cull notification method. We have to implement
     * this even though we don't care about these events since the method is
     * abstract, however, we can just make it a noop.  */
```

```cpp
    virtual void notify(vsChannel::Event, const vsChannel *,vsTraversalCull *) {}

/** inherited pre / post draw notification method.  Note here that since
 * we're using this function for both channels that we'll need to add some
 * extra logic to organize our code here.   */
    virtual void notify(vsChannel::Event event, const vsChannel *channel, vrDrawContext *context)
{
    // for the bigger channel we'll do all of our stencil stuff so
    // we can get the rounded outputs
    //为子通道设置模板效果          ***********************************
    if (channel == PublicMember::biggerChannel)
    {
        vrStencilTest::Element stencilTestElement;

        // for the pre draw we'll write our mask to the stencil buffer
        // and enable the stencil test for the channel rendering.  Note
        // that all of the operations that are being performed below wrt
        // writing to the stencil buffer technically only need to be
        // performed whenever the window is resized, but for the sake of
        // simplicity, we'll do them every frame here.  A
        // vsWindow::EVENT_RESIZE event could be used to be more optimal,
        // or if you knew your window size was not going to change, you
        // could just do these operations the first time through.

        //只在子通道的 vsChannel::EVENT_PRE_DRAW 事件中设置模板效果***************
        if (event == vsChannel::EVENT_PRE_DRAW)
        {
            // since OpenGL doesn't directly support rendering of a
            // circle, we'll create a simple triangle fan primitive to
            // approximate one the first time through.
            int i;
            static bool first = true;
            const int numVertices = 38;
            static vuVec2<float> vertex[numVertices];

            if (first) {
                float s, c, angle = 0.0f;

                vertex[0].set(0.0f, 0.0f);
                for (i=1;i<numVertices;i++,angle+=10.0f) {
                    vuSinCos(360.0f - angle, &s, &c);
                    vertex[i].set(0.25f * s, 0.5f * c);
                }
                vertex[37] = vertex[1];
                first = false;
```

```cpp
        }
        // since we're in a pre draw callback here, the channel hasn't
        // actually been configured yet (i.e. the viewport, projection
        // matrix, etc. don't get specified until the channel's draw
        // function is called), so we'll need to call
        // vrChannel::apply() here to force the update.  This update
        // also clears the frame buffer, so this is a nice opportunity
        // to reset the clear color as well.   Note that just changing
        // the clear color in the ACF wouldn't work, since the clear
        // color is being overridden by the environment.   By the action
        // we're taking here we're actually overriding the clear color
        // specified by the environment.
        channel->getVrChannel()->setClearColor(0.0f, 0.0f, 1.0f, 1.0f);
        channel->getVrChannel()->apply(context);

        // store the current graphics state and reset all of the
        // graphics state elements to their default values
        context->pushElements(true);

        // disable the depth test and set up the projection and model
        // view matrices for orthographic rendering
        glDisable(GL_DEPTH_TEST);
        glMatrixMode(GL_PROJECTION);
        glPushMatrix();
        glLoadIdentity();
        gluOrtho2D(0.0, 1.0, 0.0, 1.0);
        glMatrixMode(GL_MODELVIEW);
        glPushMatrix();
        glLoadIdentity();

        // clear the stencil buffer to all zeros
        glClearStencil(0);
        glClear(GL_STENCIL_BUFFER_BIT);

        // write ones to the stencil buffer where the circles that
        // represent the binoculars are
        stencilTestElement.m_enable = true;
        stencilTestElement.m_mode = vrStencilTest::MODE_ALWAYS;
        stencilTestElement.m_ref = 1;
        stencilTestElement.m_mask = 1;
        stencilTestElement.m_fail = vrStencilTest::OPERATION_REPLACE;
        stencilTestElement.m_dfail = vrStencilTest::OPERATION_REPLACE;
        stencilTestElement.m_dpass = vrStencilTest::OPERATION_REPLACE;
        context->setElement(vrStencilTest::Element::Id,&stencilTestElement);

        glBegin(GL_TRIANGLE_FAN);
        for (i=0;i<numVertices;i++)
            glVertex2f(PublicMember::drawX + vertex[i][0], PublicMember::drawY + vertex[i][1]);
```

```cpp
                glEnd();

                // restore the graphics state
                context->popElements(false);

                // now configure the stencil buffer to only accept fragments
                // whose stencil value is equal to one
                stencilTestElement.m_enable = true;
                stencilTestElement.m_mode = vrStencilTest::MODE_EQUAL;
                stencilTestElement.m_ref = 1;
                stencilTestElement.m_mask = 1;
                stencilTestElement.m_fail = vrStencilTest::OPERATION_KEEP;
                stencilTestElement.m_dfail = vrStencilTest::OPERATION_KEEP;
                stencilTestElement.m_dpass = vrStencilTest::OPERATION_KEEP;
                context->setElement(vrStencilTest::Element::Id,&stencilTestElement);
            }
            //子通道的 vsChannel::EVENT_PRE_DRAW 事件功能结束**************************

            // for the post draw, just disable the stencil test so it doesn't
            // effect the main channel
            else {
                context->setElement(vrStencilTest::Element::Id,&stencilTestElement);
            }
        }
    //为主通道绘制矩形框和放大倍数 ******************************************************
        // for the main channel we'll just draw some graphics overlays
        else
        {
            // store the current graphics state and reset all of the graphics
            // state elements to their default values
            context->pushElements(true);

            // disable the depth test and set up the projection and model view
            // matrices for orthographic rendering
            glDisable(GL_DEPTH_TEST);
            glMatrixMode(GL_PROJECTION);
            glPushMatrix();
            glLoadIdentity();
            gluOrtho2D(0.0, 1.0, 0.0, 1.0);
            glMatrixMode(GL_MODELVIEW);
            glPushMatrix();
            glLoadIdentity();

            // get the fields of view for the main channel.  Note that this
            // code assumes the vertical field of view was specified as -1,
            // so we'll need to determine the actual field of view by
            // computing the aspect ratio.
            int width, height;
```

```cpp
        float hfov, vfov, aspect;
        vpWindow *window = *vpWindow::begin();
        assert(window);
        window->getSize(&width, &height);
        channel->getFOVSymmetric(&hfov, &vfov);
        if (width > height)
              aspect = (float) width / (float) height;
        else aspect = (float) height / (float) width;
        vfov = hfov / aspect;

        // draw a box around what the binoculars are looking at.   We'll
        // compute the size of the box by computing the ratio between the
        // main and binocular fields of view.
        float x = 40.0f / PublicMember::m_magnification / hfov / 2.0f;
        float y = 40.0f / PublicMember::m_magnification / 2.0f / vfov / 2.0f;

        glColor4f(1.0f, 1.0f, 1.0f, 1.0f);
        glBegin(GL_LINE_LOOP);

        glVertex2f(PublicMember::drawX + -x, PublicMember::drawY +  y);
        glVertex2f(PublicMember::drawX + -x, PublicMember::drawY + -y);
        glVertex2f(PublicMember::drawX +  x, PublicMember::drawY + -y);
        glVertex2f(PublicMember::drawX +  x, PublicMember::drawY +  y);

        glEnd();

        // draw the magnification level
        char str[256];
        sprintf(str, "%dX", PublicMember::m_magnification);
        m_string->setColor(1.0f, 1.0f, 1.0f, 1.0f);
        m_string->setString(str);
        m_string->setPosition(0.02f, 0.02f, 0.0f);
        m_string->draw(context);

        // restore the graphics state
        context->popElements(false);
    } //主通道绘制矩形框和放大倍数结束 ****************************************
  }; //成员函数供结束********************************
}; //双通道模板类结束***************
```

图 9.11.9 双通道模板效果类设计

（3）双通道模板效果的配置使用。

为了完全使用双通道模板功能，还需要设计一个键盘函数。这个键盘函数的第一个效果是移动圆形模板，主要是利用上下左右键修改 PublicMember::drawX 和 PublicMember::drawY 的值，从而移动圆形模板效果的位置；第二个效果则是修改 PublicMember::m_magnification 的值，从而改变子通道的放大倍数。键盘函数的详细代码如图 9.11.10 所示。

```
void   PublicMember::CTS_Keyboard(vpWindow *window,vpWindow::Key key, int modifier,void *)
{
    switch(key)
    {
        case vpWindow::KEY_LEFT:        //圆形模板中心坐标
            PublicMember::drawX=PublicMember::drawX-0.01;break;

        case vpWindow::KEY_RIGHT:
            PublicMember::drawX=PublicMember::drawX+0.01; break;

        case vpWindow::KEY_UP:
            PublicMember::drawY=PublicMember::drawY+0.01;break;

        case vpWindow::KEY_DOWN:
            PublicMember::drawY=PublicMember::drawY-0.01;break;

        case vrWindow::KEY_m: // 缩小场景
            PublicMember::m_magnification=PublicMember::m_magnification-1;
            if( PublicMember::m_magnification=PublicMember::m_magnification<1)
                PublicMember::m_magnification=1;
PublicMember::biggerChannel->setFOVSymmetric(40.0/PublicMember::m_magnification,-1.00f);
            break;

        case vrWindow::KEY_M: // 放大场景
            PublicMember::m_magnification=PublicMember::m_magnification+1;
PublicMember::biggerChannel->setFOVSymmetric(40.0/PublicMember::m_magnification,-1.00f);
            break;
            default:   ;
    }
}
```

图 9.11.10　双通道模板效果的键盘函数

使用双通道模板效果需要在主线程中进行必要的设置。
实例化一个双通道模板效果类：

　　myChannelStencil * myChStencil=new myChannelStencil();
查找到主通道，并为其添加通道订阅事件及其事件处理对象：

　　vpChannel *channel = vpChannel::find("myChannel");
　　channel->addSubscriber(vsChannel::EVENT_POST_DRAW, myChStencil);
查找到子通道，并为其添加通道订阅事件及其事件处理对象：

　　PublicMember::biggerChannel = vpChannel::find("biggerChannel");
　　PublicMember::biggerChannel->ref();
　　PublicMember::biggerChannel->addSubscriber(vsChannel::EVENT_PRE_DRAW,
                                            myChStencil);

PublicMember::biggerChannel->addSubscriber(vsChannel::EVENT_POST_DRAW,
myChStencil);

可以看出，主通道只订阅了 vsChannel::EVENT_POST_DRAW 事件，而子通道则同时订阅了 vsChannel::EVENT_PRE_DRAW 和 vsChannel::EVENT_POST_DRAW，全部的功能都由模板对象 myChStencil 完成。其实，读者可以根据实际需要分别进行处理，以减少主通道与子通道之间的耦合性。双通道模板通道订阅类在主线程中的使用的程序代序如图 9.11.11 所示。

```
UINT PublicMember::CTS_RunBasicThread(LPVOID)
{
    vp::initialize(__argc,__argv); //初始化
    PublicMember::CTS_Define();        //定义场景
    vpKernel::instance()->configure();
    //配置双通道模板效果    *********************************************

    //实例化模板功能类
    myChannelStencil * myChStencil=new myChannelStencil();

    //查找到主通道，并为其添加通道订阅事件及其处理对象
    vpChannel *channel = vpChannel::find("myChannel");
    channel->addSubscriber(vsChannel::EVENT_POST_DRAW, myChStencil);

    //查找到子通道，并为其添加通道订阅事件及其处理对象
    PublicMember::biggerChannel = vpChannel::find("biggerChannel");
    PublicMember::biggerChannel->ref();

    PublicMember::biggerChannel->addSubscriber(vsChannel::EVENT_PRE_DRAW, myChStencil);
    PublicMember::biggerChannel->addSubscriber(vsChannel::EVENT_POST_DRAW, myChStencil);

    //模板效果配置结束*********************************************
    //设置窗体
    vpWindow * vpWin= * vpWindow::begin();
    vpWin->setParent(PublicMember::CTS_RunningWindow);
    vpWin->setBorderEnable(false);
    vpWin->setFullScreenEnable(true);

    //设置键盘
    vpWin->setInputEnable(true);
    vpWin->setKeyboardFunc((vrWindow::KeyboardFunc)PublicMember::CTS_Keyboard,NULL);

    vpWin->open();
    ::SetFocus(vpWin->getWindow());

    //帧循环
    while(vpKernel::instance()->beginFrame()!=0)
    {
        vpKernel::instance()->endFrame();
```

```
        if(!PublicMember::CTS_continueRunVP)
        {
        vpKernel::instance()->unconfigure();
        vp::shutdown();
        return 0;
            }
    }
    return 0;
}
```

图 9.1111　双通道模板通道订阅类在主线程中的使用

整个功能配置完成后，在键盘的左移键功能中，减小 PublicMember::drawX 的值使模板圆形左移，其效果如图 9.11.12（a）所示。当然，右移与此类似。

在键盘的下移键功能中，增大 PublicMember::drawY 的值使模板圆形上移，其效果如图 9.11.12（c）所示。当然，下移与此类似。

另外，可以通过 M 键修改 PublicMember::m_magnification 的值，从而减小子通道的 FOV 角度值，实现场景放大，其效果如图 9.11.12（b）所示，图中为放大 2 倍效果。图 9.11.12（d）所示为放大 4 倍效果。

当然，还可以通过 m 键修改 PublicMember::m_magnification 的值，从而增大子通道的 FOV 角度值，实现场景缩小。

（a）模板左移

（b）放大 2 倍

（c）模板上移

（d）放大 4 倍

图 9.11.12　双通道模板效果

9.12 仿真辅助线程设计

对于仿真场景的数据驱动有多种模式，如数据文件方式、网络数据通信方式等。总的来说，为了便于区别处理仿真模块和数据模块，使用多线程方式是一种比较合理可行的方式。仿真模块和数据模块作为独立的两个模块，可以利用多线程方式降低两者之间的耦合度，增加编程控制的灵活性。

9.12.1 C++中的线程

线程是操作系统调度的最小单位。线程包含在进程中，是进程中实际运行的单位。一个进程中可以同时运行多个线程，每个线程可以执行不同的任务，这就是所谓的多线程。同一进程中的多个线程将共享该进程中的全部系统资源，如虚拟地址空间、文件描述符和信号处理等，但是同一个进程中的多个线程都有各自的调用栈、寄存器环境和线程本地存储。

对于单核（单 CPU）系统来说，即便处理器一次只能运行一个线程，但是操作系统通过时间片轮转技术，在不同的线程之间进行切换，让用户产生可以同时处理多个任务的错觉，这样的程序运行机制称为软件的多线程。对于多核（多个 CPU）系统来说，这样的系统能同时进行真正的多线程多任务处理。这种运行机制可以称为硬件的多线程技术。

多线程程序作为一种多任务、并发的工作方式，它具有以下优点：

（1）提高应用程序响应。这对图形界面的程序尤其有意义，当一个操作耗时很长时，整个系统都会等待这个操作，此时程序不会响应键盘、鼠标、菜单的操作，而使用多线程技术，将耗时长的操作置于一个新的线程，可以避免这种尴尬的情况。

（2）使多 CPU 系统更加有效。操作系统会保证当线程数不大于 CPU 数目时，不同的线程运行于不同的 CPU 上。

（3）改善程序结构。一个既长又复杂的进程可以考虑分为多个线程，成为几个独立或半独立的运行部分，这样的程序会利于理解和修改。

Vega Prime 本身就是一个包含多线程的程序，灵活掌握了多线程的使用，可以更灵活方便地操控可视化仿真场景，使自己的仿真程序更为强大。

对于 C++多线程的更多内容，读者可参考更多专门讲解多线程开发的资料，对线程的同步和异步通信、线程池等内容进行更全面的掌握。

9.12.2 线程函数

线程函数必须是全局函数或是某个类的静态成员，而且线程函数使用外部的变量或函数也要求是全局变量或全局函数，或者静态成员。

为了管理方便，一般都不使用全局函数或变量，而是尽量使用静态成员，也就是本节为什么会有一个包含众多静态成员的类 PublicMember。

其实，VP 程序本身是基于多线程的设计，开发者也可以借鉴这种方式：在实际项目的开发中将程序进行更为细致的功能划分，把与 VP 场景仿真相关的程序放到 PublicMember 类中，把与文件操作相关的程序放到 FileOperate 类中，把与网络通信相关的程序放到 NetworkOperate 类中，等等。这些模块与界面没有任何直接关联，完全可以在多个项目中自由组装。这完全符合软件工程中的模块耦合性要求。模块间耦合性越低越好，模块内部内聚性越高越好，可以最大限度地提高模块的复用性。读者可以在实际项目中深入体会使用。

对于线程函数的声明，可参考以下介绍：

线程函数的声明：

  static UINT AutoDisplay(LPVOID) ;

线程函数的运行体：

```
UINT ScenceDisplay::AutoDisplay (LPVOID)
{
    while(ScenceDisplay::CTS_continueDisplay)
    {
      doSomething();
        ::Sleep(10);
    }
    return 0;
}
```

线程函数体一般采用一个循环体，反复完成一些操作，操作完成后使自己短暂休眠，释放 CPU 的使用权，以免因自己长期霸占 CPU 而影响其他线程的执行。最后，通过布尔变量的控制自动结束循环退出线程的执行。

9.12.3 线程控制

在 VC 中，有一个线程对象，可以方便"野蛮"地控制线程，那就是：

  CWinThread *pThread;

有各种文献讨论过 CWinThread 控制线程的优缺点，读者可以通过其他渠道掌握更多。笔者一直使用它，非常方便：

启动线程：

  pThread=AfxBeginThread(ScenceDisplay::AutoDisplay,this);

暂停线程执行：

  pThread->SuspendThread();

继续执行线程：

  pThread->ResumeThread();

停止线程执行：

  TerminateThread(pThread->m_hThread,1);

  CloseHandle(pThread->m_hThread);

对于线程的结束，理想状态是使用布尔变量，使线程函数结束循环退出线程。但是，对于某些线程函数，并不能简单地使用布尔变量进行控制，则使用 CWinThread 结束线程也是切实可行的。

9.12.4 数据辅助线程设计

利用线程相关知识设计了一个数据辅助线程，其完整代码如图 9.12.1 所示。

```cpp
//声明静态成员
  static UINT autoRun(LPVOID);      // 线程函数
  static bool autoRunBool;          // 线程控制布尔变量
  CWinThread *pThread;              //线程控制对象
//-------------------------------------------------------------
  bool CVPTestDialogDlg::autoRunBool=false;
  UINT CVPTestDialogDlg::autoRun(LPVOID)
  {
       while(CVPTestDialogDlg::autoRunBool)
       {
          PublicMember::CTS_pObject_observer->setTranslateX(0.01,true);
          ::Sleep(5);
       }
     return 0;
  }
//-------------------------------------------------------------
void CVPTestDialogDlg::OnBnClickedautorun()//启动辅助线程
{
    CVPTestDialogDlg::autoRunBool=true;
    pThread=AfxBeginThread(CVPTestDialogDlg::autoRun,this);
}

void CVPTestDialogDlg::OnBnClickedstoprun()//关闭线程
{
  //线程函数体自然结束返回
  //CVPTestDialogDlg::autoRunBool=false;

  //强行结束线程
  TerminateThread(pThread->m_hThread,1);
  CloseHandle(pThread->m_hThread);
}
void CVPTestDialogDlg::OnBnClickedsuspend()//暂停线程
{
  pThread->SuspendThread();
}
void CVPTestDialogDlg::OnBnClickedresume()//继续执行线程
{
  pThread->ResumeThread();
}
```

图 9.12.1 数据辅助线程设计使用

整个设计中，设计了一个静态函数 autoRun 作为数据辅助线程的函数，使汽车自动往 X 正向增加；设计了一个布尔变量 autoRunBool 来控制线程循环体的执行；设计了一个线程控制对象 pThread，从而实现线程的开始运行、暂停、继续执行和结束操作。

（1）启动数据辅助线程：

  CVPTestDialogDlg::autoRunBool=true;

  pThread=AfxBeginThread(CVPTestDialogDlg::autoRun,this);

其中，在启动线程时，AfxBeginThread 函数的返回值赋予了 pThread，以便于控制数据线程。

（2）关闭线程：

  //CVPTestDialogDlg::autoRunBool=false;

  TerminateThread(pThread->m_hThread,1);

  CloseHandle(pThread->m_hThread);

其中，给布尔变量赋 false 值的方式，使线程循环体执行完当次循环后，自动退出循环，结束线程。而后面两句则是不论线程函数执行到哪里，都会执行线程，强行结束操作。

（3）暂停线程：

  pThread->SuspendThread();

暂停操作是不论线程函数执行到哪里，都会暂停线程函数的执行，保留在当前状态。

（4）继续执行线程：

  pThread->ResumeThread();

继续执行线程操作则是从线程函数最新暂停的位置继续执行操作。

运行效果如图 9.12.2 所示。

图 9.12.2　数据辅助线程效果

第 10 章　Vega Prime 编程框架设计

依据实际的项目开发经验，在这里推荐一种简单的 Vega Prime 程序开发框架。该框架依据 MVC（Model-View-Control）设计模型，把界面、功能和控制进行分离：把 Vega Prime 的功能设计为 PublicMember 类和其他独立的功能类；界面上利用 TabControl 控件，实现多功能 MFC 对话框程序。主要内容包括 MFC 下的框架总体设计、主窗口设计、全屏幕自适应设计、窗口背景设计、功能窗口选择设计、加载进度显示设计、Vega Prime 场景操控设计等。整个过程充分展现了该框架设计的方法和步骤，开发者可以根据实际需求进行进一步设计。

【本章重点】

- MFC 下的框架总体设计；
- 主窗口设计；
- 全屏幕自适应设计；
- 窗口背景设计；
- 功能窗口选择设计；
- 加载进度显示设计；
- Vega Prime 场景操控设计。

10.1　MFC 下的框架总体设计

前面的设计集中精力于 VP 的功能，因此，界面设计非常简单，具体来说，就只是一个单对话框程序。对于具有较多功能的程序，在这里推荐一种设计框架，方便开发者设计较为复杂的功能程序。实际上，该框架也是笔者在工程中所使用的框架。该框架简单易懂，操作方便，把界面与 VP 功能完全脱离，VP 的基本功能集中在 PublicMember 类和具体的功能类中，基本符合 MVC 的设计理念。

设计 5 个窗口，其 ID 分别是：IDD_VP2DIALOGVS2003_DIALOG、IDD_VPWindow、IDD_NotifyMSG、IDD_FunctionWindowA、IDD_FunctionWindowB。其中，IDD_VP2DIALOGVS2003_DIALOG 窗口为程序的主窗口，其他窗口都覆盖在其上面；IDD_VPWindow 为 VP 线程运行的窗口；IDD_NotifyMSG 为 VP 加载进度显示窗口，只在启动 VP 线程后出现，在 VP 进入循环后结束；IDD_FunctionWindowA 和 IDD_FunctionWindowB 位于 IDD_VP2DIALOGVS 2003_DIALOG 右边的 TabControl 控件中，完成具体的操作功能。当然，可以根据需要调整功

能窗口的数目。图 10.1.1 所示为窗口布局图。在实际应用中，其效果如图 10.1.2 所示。

IDD_VP2DIALOGVS2003_DIALOG	A	B
IDD_VPWindow	IDD_FunctionWindowA	

图 10.1.1 窗口布局图

 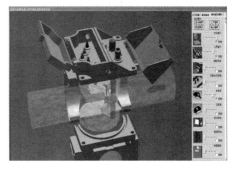

图 10.1.2 实际项目窗口布局效果图

10.2 具体窗口功能设计实现

10.2.1 预备设计

首先，在 VS 2003 下，建立一个 MFC 对话框程序。建立过程依次为：① 启动 VS 2003 后，如图 10.2.1(a)所示，打开"文件"→"新建"→"项目"；② 出现图 10.2.1 (b)所示的界面后，选择"Visual C++项目"下的"MFC"，然后选中"MFC 应用程序"图标，输入名称；③ 在图 10.2.1 (c)中，需要选择"应用程序类型"为"基于对话框"；④ 图 10.2.1 (d)为完成项目建立后的界面。对于主对话框，系统会自动建立对应的类。在该项目中，主对话框的 ID 为 IDD_VP2DIALOGVS2003_DIALOG，对应的类名为 CVP2DialogVS2003Dlg。

（a）新建项目

第 10 章　Vega Prime 编程框架设计

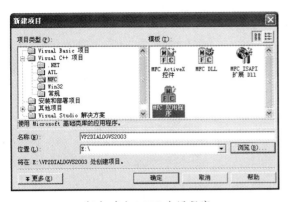
（b）建立 MFC 应用程序

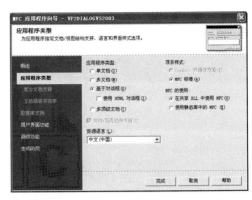
（c）选择基于对话框

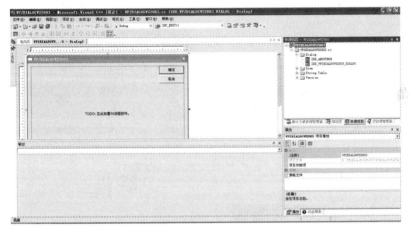
（d）完成项目建立

图 10.2.1　建立 MFC 对话框项目

对于其他窗口，需要在项目中依次添加。如图 10.2.2 (a)所示，首先在窗口左下角，切换到"资源视图"，然后右键单击"Dialog",在右键菜单中选择"插入 Dialog"。将会出现如图 10.2.2 (b)左图所示的界面,单击资源视图中新插入窗口的 IDD_DIALOG1,将会出现如图 10.2.2 (b)右图所示的界面，可以在右上角输入自己需要的窗口 ID。

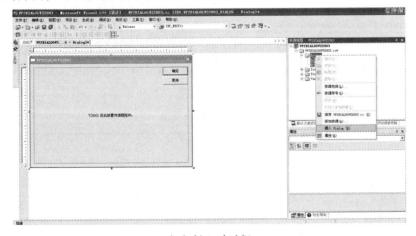
（a）插入对话框

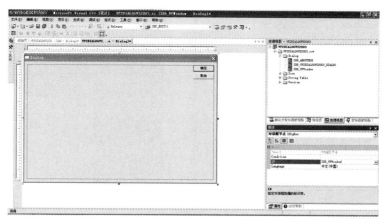

(b)修改窗口ID

图 10.2.2 插入其他对话框

插入对话框资源后，如果要实现功能，还需要建立对应的类。如图 10.2.3 所示，双击图中刚刚插入的对话框，将弹出建立类的向导窗口，一般只需要输入自己设定的类名即可。但是，需要注意类名下面的基类一定要选择 CDialog。

图 10.2.3 建立对话框类

整个框架的主要功能需要 4 个窗口：IDD_VP2DIALOGVS2003_DIALOG、IDD_VPWindow、IDD_FunctionWindowA 和 IDD_FunctionWindowB。为了便于操作，给 4 个窗口定义了相应的窗口指针，对应的窗口指针依次为：PublicMember::pMainWindow、PublicMember::pVPWindow、PublicMember:: pFunctionWindowA 和 PublicMember::pFunction WindowB，类型均为 HWND，而且都为 PublicMember 类的静态成员，具体清单如表 10.2.1 所示。

表 10.2.1 窗口清单

窗口名称	主窗口	VP 窗口	功能 A 窗口	功能 B 窗口
窗口 ID	IDD_VP2DIALOGVS2003_DIALOG	IDD_VPWindow	IDD_FunctionWindowA	IDD_FunctionWindowA
窗口指针	PublicMember::pMainWindow	PublicMember::pVPWindow	PublicMember::pFunctionWindowA	PublicMember::pFunctionWindowB
窗口类	CVP2DialogVS2003Dlg	VPWindow	FunctionWindowA	FunctionWindowB

第 10 章 Vega Prime 编程框架设计

整个框架的基本窗口资源如图 10.2.4 所示，类视图如图 10.2.5 所示。

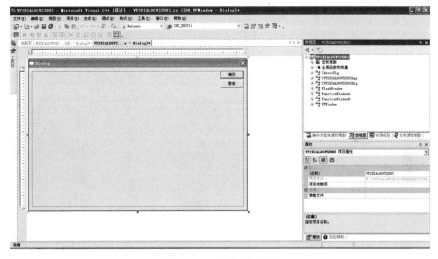

图 10.2.4 框架窗体视图

图 10.2.5 框架类视图

主窗口是整个应用程序的主舞台，包含相对较多的功能，主要包含窗口背景色彩控制、背景图片布局、自适应屏幕设计、TabControl 功能窗口设计。

10.2.2 主窗口背景色彩控制

MFC 提供的窗口背景是单一的颜色。虽然有很多的方法控制窗口背景颜色，由于本书篇幅有限，在这里仅介绍一种常用的方法——采用窗口事件 OnCtlColor(CDC* pDC, CWnd* pWnd, UINT nCtlColor)。它不仅可以控制窗口的背景颜色，还可以控制窗口控件的字体颜色和背景等。如图 10.2.6 所示，切换到资源视图，选中主窗口 IDD_VP2DIALOGVS2003_DIALOG，选中属性栏右上角的"消息"图标，选中其中的 OnCtlColor 消息，添加相关的代码即可。

- 339 -

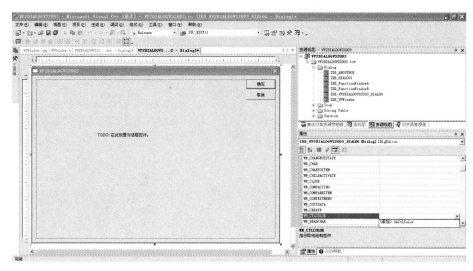

图 10.2.6 添加 OnCtlColor 消息

关于 OnCtlColor 消息的更多应用，可以自行查询。在这个设计中，只是简单设置了静态文本的颜色为蓝色，设置窗体背景颜色为一个 RGB 值。改变颜色的功能，主要改变窗体返回的画刷 Brush 的默认参数。具体代码如图 10.2.7 所示。这里，画刷颜色采用一个 RGB 值，具体的 RGB 值可以自行调制，或者用工具取得，直到取得满意的效果位置。

```
HBRUSH CVP2DialogVS2003Dlg::OnCtlColor(CDC* pDC, CWnd* pWnd, UINT nCtlColor){
    HBRUSH hbr = CDialog::OnCtlColor(pDC, pWnd, nCtlColor);
    //改变静态文本颜色
    if(nCtlColor==CTLCOLOR_STATIC)
    {
        pDC->SetTextColor(RGB(0, 0,250));
        pDC->SetBkMode(TRANSPARENT);
    }
    //设置窗体背景色彩
    hbr = CreateSolidBrush(RGB(200,224,251)); //创建新画刷
    return hbr;
}
```

图 10.2.7 窗口背景色彩控制代码

代码颜色的具体效果如图 10.2.8 所示，窗口背景色已经改变，静态文本的字体颜色也改变为设置的蓝色。

图 10.2.8 窗口背景色彩控制

10.2.3 主窗口背景图片布局

背景颜色的设置步骤和代码相对简单，效果也非常有限。因此，可以使用图片对窗口进行装饰，以达到更好的效果。

首先，需要导入一张合适的位图图片，操作步骤如图 10.2.9 所示，在资源视图中，右键单击"VP2DialogVS2003.rc"，在弹出菜单中选择"添加资源"，就能得到背景位图，一般的 24 色位图都可以导入，使用默认 ID 为 IDB_BITMAP1。

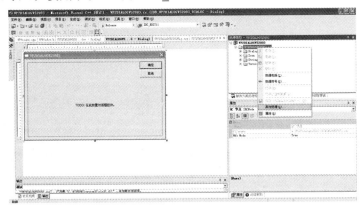

（a）右键导入资源

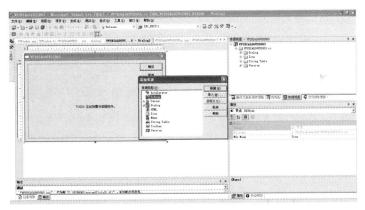

（b）导入图片资源

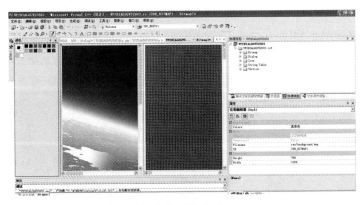

（c）位图导入后

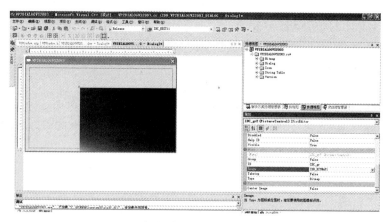

(d)使用位图资源

图 10.2.9　位图背景设计

在图 10.207(d)中，首先在主窗口中添加一个 picture control 控件。在图片控件属性栏中，ID 设置为 IDC_gr，type 设置为 Bitmap，image 设置为刚才导入的位图 IDB_BITMAP1。运行程序，就能看到具有图片背景的窗口。但是，效果不会太理想，需要完成下一步设计后才能完整呈现理想的效果。

10.2.4　主窗口全屏幕自适应设计

在这里就需要建立公共类 PublicMember，用于储存窗口的指针。整个框架的主要功能需要 4 个窗口：IDD_VP2DIALOGVS2003_DIALOG、IDD_VPWindow、IDD_FunctionWindowA 和 IDD_FunctionWindowB，给 4 个窗口定义了相应的窗口指针，依次为：PublicMember::pMainWindow、PublicMember::pVPWindow、PublicMember:: pFunctionWindowA 和 PublicMember::pFunctionWindowB，类型均为 HWND，而且都为 PublicMember 类的静态成员。

全屏幕的自适应设计包含两个方面的内容：一是主窗口的全屏幕自适应设计；二是控件适应窗口的相应变化。

在此框架设计中，把主窗口分为左右两大部分。左边为主要部分，未启动 VP 仿真之前，就显示图片；启动后，就显示 VP 的仿真场景。右边为功能区，为了能容纳多个功能窗口，使用 TabControl 控件，在这里，只先实现 TabControl 的布局。从工具箱中添加一个 TabControl 控件到主窗口，其 ID 为 IDC_tabControl。当然，主窗口和控件的布局实现，都应该在主窗口的初始化函数里实现，也就是函数"BOOL CVP2DialogVS2003Dlg::OnInitDialog();"，该函数的系统代码应该保留，自己的代码一般都添加到最后，具体代码如图 10.2.10 所示。

```
BOOL CVP2DialogVS2003Dlg::OnInitDialog()
{
//系统默认的代码应该保留
// TODO: 在此添加额外的初始化代码
// 窗口布局代码******************************
//获取主窗口句柄
 PublicMember:: pMainWindow =theApp.GetMainWnd()->GetSafeHwnd();
```

```
//获取屏幕的尺寸
 int width=::GetSystemMetrics(SM_CXSCREEN);
 int height=::GetSystemMetrics(SM_CYSCREEN);

//主窗体布局
 ::MoveWindow(PublicMember::pMainWindow,0,0,width,height+5,true);

 //控件布局
 GetDlgItem(IDC_gr)->SetWindowPos(NULL,0,0,width-180,height,SWP_SHOWWINDOW);
 GetDlgItem(IDC_tabControl)->SetWindowPos(NULL,width-180,0,180, height,SWP_SHOWWINDOW);

// 主窗口和控件布局结束******************************
return TRUE;
}
```

图 10.2.10　全屏幕自适应设计

主窗口的全屏幕自适应设计，关键是获取屏幕的宽度和高度，然后让主窗口布满整个屏幕。获取屏幕尺寸可以使用 MFC 的全局函数"::GetSystemMetrics()"进行实现，重新布局窗口可以使用 MFC 的全局函数"::MoveWindow()"进行实现。为了便于控制主窗口，需要获取主窗口的窗口句柄。这里，在 PublicMember 类里定义了一个静态句柄变量"pMainWindow;"，格式为 static HWND pMainWindow。

在其他窗口中使用 PublicMember 类需要先添加其头文件。

在图 10.2.10 中，下面一句代码获取主窗口句柄：

　　PublicMember:: pMainWindow =theApp.GetMainWnd()->GetSafeHwnd();

下面两句分别获取屏幕的宽度和高度：

　　int width=::GetSystemMetrics(SM_CXSCREEN);

　　int height=::GetSystemMetrics(SM_CYSCREEN);

下面一句使主窗口布满全屏：

　　::MoveWindow(PublicMember::pMainWindow,0,0,width,height+5,true);

其中，高度加 5 的目的是覆盖屏幕底端的任务栏，可根据需要自行调整。

对于控件的布局，主要是根据需要重新布局图片控件和 TabContol 控件。图片控件的高度为屏幕高度，其宽度为屏幕宽度减去 180 个像素，这 180 个像素的宽度也就是右边 TabContol 的宽度。其具体代码为：

　　GetDlgItem(IDC_bgPIC)->SetWindowPos(NULL,0,0,width-180,height,
　　　　　　　　　　　　　　　　　SWP_SHOWWINDOW);

TabContol 控件的布局代码为：

　　GetDlgItem(IDC_tabControl)->SetWindowPos(NULL,width-180,0,180,
　　　　　　　　　　　　　　　　　height,SWP_SHOWWINDOW);

其宽度为 180 个像素，高度为 height 个像素。当然，这些值都可以根据项目的需要自行调整。

运行主窗口的初始化函数"BOOL CVP2DialogVS2003Dlg::OnInitDialog();"，其效果如图 10.2.11 所示。

图 10.2.11　自适应屏幕设计

10.2.5　功能窗口初步设计

整体布局完成后，需要对功能窗口进行初次设计。因为功能窗口都是位于 TabControl 控件之内的，因此，它们都需要设计成没有标题栏的窗口。对 IDD_FunctionWindowA 窗口添加 2 个按钮，分别为"开始仿真"和"停止仿真"，用于启动窗口和停止仿真窗口，如图 10.2.12 所示；对 IDD_FunctionWindowB 窗口，添加 4 个单选按钮，分别完成"前视""后视""左视"和"右视"功能，如图 10.2.13 所示。另外，两个功能窗口的尺寸需要满足要求，也就是宽度为 180 个像素，高度不能超过屏幕高度。两个功能窗口的背景颜色，也是利用 OnCtlColor 函数设计为"RGB(200,224,251);"，窗口属性栏中窗口类型（Style）需要从默认的 Popup 修改为 Child 类型。

图 10.2.12　设置功能窗口 A 的属性

图 10.2.13　设置功能窗口 B 的属性

10.2.6 TabControl 初始化设计

对于主对话框的 TabControl 控件的初始化,首先需要为主窗口类建立三个成员变量。

第一个成员变量就是 TabControl 控件成员变量,在主窗口中,右键单击 TabControl 控件,从右键菜单中选择"添加成员变量",将会弹出如图 10.2.14 所示的设置界面,设置 TabControl 控件变量的名称为 tabControl。

图 10.2.14　设置 tabControl 成员变量

第二个和第三个成员变量为功能窗口的变量。首先把 IDD_FunctionWindowA 和 IDD_FunctionWindowB 对应类的头文件放入主对话框类的头文件中,也就是在 VP2DialogVS2003Dlg.h 的最开头放入:

　　#include "FunctionWindowA.h"
　　#include "FunctionWindowB.h"

然后,分别定义:

　　FunctionWindowA *FunctionA;

　　FunctionWindowB *FunctionB;

分别为主窗口类的私有成员,如图 10.2.15 所示,子窗口 A 和子窗口 B 成为主窗口类的成员。

图 10.2.15　添加窗口类成员

窗口类成员,还需要在构造函数里赋初值,在析构函数里销毁指针成员变量。其代码如图 10.2.10 所示。其中,构造函数 CVP2DialogVS2003Dlg 中的赋初值操作代码为:

　　FunctionA =new FunctionWindowA();

FunctionB =new FunctionWindowB();
另外，OnDestroy()函数需要在窗口消息里添加，并加入代码
FunctionA =NULL;
FunctionB =NULL;
完成销毁功能。

```
CVP2DialogVS2003Dlg::CVP2DialogVS2003Dlg(CWnd* pParent /*=NULL*/)
 : CDialog(CVP2DialogVS2003Dlg::IDD, pParent)
 , FunctionA(NULL)
 , FunctionB(NULL)
{
//初始化窗口成员变量
FunctionA =new FunctionWindowA();
FunctionB =new FunctionWindowB();
m_hIcon = AfxGetApp()->LoadIcon(IDR_MAINFRAME);
}
void CVP2DialogVS2003Dlg::OnDestroy()
{
//释放指针成员
FunctionA =NULL;
FunctionB =NULL;
CDialog::OnDestroy();
}
```

图 10.2.16　窗口成员初始化和销毁

准备工作完成后，即可在 OnInitDialog()函数中完成 TabControl 的初始化工作。具体代码如图 10.2.17 所示，其中的主要功能依次是：设置 TabControl 控件的文字标签；以非模式对话框方式创建功能窗口；设置功能窗口 A 为默认显示窗口。

```
BOOL CVP2DialogVS2003Dlg::OnInitDialog()
{
//系统默认的代码应该保留
// tabControl 初始化代码****************************
    TCITEM tc1,tc2;
    tc1.mask = TCIF_TEXT;
    tc1.pszText = "功能 A";

    tc2.mask = TCIF_TEXT;
    tc2.pszText = "功能 B";

    tabControl.InsertItem(0,&tc1);
    tabControl.InsertItem(1,&tc2);

    CRect rec;
    tabControl.GetClientRect(&rec);//获得 Tab 控件的用户区

    FunctionA->Create(IDD_FunctionWindowA,GetDlgItem(IDC_tabControl));
```

```
    FunctionB->Create(IDD_FunctionWindowB,GetDlgItem(IDC_tabControl));

    //初始化时，显示功能 A,隐藏共 B
    //将功能窗口 A 移动到 TabControl 的位置
    FunctionA->MoveWindow(2,20,rec.right,rec.bottom);

    FunctionB->ShowWindow(SW_HIDE);
    FunctionA->ShowWindow(SW_RESTORE);

    tabControl.SetCurSel(0);

    PublicMember:: pFunctionWindowA = FunctionA->GetSafeHwnd();
    PublicMember:: pFunctionWindowB = FunctionB->GetSafeHwnd();

    // 初始化结束***************************
    return TRUE;
}
```

图 10.2.17 TabControl 初始化

运行效果如图 10.2.18 所示。

图 10.2.18 TabControl 初始化效果图

10.2.7 TabControl 功能切换设计

对于 TabControl 控件，还需要为其标签选择事件编写相关代码，双击主窗口上的 TabControl 控件，就会在主窗口中生成 OnTcnSelchangetabcontrol 事件，如图 10.2.19 所示。图中代码就能完成功能窗口 A 和功能窗口 B 的切换功能。

```
void CVP2DialogVS2003Dlg::OnTcnSelchangetabcontrol(NMHDR *pNMHDR, LRESULT *pResult)
{
    CRect rec;
    tabControl.GetClientRect(&rec);//获得 TAB 控件的用户区
    switch (tabControl.GetCurSel())
        {
        case 0: //显示功能 A
                FunctionA->MoveWindow(2,20,rec.right,rec.bottom);
                FunctionA->ShowWindow(SW_RESTORE);
```

```
                FunctionB->ShowWindow(SW_HIDE);
                break;
        case 1://显示功能 B
                FunctionB->MoveWindow(2,20,rec.right,rec.bottom);
                FunctionB->ShowWindow(SW_RESTORE);
                FunctionA->ShowWindow(SW_HIDE);
                break;
        default:   break;
    }
    ::SetFocus(PublicMember:: pVPWindow);
    *pResult = 0;
}
```

图 10.2.19   TabControl 标签切换代码

运行代码，分别点击"功能 A"和"功能 B"标签，就能完成不同功能窗口的切换，运行效果如图 10.2.20 所示。

图 10.2.20   TabControl 标签切换效果

10.2.8 功能窗口再次设计

功能窗口 A 需要进一步设计。这个设计主要是按钮"启动仿真"和"停止仿真"的功能实现。

实现"启动仿真"按钮的功能需要几个步骤：第一步，启动 VP 展示窗口，并放置到合适位置，同时启动进度显示窗口；第二步，在 VP 展示窗口的初始化事件里启动 VP 线程；第三步，在启动进度显示窗口中显示加载进度，并判断 VP 线程是否加载完毕，完毕后，关闭计时窗口。

功能窗口 A 中的按钮"启动仿真"涉及另外两个窗口，一个是 VP 场景仿真窗口，另一个是仿真场景加载进度显示窗口。首先，需要完成这两个窗口的设计。

对于 VP 场景仿真窗口，在预备设计中，只完成了最基本的设计，得到 ID 为 IDD_VPWindow，类名为 VPWindow，窗口指针为 PublicMember::VPWindow。对于该窗口的外观，需要在属性栏中进行设置。把标题栏（Title Bar）设置为 False，把边界(Border)设置为 None，删除掉原来的"确定"和"取消"按钮，效果如图 10.2.21 所示。当然，也可以根据需求，自行设计。例如，可以按照前面的方法，为该窗口设计背景颜色。

对于 VP 场景仿真窗口，另外一个重要的设计是为其添加初始化函数 OnInitDialog。在所有窗口中，只有主窗口 IDD_VP2DIALOGVS2003_DIALOG 一开始就有初始化函数，其他窗口都需要手动添加。参考主窗口初始化函数 OnInitDialog，其返回类型为 BOOL 类型，函数最开始有一个系统的对话框初始化函数"CDialog::OnInitDialog();"，然后才是加入自己需要完成的功能代码。其添加过程如图 10.2.22 所示，切换到类视图，然后右键单击 VP 场景仿真窗口类 VPWindow，在弹出菜单中选择"添加"→"添加函数"，在弹出窗口中输入函数范围类型及名称，如图 10.2.23 所示。

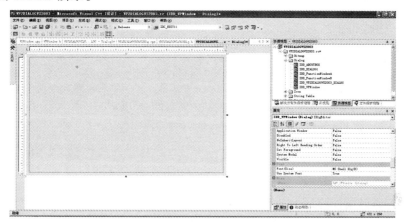

图 10.221　设置 VPWindow

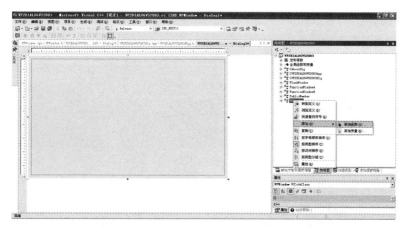

图 10.2.23　添加初始化函数

图 10.2.23　添加初始化函数

输入完成后，需要在 BOOL VPWindow::OnInitDialog(void)函数中输入功能代码，具体代码如图 10.2.24 所示。这里，首先调用系统的对话框初始化函数。然后是设置一个静态变量，在仿真场景加载前设置为 false，进入帧循环后设置为 true，加载进度显示窗口可依据此变量进行相关操作。另外一个功能就是启动 VP 仿真线程。

```
BOOL VPWindow::OnInitDialog(void)
{
    CDialog::OnInitDialog();              //调用系统对话初始化函数

    PublicMember::vpContinue=true;        //允许进入帧循环

    PublicMember::LoadVPSuccess=false;    //设置加载进度变量
    AfxBeginThread(PublicMember::vp2MainThread ,this ); //启动 VP 线程

    return 0;
}
```

图 10.2.24　VPWindow 初始化函数代码

仿真场景加载进度控制变量 PublicMember::LoadVPSuccess 在 VP 的主线程中还有一个修改，具体代码如图 10.2.25 所示的 VP 主线程代码。该线程非常简单，加载一个 acf 文件，设置窗口属性，在进入帧循环后，修改加载进度控制变量的值为 true。

```
//主线程
UINT PublicMember::vp2MainThread(LPVOID)
{
    vp::initialize(__argc,__argv);
    vpKernel::instance()->define("vp_mfc.acf");
    vpKernel::instance()->configure();

    //设置窗口属性
    vpWindow * vpWin=*vpWindow::begin();

    vpWin->setParent(RunningWindow);
    vpWin->setBorderEnable(false);
    vpWin->setFullScreenEnable(true);

    //帧循环
    while(vpContinue)
    {
        PublicMember::LoadVPSuccess=true; //场景加载成功
        vpKernel::instance()->beginFrame();
        vpKernel::instance()->endFrame();
    }
    //释放资源并关闭 VP
    vpKernel::instance()->unconfigure();
    vp::shutdown();

    return 0;
}
```

图 10.2.25　TabControl 标签切换代码

对于加载进度显示窗口，在前面的设计中没有过多提及，因为该窗口寿命本身很短，只在 VP 线程加载时出现，在加载过程中显示加载进度，在 VP 线程加载结束后自动退出。在进一步设计之前，需要确认前面的设计，该窗口 ID 为 IDD_NotifyMSG，类名为 NotifyWindow。

对于加载进度显示窗口，删掉原来的"确认"和"取消"按钮，添加一个标签控件，设置其 ID 为 IDC_Notify，添加一个进度条控件，界面如图 10.2.26 所示。另外，在窗口属性栏设置该窗口的标题栏（Title Bar）为 False，边界(Border)为 None，设置透明属性（Transparen）为 True。

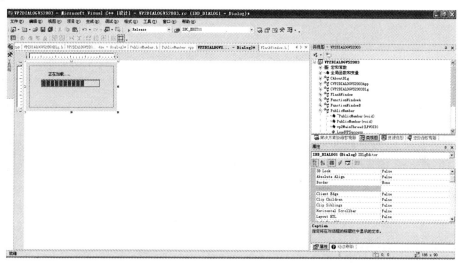

图 10.2.26　设置加载进度显示窗口

对于仿真场景进度的显示，还需要一些辅助设计：第一，为该窗口类添加一个整数 int 数据成员"int proress;"，用于记录加载进度；第二，为进度控制条设置一个控件成员"CProgressCtrl m_p;"，其过程如图 10.2.27 所示。在资源视图中，右键单击进度条按钮，在弹出菜单中选择"添加变量"，在弹出的窗口中输入名称 m_p，然后确认完成，如图 10.2.28 所示。另外的工作，就是为该窗口添加初始化函数和窗口定时器相应函数。

图 10.2.27　添加进度条成员变量

图 10.2.28 设置进度条成员

对于初始化函数,添加和设置过程与 VP 场景运行窗口类似,其代码如图 10.2.29 所示。其主要功能是设置进度值 progress 为 0,启动一个窗口定时器,其每秒执行一次功能代码。

```
BOOL NotifyWindow::OnInitDialog(void)
{
    CDialog::OnInitDialog();//调用系统对话初始化函数

    progress=0;
    SetTimer(1,1000,NULL);

    return 0;
}
```

图 10.2.29 进度显示窗口的初始化函数

初始化函数启动了一个 ID 为 1 的定时器,但定时器的功能代码还需要额外的添加设置。对于窗口定时器,需要在窗口消息中,启用窗口的时钟消息,如图 10.2.30 所示。选择窗口定时器事件 OnTimer,其功能代码如图 10.2.31 所示。

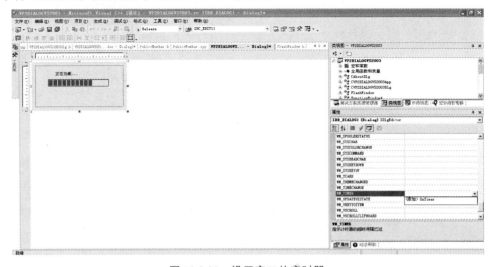

图 10.2.30 设置窗口的定时器

```
void NotifyWindow::OnTimer(UINT_PTR nIDEvent)
{  // TODO: 在此添加消息处理程序代码和/或调用默认值
   if(PublicMember::LoadVPSuccess)
     {   KillTimer(1);                                      //停止定时器
         ::SetFocus(PublicMember:: pVPWindow);              //vp 窗口获得焦点
         //获取屏幕的尺寸
         int width=::GetSystemMetrics(SM_CXSCREEN);
         int height=::GetSystemMetrics(SM_CYSCREEN);
         ::MoveWindow(PublicMember:: pVPWindow,4,25,width-180,height-10,true);
         OnOK();                                            //关闭进度显示窗口
     }
//进度值超过 90 以后就不再增加,一般可以提示强行退出计时。
   if(progress>90)
         ;
   else
         progress+=2;
   m_p.SetPos(progress);                                    //推进 进度控件
   char tt[200];
   sprintf(tt,"用时%d 秒  正在加载......",progress/2); //显示耗费时间
   SetDlgItemText(IDC_Notify,tt);
   CDialog::OnTimer(nIDEvent);
}
```

图 10.2.31　进度显示窗口定时器相应函数

定时器函数的功能,主要是判断进度控制变量 PublicMember::LoadVPSuccess。如果加载成功,就先停止定时器,移动刷新 VP 仿真窗口,最后关闭进度显示窗口。如果还在加载过程中,就判断进度是否超过 90,如果超过,一般让用户无法接受,则可以强行停止加载,这里,没有做任何处理。如果进度值没有超过 90,则进度值增加 2,设置进度条进度,显示已经耗费的时间。

完成前面的设计以后,回到功能窗口 A 的按钮功能设计。对于"启动仿真"按钮,其代码如图 10.2.32 所示。第一步是判断仿真是否已经开始,开始了就直接返回。第二步是修改加载进度显示控制变量为 false,表示加载没有完成。第三步是以非模态方式启动 VP 仿真窗口,并自适应屏幕尺寸。第四步是以模态方式启动加载进度显示窗口。

```
void FunctionWindowA::OnBnClickedbtbeginvr()
{
   // TODO: 在此添加控件通知处理程序代码
       if(PublicMember::vpContinue)
         {
              AfxMessageBox("仿真程序已经开始运行了!");
              return ;
         }


       PublicMember::LoadVPSuccess =false;
```

```
            //以非模态方式启动 VP 运行窗口
            VPWindow * dlg;
            dlg=new VPWindow;
            dlg->Create(IDD_VPWindow);
            dlg->ShowWindow(SW_RESTORE);

            //设置 VP 窗口的位置
            PublicMember:: pVPWindow=dlg->m_hWnd;
             int width=::GetSystemMetrics(SM_CXSCREEN);
             int height=::GetSystemMetrics(SM_CYSCREEN);
             ::MoveWindow(PublicMember::RunningWindow,4,25,width-180,height-10,true);

            //以模态方式启动进度显示窗口
            NotifyWindow dlgFlash;
            dlgFlash.DoModal();
        }
```

图 10.2.32 启动仿真函数

对于"停止仿真"按钮，首先是停止 VP 的仿真，直接修改变量 PublicMember::vpContinue 的值，然后是销毁 VP 显示窗口，其代码为：

　　　　PublicMember::vpContinue=false;
　　　　::DestroyWindow(PublicMember:: pVPWindow);

10.3 运行效果设计

启动程序后，点击"运行仿真"就会出现图 10.3.1 所示的加载进度显示窗口。加载完成后，将呈现图 10.3.2 所示的运行仿真界面。当然，也可以切换 TabControl 按钮，使用不同功能窗口的功能操控仿真效果。

图 10.3.1 仿真场景加载进度窗口

第 10 章 Vega Prime 编程框架设计

图 10.3.2　仿真场景加载完毕

对于程序框架的设计，是一个见仁见智的问题。这里的设计，只是一个抛砖引玉过程，有很多细节不够完善。本书的整个框架基本上能满足部分实际工程应用，满足 MVC 的要求，能够把功能与界面分开，仿真功能集中于 PublicMember 类和其他功能类，如鼠标操作类、文件操作类、网络通信类等，界面就完全交由 MFC 来完成。

参考文献

[1] 王孝平，董秀成，郑海春，等. Vega Prime 实时三维虚拟现实开发技术[M]. 成都：西南交通大学出版社，2012.

[2] Wang Xiaoping, Dong Xiucheng, Jiang Wenbo.Research on oriented-object bounding box algorithm and its application in collision detection of virtual reality[J]. ICIC Express Letters, 2015, 9(2):447-452.

[3] 曹旻. Visual C++ Windows 编程技术[M]. 北京：清华大学出版社，2011.

[4] 潘爱民，陈铭，邹开红. Effective STL 中文版[M]. 北京：清华大学出版社，2006.

[5] 周靖，葛子昂. Windows 核心编程[M]. 5 版. 北京：清华大学出版社，2008.

附件　虚拟现实开发实例

源代码下载

学习论坛

作者博客